It is now well accepted that microbial life followed very quickly after the formation of liquid water on the early earth, and that, for the next 3000 million years, life was a unicellular phenomenon. During this immense period of time, all the major types of microbes appeared, as did all the fundamental chemical pathways necessary for life. This diversification was not well appreciated until the techniques of molecular biology provided a means to examine the relationships between microorganisms which share few, if any, phenotypic characters. This volume reviews the current understanding of the evolution of microbial life during that time. The chapters draw together the various threads of the story to uncover what has been learned about the process of evolution itself, and what this knowledge can contribute to the understanding of biodiversity.

EVOLUTION OF MICROBIAL LIFE

SYMPOSIA OF THE SOCIETY FOR GENERAL MICROBIOLOGY

Series editor (1996–2001): Dr D. Roberts, Zoology Department, The Natural History Museum, London

Volumes currently available:

28 STRUCTURE AND FUNCTION OF PROKARYOTIC MEMBRANES
29 MICROBIAL TECHNOLOGY: CURRENT STATE, FUTURE PROSPECTS
30 THE EUKARYOTIC MICROBIAL CELL
32 MOLECULAR AND CELLULAR ASPECTS OF MICROBIAL EVOLUTION
33 VIRUS PERSISTENCE
35 INTERFERONS FROM MOLECULAR BIOLOGY TO CLINICAL APPLICATIONS
36 THE MICROBE 1984 PART I: VIRUSES
PART II: PROKARYOTES AND EUKARYOTES
37 VIRUSES AND CANCER
38 THE SCIENTIFIC BASIS OF ANTIMICROBIAL CHEMOTHERAPY
39 REGULATION OF GENE EXPRESSION – 25 YEARS ON
41 ECOLOGY OF MICROBIAL COMMUNITIES
42 THE NITROGEN AND SULPHUR CYCLES
43 TRANSPOSITION
44 MICROBIAL PRODUCTS: NEW APPROACHES
45 CONTROL OF VIRUS DISEASES
46 THE BIOLOGY OF THE CHEMOTACTIC RESPONSE
47 PROKARYOTIC STRUCTURE AND FUNCTION: A NEW PERSPECTIVE
49 MOLECULAR BIOLOGY OF BACTERIAL INFECTION: CURRENT STATUS AND FUTURE PERSPECTIVES
50 THE EUKARYOTIC GENOME: ORGANISATION AND REGULATION
51 VIRUSES AND CANCER
52 POPULATION GENETICS OF BACTERIA
53 FIFTY YEARS OF ANTIMICROBIALS: PAST PERSPECTIVES AND FUTURE TRENDS

EVOLUTION OF MICROBIAL LIFE

EDITED BY

D. McL. ROBERTS, P. SHARP, G. ALDERSON
AND M. A. COLLINS

FIFTY-FOURTH SYMPOSIUM OF THE
SOCIETY FOR GENERAL MICROBIOLOGY
HELD AT THE UNIVERSITY OF WARWICK
MARCH 1996

Published for the Society for General Microbiology

CAMBRIDGE
UNIVERSITY PRESS

Published by the Press Syndicate of the University of Cambridge
The Pitt Building, Trumpington Street, Cambridge CB2 1RP
40 West 20th Street, New York, NY 10011-4211, USA
10 Stamford Road, Oakleigh, Melbourne 3166, Australia

First published 1996

Printed in Great Britain at the University Press, Cambridge

A catalogue record for this book is available from the British Library

Library of Congress cataloguing in publication data

Society for General Microbiology. Symposium (54th : 1996 : University of Warwick)
Evolution of microbial life: fifty-fourth Symposium of the Society for General Microbiology held at the University of Warwick, March 1996/ edited by D. McL. Roberts . . . [et al.].
p. cm.
'Published for the Society of General Microbiology'
Includes index.
ISBN 0 521 56432 8
1. Microorganisms – Evolution – Congresses. I. Roberts, D. McL. [David McLean) II. Title.
QR13.S63 1996
576'.138 dc20 96-9637 CIP

ISBN 0 521 56432 8 hardback

PN

CONTENTS

CONTRIBUTORS

De A. Zanotto, P. M. NERC Institute of Virology and Environmental Microbiology, Mansfield Road, Oxford OX1 3SR, UK

Dijkhuizen, L. Department of Microbiology, Groningen Biomolecular Sciences and Biotechnology Institute (GBB), University of Groningen, Kerklaan 30, 9751 NN Haren, The Netherlands

Doolittle, W. F. Canadian Institute for Advanced Research and Department of Biochemistry, Dalhousie University, Halifax, Nova Scotia, Canada

Douglas, A. E. Department of Biology, University of York, PO Box 373, York YO1 5YW, UK

Fenchel, T. Marine Biological Laboratory (University of Copenhagen), Strandpromenaden 5, DK-3000 Helsingor, Denmark

Field, K. G. Department of Microbiology, 220 Nash Hall, Oregon State University, Corvallis, OR 97331-3804, USA

Giovannoni, S. J. Department of Microbiology, 220 Nash Hall, Oregon State University, Corvallis, OR 97331-3804, USA

Gogarten, J. P. Department of Molecular and Cell Biology, University of Connecticut, 75 North Eagleville Road, Storrs, CT 06269-3044, USA

Gordon, D. Department of Microbiology, 220 Nash Hall, Oregon State University, Corvallis, OR 97331-3804, USA

Gould, E. A. NERC Institute of Virology and Environmental Microbiology, Mansfield Road, Oxford OX1 3SR, UK

Gray, M. W. Program in Evolutionary Biology, Canadian Institute for Advanced Research and Department of Biochemistry, Dalhousie University, Halifax, Nova Scotia B3H 4H7, Canada

Hilario, E. Department of Molecular and Cell Biology, University of Connecticut, 75 North Eagleville Road, Storrs, CT 06269-3044, USA

Hinkle, G. Program in Molecular Evolution, Marine Biological Laboratory, Woods Hole, MA 02543, USA

Holmes, E. C. Wellcome Centre for the Epidemiology of Infectious Disease, Department of Zoology, University of Oxford, South Parks Road, Oxford, OX1 3PS, UK

LAKE, J. A. Molecular Biology Institute and MCD Biology, University of California, Los Angeles, CA 90095, USA

MOORE, R. T. School of Applied Biological and Chemical Sciences, University of Ulster at Coleraine, Northern Ireland BT52 1SA, UK

MORRISON, H. G. Program in Molecular Evolution, Marine Biological Laboratory, Woods Hole, MA 02543, USA

OLENDZENSKI, L. Department of Molecular and Cell Biology, University of Connecticut, 75 North Eagleville Road, Storrs, CT 06269-3044, USA

RAPPÉ, M. S. Department of Microbiology, 220 Nash Hall, Oregon State University, Corvallis, OR 97331-3804, USA

RIVERA, M. C. Molecular Biology Institute and MCD Biology, University of California, Los Angeles, CA 90095, USA

SCHOPF, J. W. IGPP Center for the Study of Evolution and the Origin of Life, Department of Earth and Space Sciences, and Molecular Biology Institute, University of California, Los Angeles, CA 90095, USA

SILBERMAN, J. D. Program in Molecular Evolution, Marine Biological Laboratory, Woods Hole, MA 02543, USA

SIMMONDS, P. Department of Medical Microbiology, University of Edinburgh, Teviot Place, Edinburgh EH8 9AG, UK

SOGIN, M. L. Program in Molecular Evolution, Marine Biological Laboratory, Woods Hole, MA 02543, USA

SPENCER, D. F. Program in Evolutionary Biology, Canadian Institute for Advanced Research and Department of Biochemistry, Dalhousie University, Halifax, Nova Scotia B3H 4H7, Canada

SUZUKI, M. Department of Microbiology, 220 Nash Hall, Oregon State University, Corvallis, OR 97331-3804, USA

URBACH, E. Department of Microbiology, 220 Nash Hall, Oregon State University, Corvallis, OR 97331-3804, USA

EDITORS' PREFACE

As the millenium approaches, it is becoming safe to make assertions about the most significant developments of the twentieth century. In microbial systematics, the use of DNA sequence comparisons culminating in Woese's revelation of three domains of cellular life must surely rank high on any list. Considering that, at the beginning of the century, DNA had not been demonstrated to be the repository of genetic information and that, as the century draws to a close, we have complete genomic sequences for a number of organisms shows just how far we have come.

Evidence for the three domains sprang from the still young field of molecular biology and, although it is not universally accepted (see Lake's chapter, for example) it has dramatically changed our way of thinking about the evolution and relationships of cellular life (see Doolittle's introductory chapter).

Another young field of study, palaeoecology, has contributed to our appreciation of the antiquity of cellular life on this planet. The world formed some 4.5 GYr (4.5×10^9 years) ago. The surface cooled and liquid water formed during those first 0.5 GYr, but it was widely believed that the extensive meteoric bombardment, which also resulted in the visible cratering on the moon, was of such ferocity that liquid water would have been evaporated to condense again later. However, cyanobacteria seem to have been active earlier than 3.5 GYr (see Schopf's chapter). Cyanobacteria are not simple or 'primitive' cells and our understanding of bacterial evolution does not place their appearance early. This is, for many, an astonishingly short time period for life to have begun and evolved so far and it does present paradoxes that have not yet been satisfactorily resolved. It does, at least, cast doubt on the idea that liquid water was completely removed during the meteoric bombardment. Evidence for the first multicellular organisms appears in the geological record at about 0.7 GYr, so for the majority of the Earth's history life was an unicellular phenomenon. All the major chemical cycles must have been in place during this time, so macroscopic life evolved into a functioning ecosystem. It should not be a great surprise to learn that the microbial cycles are those which drive life on the planet.

The techniques of molecular biology, particularly DNA sequencing, are providing data in ever greater quantities with which diverse organisms can be compared. Historically, taxonomy has been developed by comparing very similar organisms with very small (usually phenotypic) differences in order to deduce their relationships. It becomes exponentially more difficult to deduce relationships as the similarity decreases and the number of characters in common falls to a tiny fraction of the total number of characters available. The phrase 'cellular life' has been used here principally to avoid a discussion of what, exactly, constitutes life. The origin of the viruses is still very much an open question and suffers greatly from the paucity of characters in common, not just with cellular life, but even between the major groups of viruses (see the chapters by Holmes and Simmonds).

Molecular biology has allowed us to recover vast numbers of characters, many of which are shared by all cellular life. By far and away the most studied gene has been that encoding the small-subunit (16S) ribosomal RNA. If Darwinian evolution (descent with modification) is the dominant evolutionary model through the history of life, then sufficient quantity of sequence data of this type will allow us to build bifurcating phylogenetic trees establishing the relationship of one taxon with another. The tantalizing prospect is that there is another model of evolution visible in the predominantly asexual microbial world that owes much to lateral transfer and promiscuous genetic exchange (i.e. between quite unrelated partners) which would render such a tree an artefact of the gene chosen to build it, and in which the evolutionary relationships are a network (allowing for anastomization) rather than a simple tree. Issues of this kind were addressed in a previous symposium volume (**52**: *Population genetics of bacteria*) to which the reader's attention is drawn. Such models will not be resolved until large sections of genomes of a taxonomically diverse set of organisms are available for analysis.

Unicellular eukaryotes are a poorly understood group of microbes; for those of us who work with them it is depressingly common to find 'microbe' and 'bacteria' being used as synonyms. There is now no doubt that the eukaryotic cell evolved by successive mergers of cells (see chapters by Sogin, Gray, Douglas and Gogarten). The resulting organisms are capable of staggering morphological richness (see Moore's chapter on the fungi, for example), but also of 'reverting' to a phenotype regarded as primitive in order to exploit an available niche (see Fenchel's chapter on eukaryotic anaerobes, for example). It is going to be fascinating to see how these intertwined genetic histories resolve themselves into an evolutionary history.

Biodiversity remains high on the political agenda, especially following the 1992 Earth Summit in Rio de Janeiro. There are comparatively few microbial taxa named but their diversity cannot be doubted. Indeed, our appreciation of what is meant by the words 'microbial biodiversity' is still developing (see Giovannoni's chapter). Microbes are capable of almost any physiological process (see Dijkuizen's chapter), many of which are beyond the capacity of their multicellular descendants. The simple system of counting 'species' is only a single facet of the jewel of biodiversity. A greater appreciation of microbial evolution will surely help to understand just how rich and varied life truly is.

Modern systematics, through the study of evolution and the use of phylogenetic principles, is an exciting and vibrant field. It is the cornerstone of modern biology and will be of interest to both the specialist and general reader. In this fast moving field, the editors have placed the authors under a great deal of time pressure to deliver state-of-the-art chapters. We would like to thank them for producing such current and readable accounts so quickly.

D. McL. Roberts, P. Sharp, G. Alderson
and M. A. Collins August 1996

SOME ASPECTS OF THE BIOLOGY OF CELLS AND THEIR POSSIBLE EVOLUTIONARY SIGNIFICANCE

W. FORD DOOLITTLE

Canadian Institute for Advanced Research and Department of Biochemistry, Dalhousie University, Halifax, Nova Scotia, Canada

INTRODUCTION

Readers with a memory of earlier Society for General Microbiology symposium volumes may recognize the title of this essay. It is the same as that used by Roger Stanier, for his erudite but discursive introductory chapter in the 1970 (20th) volume (Stanier, 1970). That symposium: 'Organization and Control in Prokaryotic and Eukaryotic Cells' was about comparing and contrasting cells of the two sorts, and about how the one might have given rise to the other – issues still very much with us and central to this meeting of the Society, a quarter of a century later.

Stanier wrote at a crucial juncture in the development of our thinking about cell evolution. One 'fact', the existence of two and only two basic types of cellular organism, prokaryotes and eukaryotes, seemed solidly established. A second central evolutionary concept, that eukaryotic organelles of photosynthesis and respiration are the degenerate descendants of prokaryotic endosymbionts, anciently established within the eukaryotic cytoplasm, was surging towards acceptance and 'proof'. In addition, molecular biology (including molecular phylogenetic reconstruction) was about to replace electron microscopy and comparative biochemistry on the research agendas of cellular evolutionists.

We are again, I think, at a crucial juncture in the development of our field. Although little of what Stanier believed 25 years ago was actually wrong, we can no longer accept a simple prokaryote/eukaryote dichotomy. Although we do not doubt the bacterial origins of chloroplasts and mitochondria, we may be on the verge of subsuming the endosymbiont hypothesis within a broader chimaeric theory for the origin of eukaryotic cells. Once more a powerful new analytical methodology, this time based on the comparison of whole genome sequences, is about to overwhelm us all.

Here I will revisit the issues which Stanier addressed, presenting what I think to be the current consensus about them. I will then highlight the weak points in that consensus and make some suggestions about how the next synthesis might look. Many of the more contentious issues will be discussed with more detail by other contributors to this volume.

ISSUES FOR STANIER

Stanier (1970) concerned himself primarily with (i) defining sets of prokaryote-specific and eukaryote-specific 'least common denominators', traits by which organisms of one type could always be distinguished from those of the other, and (ii) elaborating schemes by which we might imagine cells of the first type had evolved into cells of the second.

Least common denominators

For prokaryote-specific traits, he saw the first task as well in hand, especially since the so-called 'blue-green algae', which formerly seemed like missing links between the two major cell types, had been brought firmly into the prokaryote camp (as 'cyanobacteria') by ultrastructural studies. Prokaryotes had anucleate cells without membrane-enclosed organelles of respiration or photosynthesis, divided by fission, not mitosis, and used peptidoglycan to strengthen their cell walls – unifying criteria he and van Niel had already identified in 1962 (Stanier & van Niel, 1962).

For the eukaryotic cell, Stanier demurred, common denominators were generally less certain, because of extremes of specialization between species, and between tissues and cell types in differentiated forms. However, recent developments in cell and molecular biology had identified in plastids and mitochondria 'a major new common denominator of eukaryotic cells which can be stated as follows. The eukaryotic cell contains two (or more) genetic systems, each housed and replicated in a separate intracellular structure, and each associated with a machinery of replication, transcription and translation, some elements of which are specific to that genetic system'.

He also saw in the complex cytoplasmic membrane and cytoskeletal systems of eukaryotes a host of eukaryote-specific common features. The former introduced complications into eukaryotic intracellular transport that are unknown to prokaryotes. The latter, with its special component actin, tubulin and intermediate filament proteins (also unknown from prokaryotes) conferred on eukaryotes the ability to effect directed intracellular translocation. Together, they made endocytosis (phagocytosis) possible.

Stanier believed it was the pressure to improve endocytotic ability which forged the complex internal workings of the eukaryotic cell, while endocytosis itself relieved eukaryotes of the need to evolve beyond simple forms of glycolysis for energy generation. They could instead eat bacteria. Endocytosis not only allowed eukaryotes to engulf bacteria as food but also on many occasions to engulf beneficial energy-producing prokaryotes which, escaping digestion, could make a more enduring contribution as endosymbionts.

How prokaryotes produced eukaryotes

Stanier believed that eukaryotes arose from prokaryotes because 'completely separate origins are improbable, even though most of the evidence of [prokaryote: eukaryote] homology are to be found only at the deepest level (i.e. the molecular one)'. He presented two scenarios.

The first scenario (autogenous origin or direct filiation), which had enjoyed general favour for most of the twentieth century, posited that mitochondria and plastids were (like the endomembrane and cytoskeletal system) further steps in the internally driven complexification process which produced primitive eukaryotes from advanced prokaryotes. The second scenario (the xenogenous origin, or serial endosymbiosis hypothesis) was an obvious extension of, and stimulus for, Stanier's views on the importance of endocytosis in eukaryote evolution. It was also informed by a review (Sagan, 1967 – now Margulis) published three years earlier – a review that became a book, *Origin of Eukaryotic Cells* – the book that turned many molecular biologists of my generation on to cell evolution.

Stanier (1970) preferred the second scenario, but in part for the wrong reasons – the apparent multiplicity of organellar (especially plastid) origins. Major groups of algae have differently pigmented oxygen-producing plastids which Stanier saw as the descendants of different cyanobacteria-like endosymbionts. He imagined an 'evolutionary shift from predation pure and simple, to predation combined with (or replaced by) endosymbiosis' taking place 'repeatedly, in many different lines of eukaryotes, and over a long evolutionary period'. Stanier also espoused a polyphyletic origin for mitochondria, arguing that respiration was not needed until long after the emergence of oxygenic photosynthesis and the divergence of eukaryotic lineages; thus it was 'probable that mitochondria were the last components established in the eukaryotic cell'.

The current view (see Gray & Spencer, this volume) is that modern mitochondria, and plastids, had but one primary origin each (with mitochondria first). This is not to say, however, that many early eukaryotic lines did not harbour many prokaryotic symbionts, any number of which might have made lasting (genetic) contributions to eukaryotic biology.

Stanier concluded his essay with an often-quoted disclaimer. He wrote 'it may have happened thus; but surely we will never know with certainty. Evolutionary speculation constitutes a kind of metascience, which has the same intellectual fascination for some biologists that metaphysical speculation possessed for some mediaeval scholastics. It should be considered a relatively harmless habit, like eating peanuts, unless it assumes the form of an obsession: then it becomes a vice'.

Nowadays, evolutionary speculations are as common in otherwise hard-nosed molecular biological manuscripts as peanut butter sandwiches in school lunches. Some are stale and some soggy, but as the chapters in this symposium volume will show, there is real progress and real science too.

THE CURRENT CONSENSUS

Our present beliefs about prokaryotes, eukaryotes and the evolutionary relationships between them are sketched in Fig. 1. These notions derive in very large part from comparing sequences of a single universal macromolecule, SSU rRNA (small subunit, or 16S/18S ribosomal RNA), which Woese began promoting as the 'ultimate molecular chronometer' in the mid-1970s (Woese, 1987).

Early results with 16S

This molecule first proved its worth to the rest of us in affirming the endosymbiont hypothesis: partial sequence data (T1 oligonucleotide 'catalogues') showed that cyanobacteria and proteobacteria were indeed closer to plastids and mitochondria than these were to the nuclei of the cells they inhabit (Gray & Doolittle, 1982). Sequence studies with cytochromes and ferredoxins had already pointed to this conclusion (Schwartz & Dayhoff, 1978), but provided incomplete proof: organellar proteins could track only organellar evolutionary history, and told us nothing about the nucleus.

This result would have been welcomed heartily by Stanier, but the next major conclusion from SSU rRNA analyses would have troubled him. Prokaryotes were, at least on the basis of rRNA sequence, of two fundamentally different types. The first type, the eubacteria, were what he and van Niel had in mind when they sought defining characters of prokaryotes. The second type, the archaebacteria, included cells which certainly lacked nuclei, but about which little else was known, either in 1970 or in 1977, when Woese and Fox (1977*a*) presented the first results of their massive oligonucleotide catalogue survey of SSU rRNAs.

Acceptance of the archaebacteria

There was widespread resistance to the claim of Woese that prokaryotes were of two fundamentally different kinds. Some argued that rRNA sequences constituted but a single character, not 1500 different ones – in no way enough to outweigh eubacterial/archaebacterial similarities in cell size and internal simplicity. Others did not accept (and still do not accept, see Margulis, 1992) the basic tenet of most molecular biologists studying evolution – that gene and gene product sequences are closer to the phylogenetic bone and the ultimate guide to evolutionary relationships, phenotypic differences traditionally used by taxonomists being of value only in that they can (in principle) be traced back to sequence differences.

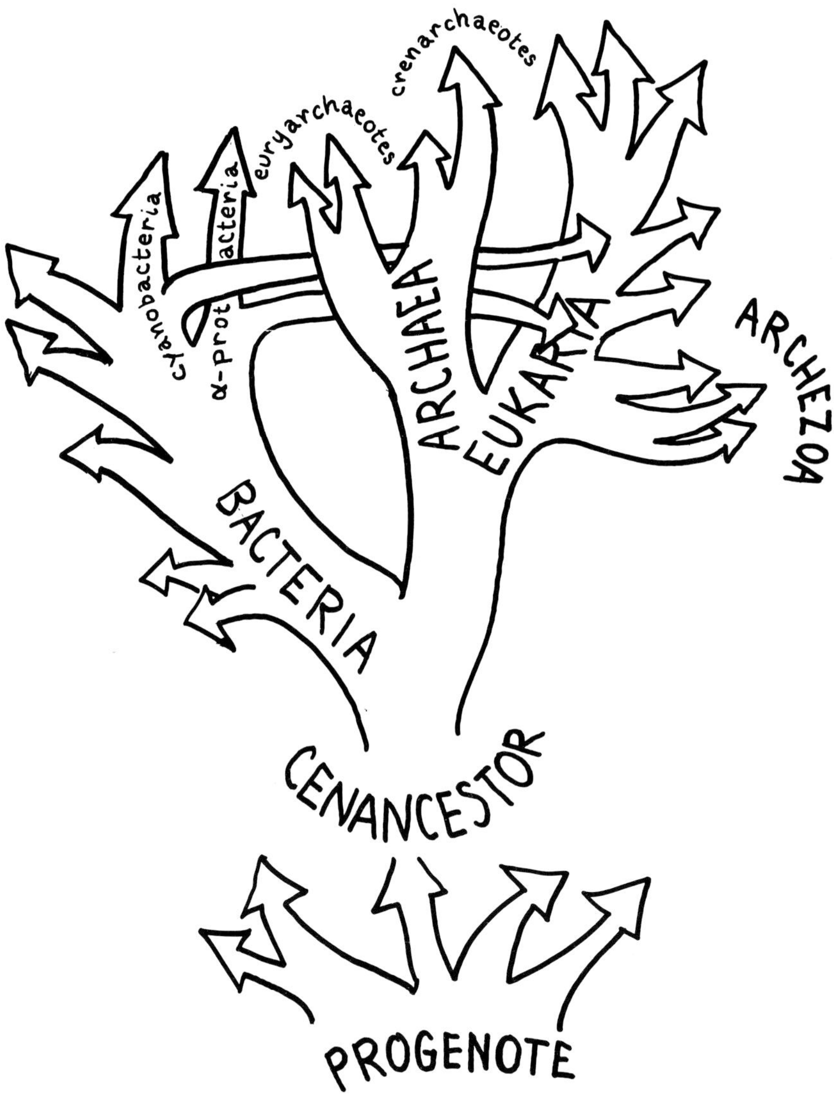

Fig. 1. The current consensus about cell evolution (Doolittle & Brown, 1994). The last common ancestor of all Life (the cenancestor) was a fairly sophisticated 'prokaryotic' cell, with much modern biochemistry and molecular biology. It descended from a more primitive, inefficiently organized ancestor (or series of ancestral quasi-cellular entities) called 'the progenote'. Other descendants of the progenote, indeed all cellular lineages but that of the cenancestor, went extinct, probably more than 3.5 billion years ago. Eubacteria diverged from the archaebacterial/eukaryotic clade at the base of the tree, but re-entered the eukaryotic cell as endosymbionts (mitochondria from α-proteobacteria, plastids from cyanobacteria) later, after the divergence from the main eukaryotic trunk of the Archezoa.

Work in the decade after Woese and Fox's announcement fleshed out the archaebacterial concept, providing us with lists of archaebacteria-specific traits, often displayed in articles like this one as evidence for the coherence of the group (Kates, Kushner & Matheson, 1993). Such traits include: the use of substances other than peptidoglycan (pseudomurein, form-stabilizing protein or glycoprotein S-layers) in cell walls, the use of isoprenoid ether-linked lipids (rather than fatty acid ester-linked lipids) in membranes, many 'signature' sequences and structures in rRNA and unique modification in tRNA. There are also some ways in which archaebacteria are 'just like' eubacteria (circular genomes, operons, several metabolic pathways), and still others in which they have a distinctly eukaryotic flavour (most notably characteristics of the transcriptional and translational apparatus: Keeling, Charlebois & Doolittle, 1994; Klenk & Doolittle, 1994; Keeling and Doolittle, 1995).

At times, this mix of features has itself been seen as somehow defining archaebacteria, or as indicating that these organisms are in some way transitional between eubacteria and eukaryotes. Posters advertising the second or third international gathering of molecular archaebacteriologists showed the archaebacteria splayed out to span the distance between eubacteria and eukaryotes, with halophiles and methanogens (now euryarchaeotes) next to the eubacteria and what were then called thermoacidophiles (now crenarchaeotes) cosied up to the eukaryotes. In a similar vein, Lake proposed in the early 1980s that some or all archaebacteria be called 'eocytes', and recognized as immediate ancestors of eukaryotes (Lake *et al.*, 1984).

There is no way that any living group is transitional between or ancestral to other living groups. The mixture of features does nevertheless begin to make sense in the context of the rooted universal tree (Fig. 1) – eubacteria-like features of archaebacteria are generally primitive, eukaryote-like features often derived.

Rooting the universal tree with deep paralogy

However, we can draw conclusions about which archaebacterial features are primitive and which derived only when we have a rooted tree. With a tree rooted as in Fig. 1, primitive features are homologous traits shared by (i) eubacteria and either archaebacteria or eukaryotes, or (ii) eubacteria and both archaebacteria and eukaryotes. In either of these situations reasoning by parsimony favours the presence of the shared character in the last common ancestor of all three lineages (the cenancestor). Derived features (not present in the cenancestor) are those found in only archaebacteria or only eukaryotes. Traits shared by archaebacteria and eukaryotes but not eubacteria – or found only in eubacteria – can be either primitive or derived: parsimony does not tell us.

Thus, rooting of the universal tree is very important to our understanding of the course of cellular evolution. And yet, in principle, no universal tree can be rooted, because roots can only be defined in terms of 'outgroups'. Rooting a tree of mammals, for instance, could be done with fish as an outgroup, but knowledge gained independently from the data being treed is needed to indicate that fish did indeed split from mammals before any internal mammalian divergence.

This conundrum allowed many evolutionary biologists to make assumptions about the nature of the root organism, the cenancestor, which seem ingenuous now. For instance, I suggested in 1978 (Doolittle, 1978) that this last common ancestor of all Life might have been more like modern eukaryotes than prokaryotes in gene and genome structure (introns, no operons, much 'junk') while Woese and Fox (1977*b*) expatiated on 'the progenote', a common ancestral state in which the machineries of replication, transcription and translation were still crudely fashioned, and promiscuous interlineage genetic exchange was the order of the day.

In 1989 Gogarten *et al.* (1989) and Iwabe *et al.* (1989) managed to overcome the rooting problem. Although no universal tree based on orthologues of a single gene in all taxa can be rooted, paired trees based on paralogues which are the product of gene duplications occurring in the lineage leading to the last common ancestor can be used to root each other. Thus, for example, a tree for orthologous vertebrate haemoglobin gene sequences (and thus an organism tree for all vertebrates) could be rooted using any vertebrate myoglobin gene, since myoglobin genes are paralogues of haemoglobin genes, produced by duplication before the divergence of vertebrates. This realization seems to have been original with Iwabe *et al.* (1989) and Gogarten *et al.* (1989), although Schwartz and Dayhoff (1978) had used an internal gene duplication, and a rather different logic, in rooting their now forgotten composite cytochrome/ferredoxin/5S rRNA universal tree (Schwartz & Dayhoff, 1978).

There were only two sets of paralogues with sequenced examples from at least one archaebacterium, one eubacterium and one eukaryote available in 1989: elongation factors EF-Tu and EF-G (all genomes have at least one each) and α- and β-subunits of F_1-ATPases. Both showed the archaebacteria to be the sister group to eukaryotes (Fig. 1).

Perhaps because it validated earlier feelings that archaebacteria were somehow missing links, the Gogarten–Iwabe rooting found quick acceptance in the archaebacterial research community. Woese, Kandler and Wheelis (1990), in particular, welcomed this conclusion, and incorporated it into an extended argument for elevating (from 'kingdom' to 'domain') the taxonomic status of eubacteria, archaebacteria and eukaryotes, and assigning them new names: Bacteria, Archaea and Eukarya. In this essay, a mixed naming strategy has been adopted. Thus 'Bacteria', 'Archaea' and 'Eukarya' are used as formal domain names, emphasizing the profound differences between members of the three groups, but 'archaebacteria', 'eubacteria' and 'eukaryotes' are used as informal names because they have become familiar, and because 'bacteria' would have denoted both prokaryotic domains in Stanier's day.

More important issues were beneath the war of words which followed the publication of Woese *et al.*, 1990 (Doolittle & Brown, 1994). The word prokaryote had come to be used necessarily to describe both a clade (group made up of a single ancestral species and all its descendent species) and a grade (level of cellular or organismal organization). It had been assumed by many that because archaebacteria lacked nuclei (and thus were of prokaryotic grade) they must necessarily share with eubacteria many features (peptidoglycan, no cytoskeleton, simple genomes, Shine/Dalgarno recognition sequences for ribosome/mRNA recognition, 'simple' (four subunit) RNA polymerases which function without transcription factors) which have traditionally been seen as prokaryotic rather than eukaryotic. However, many of these features could in principle be the products of contingent events occurring in the line leading to the first eubacterium, characteristics only of the eubacterial clade. If the tree is as in Fig. 1, actual experiments need to be done to find out which eubacterial features, because they were present in the cenancestor and not lost, also characterize the archaebacteria.

Confirming the root

The root also must be confirmed. As Gogarten discusses elsewhere in this volume, the ATPase genes used to root the tree in 1989 exhibit a complex family history, with both lateral transfer and deep paralogy. There is no certainty that the paired trees for α and β subunits (Iwabe *et al.*, 1989) truly track organismal lineages. The elongation factor genes seem safe in this regard, but in 1989 only two archaebacterial sequences were available, surely not enough on which to base nearly a decade of theorizing about cellular evolution.

Baldauf, currently in my laboratory, has recently constructed new elongation factor trees with more than 60 gene sequences, including nine archaebacterial EF-Tu's and 7 archaebacterial EF-G's. Gratifyingly, these trees support, with very high bootstrap values, the sisterhood of Archaea and Eukarya (S. Baldauf, W. F. Doolittle & J. Palmer, unpublished observations).

Brown, also in my laboratory, has obtained a similar result using a quite different set of data – sequences of the genes from aminoacyl-tRNA synthetases specific for isoleucine, valine and leucine (Brown & Doolittle, 1994). The aminoacyl-tRNA synthetases are particularly appealing for this purpose. Nagel and Doolittle (1991) had shown that these essential components of the translation system could be grouped by sequence into two classes (I and II, the same classes which can be defined by three-dimensional structure and reaction mechanisms). Within each class, all enzymes charging a given amino acid (e.g. isoleucine) group together, regardless of their source within the Archaea, Bacteria or Eukarya. Thus the gene duplication events which produced paralogous isoleucine and, for example, valine tRNA synthetase genes predated the cenancestor. And, thus, valine sequences can root universal trees for isoleucine, and vice versa. There will be many possible rootings involving pairs of aminoacyl-tRNA synthetases, as the data accumulate from genome projects.

Although data from both Brown and Baldauf are very clear on the root of the universal tree, with the deepest branching separating eubacteria from an Archaeal/Eukaryal clade, they differ in an important respect. The tRNA synthetase analysis indicates a holophyletic Archaea (all descendents of the ancestral archaebacterium being still called archaebacteria (Ashlock, 1971)). But the elongation factor data support, albeit weakly, a paraphyletic Archaea, with another group not called archaebacteria (i.e. the eukaryotes) arising from within the archaebacteria (specifically as sisters to crenarchaeotes). There are several things to be said about this latter result.

Paraphyletic or holophyletic Archaea

First, although the result appears to be friendly to the 'eocyte hypothesis' of Lake *et al.* (1984) this is partly because that hypothesis remains ill-defined. Lake proposed that archaebacterial 70S ribosomes shared structural features with eukaryotic 80S ribosomes, and then more specifically by 1984 that those of crenarchaeotes like *Sulfolobus* were especially eukaryotic in character (Lake *et al.*, 1984). He argued that eukaryotes might thus have arisen from within the sulphur-dependent thermoacidophilic archaebacteria, which he called 'eocytes' to honour this special relationship. At the same time, from similar data, he proposed that halophilic archaebacteria were sisters to eubacteria, and since both contain photosynthetic members, that this clade be named 'photocytes', although the underlying photosynthetic machineries are clearly homoplaseous (independently evolved but functionally convergent). The ribosome structure data have not held up, but in 1988 Lake drew a similar tree using rRNA sequences (Lake, 1988). This 'eocyte tree' showed a root within the archaebacteria, so that the deepest division separated methanogens and photocytes on the one hand from eocytes and eukaryotes on the other. The eocyte–eukaryote relationship was further endorsed by Rivera and Lake in 1992, because of a shared insertion in EF-1a (EF-Tu) sequences, but in the same analysis these authors abandoned photocytes and adopted a paraphyletic eukaryarchaeotes (Rivera & Lake, 1992).

Second, although only rooted trees can in principle decide the holophyly/paraphyly issue, the most data-rich unrooted trees (for ribosomal RNA and RNA polymerase) do usually show euryarchaeotes and crenarchaeotes as sisters (Woese, 1993; Barnes *et al.* personal communication; Klenk & Zillig, 1994; Iwabe *et al.*, 1991). A paraphyletic Archaea could only be created placing the cenancestor (the root) at the base of one of these two groups, but the other would then be seriously paraphyletic, spawning both Bacteria and Eukarya as sisters, something no one has proposed on the basis of any other data. Furthermore, Barns and collaborators (S. Barns, C. Delwiche, J. Palmer & N. Pace, personal communication) have identified (through cloned rDNA) new hyperthermophilic archaebacteria branching below the euryarchaeote/crenarchaeote split. To restore the 'eocyte' tree, several nodes would need to be moved, not just the single attachment point of the existing crenarchaeotes.

Nevertheless (thirdly) much of what has been said about the phenotypic 'unity' or 'coherence' of the archaebacteria is not actually germane to the paraphyly/holophyly question. Several of the 'unique' features of Archaea and the general similarity of sequences of homologous proteins within the Archaea can be interpreted as shared primitive traits, no more relevant to the issue of whether Eukarya arose from within crenarchaeotes (and then diverged very rapidly) than the shared scaly cold-bloodedness of the reptiles is relevant to the origin (and rapid divergence) from within them of birds.

Finally (fourthly), incongruent trees have to be accepted until the causes of incongruency are known, and patience about final answers will be required. Too many have proposed radical revisions of existing views on organismal phylogeny from limited studies of single molecules or, conversely, refrained from publishing contradictory conclusions because they do not fit existing views. Trees will differ because (i) alignments will always require judgment, (ii) phylogenetic algorithms are imperfect, and have been pushed to the limits of their resolution, and (iii) some genes do have incongruent histories. Trees are often presented as if they were themselves facts, but in truth they are only summaries and analyses of data, on which to base inferences and future speculation.

Implications for the cenancestor

Certainly the last has not been written of the Archaeal paraphyly/holophyly issue, and there are variant views concerning the nature of the eukaryotic genome and on the extent of its chimaerism, in particular. However, it is unlikely that there will be a return to the belief espoused in an effort to rationalize the phylogenetic distribution of introns, that the deepest division in the tree of Life separates all prokaryotes from all eukaryotes (Doolittle, 1978; Darnell, 1978). Thus, even if the position of Cavalier-Smith (1987), who argued on biochemical grounds that the universal root lies in the eubacteria is accepted (Cavalier-Smith, 1987), we are obliged by parsimony to believe that homologous features shared by Archaea and Bacteria are primitive ones, present in the cenancestor.

The cenancestor, then, had a circular genome, no introns, genes linked into operons (often the same genes in the same order as they are linked today), 70S ribosomes with most of the modern suite of ribosomal proteins, the 'universal' genetic code, modern replication and recombination functions and many enzymes of modern catabolic and anabolic pathways (Gogarten & Kibak, 1992; Doolittle & Brown, 1994). It was in no way the primitive intron-ridden ur-eukaryote that I had imagined nor the loose consortium of disparate genes of varied provenance and inefficient molecular machineries that Woese (1987) described as the progenote.

There must have been a progenote somewhere at some time, however; modern biochemistry and molecular biology did not arise full-blown from the first self-replicating ribozyme. There must have as well been many lineages which diverged from that which produced the cenancestor. It may seem surprising, given the 'advanced' character of the cenancestor, that none of the other lineages left survivors, but there are no data from other planets by which we might judge how unexpected this really is.

There is also no easy way to evaluate the claim that the cenancestor was a hyperthermophile, a romantic notion consistent with our beliefs about Earth's turbulent early years, and popular with the press. Parsimony makes this seem the best guess (Woese, 1987), given that thermophiles comprise the deepest few branches within both Bacteria and Archaea. However, parsimony is no infallible guide. Evolution has not followed the shortest path, and more importantly, high temperature adaptations shown by Bacteria and Eukarya may not be homologous, in the way that identically ordered operons surely are. There are performable but difficult tests of the thermophilic ancestor hypothesis. If the cenancestor were thermophilic, then enzymes in mesophilic bacteria and archaea would have achieved their new temperature optima by independent evolutionary paths, and be less similar in important structural features than enzymes in primitively thermophilic species of the two domains; the opposite is expected if the cenancestor preferred low temperatures.

There is also no certainty about when the cenancestor lived. Schopf will have described elsewhere in this volume the unarguable evidence for sophisticated cellular life on this planet more than 3.5 billion years ago, and the strong evidence that the first fossils were cyanobacteria. This has always posed a challenge to those of us who rely on the molecular data, because cyanobacteria do not branch deeply within the Bacteria. There must have been quite a lot of cellular evolution (the divergence leading to Archaea, *Aquifex*, *Thermotoga*, and several nonthermophilic eubacterial lineages) before the appearance of these first fossils.

Archaebacteria looking forward

Accepting the consensus of the tree in Fig. 1, it would be expected that some features which we previously thought eukaryote-specific will in fact be more generally characteristic of the Archaeal/Eukaryal clade, having developed before the separation of these two domains. From our own phylogenetic position these will look like prokaryotic 'pre-adaptations' to the eukaryotic condition. There is already an impressive list, including TATA-box-like promoters, complex multi-subunit RNA polymerases (with many subunits not found at all in eubacteria), homologues of transcription factors TBP and TFIIB and translation factors eIF-5A, eIF-2, eIF-1A and eIF-2B, and several eukaryotic ribosomal proteins which eubacterial ribosomes altogether lack (see Keeling & Doolittle, 1995, for references).

We should also anticipate the discovery of archaeal homologues of proteins of the eukaryotic cytoskeleton. For actin and tubulin, eubacterial candidates have already been nominated (Bork, Sander & Valencia, 1992; Erickson, 1995), so we would expect the archaebacterial versions of these proteins, when found, to look more convincingly eukaryotic. Unless we believe that eukaryote-specific proteins arose from random sequences (well after the time when prokaryotic genes and genomes had achieved their current complexity and efficiency of function) we must accept that most important eukaryotic proteins will have homologues in prokaryotes, and if Archaea and Eukarya are truly sisters, show especial closeness to their archaebacterial counterparts. The important task is not looking for homologues, but tracking events of gene duplication and changes in function, and associating these with the origins of major groups.

Archezoa looking back

From the other direction, comparative analyses of molecular and cellular form and function in the most disparate Eukarya should reveal intimations of their common ancestral 'proto-eukaryotic' state. Cavalier-Smith (1983) proposed that three or four protist phyla which now lack mitochondria never had them, but represented a stage in cell evolution implicit (but largely glossed-over) in the endosymbiosis scenarios of Margulis (1970). Endorsing arguments of Stanier (1970) for the crucial role for the eukaryotic endomembrane and cytoskeletal systems, he ventured that eukaryotes had already evolved a nucleus, 9+2 flagella and the capacity for phagocytosis and intracellular digestion by lysosomes, before acquiring endosymbionts. Thus the amitochondriate archamoebae, metamonads, microsporidia and (possibly) parabasalia (together Archezoa) diverged from the 'main line' of eukaryotic evolution (the line leading to us) before the endosymbiotic events whose products Stanier (and Margulis) had seen as vital to the definition of the eukaryotic condition.

It must have been especially gratifying to Cavalier-Smith (and it was exciting for the rest of us) when, six years later, SSU rRNA analyses (Vossbrinck *et al.*, 1989) showed the microsporidian *Vairimorpha necatrix* to define the earliest branch in the Eukarya, and, shortly thereafter Sogin *et al.* (1989) placed the metamonad *Giardia lamblia* even deeper in the tree. These two groups, together with parabasalians (*Trichomonas*) and (sometimes) the entamoebid *Entamoeba histolytica* usually diverge before all others in eukaryotic trees based on rRNA or proteins (elongation factors, triose-phosphate isomerase, tubulins, glyceraldehyde phosphate dehydrogenase; A. Roger and P. J. Keeling, personal communication), diplomonads and microsporidia vying for low position (there are as yet no data from archamoebae). *Giardia* and *Vairimorpha* rDNAs show unusually high (71 mol%) and low (41 mol%) G+C compositions, conditions thought to mislead treeing algorithms. Very recently several authors (for example, Galtier & Guoy, 1995) have described methods for countering this effect, methods which show *Vairimorpha* as the deepest eukaryotic branch.

The lack of mitochondria is not the only seemingly primitive archezoal feature. Archamoebae, metamonads and microsporidia lack peroxisomes, hydrogenosomes and well defined Golgi dictyosomes (Cavalier-Smith, 1993). As far as is known they (and parabasalia) sport 70S ribosomes and have no spliceosomal introns in their DNA, like their presumed archaebacterial ancestors. One of the clearest messages that the current phylogenetic consensus has for experimentalists is that we need to know much more about archezoa, their physiology, biochemistry and molecular biology.

CHINKS IN THE ARMOUR OF THE CONSENSUS

Figure 1 represents the current consensus: the general understanding against and around which each of us who holds a variant view will typically construct his/her arguments – the position which each of us who hopes to establish a new conceptual framework feels he/she must overthrow. There are many variant views, several of which will be presented in this volume. Although most heat has been generated by the Archaeal paraphyly/holophyly issue, proposals for what I call radical chimaerism are by far the more threatening, since at bottom they question our ability to construct organism trees from gene trees.

Chimaerism

Chimaeric schemes have been most articulately formulated by Sogin (1991); Patterson and Sogin (1992); Zillig, Palm and Klenk (1992); and Golding and Gupta (1995). In each case, incongruence of trees constructed from sequences of different macromolecules was the stimulus. Sogin was perturbed principally by the fact that rooted EF and ATPase trees (and distance relationships for many other proteins) supported the sisterhood of Archaea and Eukarya, but archaebacterial and eubacterial rRNAs are the more similar in pairwise comparisons across the three domains. Zillig's doubts sprang from phylogenetic reconstructions of RNA polymerase, which (at that time, but see Iwabe *et al.*, 1991) showed eukaryotic RNAPII and III as immediate relatives of Archaeal RNAP, but placed eukaryotic RNAPI with the eubacterial polymerase. They deepened with the realization that several eukaryotic genes of intermediary metabolism also show stronger similarity to eubacterial homologues than to archaebacterial homologues – specifically glyceraldehyde phosphate dehydrogenase, L-malate dehydrogenase, phosphoglycerate kinase and aspartate amino transferase. The list of Golding and Gupta includes this last enzyme, plus the hsp70 protein, ferredoxin, glutamate dehydrogenase, glutamine synthetase, deoxyribodipyrimidine photolyase, and pyrolline-5-carboxylate reductase. Although in some cases such results can be explained by isolated instances of relatively recent lateral transfer, or do not hold up when taxa other than those they selected are included in the analysis, there does seem to be some sort of signal here.

The solution proposed by Patterson and Sogin (1992) is the most radical – that eukaryotes arose through the capture of a more-or-less modern archaebacterium by a '*proto-eukaryote*' of a quite unusual sort. This 'host' had developed at least a rudimentary version of the eukaryotic cytoskeleton, but retained a primitive ('progenote-like') fragmented genome made of RNA, and made extensive use of splicing and other RNA-catalysed processes which still dominate eukaryotic information transfer. These processes (and the underlying genes), the rRNAs themselves, and genes for cytoskeletal components are the proto-eukaryotes' legacy in modern eukaryotes, while most enzymes (of metabolism and information handling) are of archaebacterial provenance.

The conjectures of Zillig *et al.* (1992) and those of Golding and Gupta (1995) are similar in form. Both see the eukaryotic nucleus as having been produced by the fusion of an already advanced (contemporary form) archaebacterial cell with an already advanced eubacterium. Zillig *et al.* (1992) make no specific claims about the eubacterial lineage involved, but Golding and Gupta (1995) described it as 'Gram negative', while detecting a special affinity between 'Gram positives' and certain archaebacteria (although 'Gram negatives' are not monophyletic in rRNA-based trees, and 'Gram positives' are a restricted group). Zillig *et al.* (1992) and Golding and Gupta claimed that the kind of genomic chimaerism needed to explain such data is different from that we already accept within the consensus as shown in Fig. 1, under the aegis of the endosymbiont hypothesis. Zillig *et al.* (1992) noted the presence of a eubacterial type GAPDH in the archezoan *Giardia lamblia*, which by the consensus view never had mitochondria. Golding and Gupta (1995), who based their case in large part on sequences of hsp70, noted that *Giardia* also has this protein, while other eukaryotes show both cytosolic and organellar forms, only the latter exhibiting clear specific affinity to proteobacteria or cyanobacterial proteins. There must have been an earlier influx of eubacterial genes.

Such a two-step scheme, with the mitochondrial endosymbiont finding itself in a host whose genome was already chimaeric, thanks to some earlier and possibly altogether different type of event, is not illogical or unappealing. One feature of the consensus in particular seems to drive to this conclusion – the divergence of archezoal lineages from the main trunk of the eukaryotic tree prior to the acquisition of mitochondria. If this were not so, if archezoa once had mitochondria, then other less radical (in terms of evoking multiple events of chimaerism) schemes might still get many votes.

Recasting the endosymbiont hypothesis

Entamoeba histolytica is an amitochondriate protist lacking peroxisomes, rough endoplasmic reticulum and Golgi dictyosomes. Some analyses have placed it at the base of the eukaryotes (in the Archezoa), although rRNA trees have *Entamoeba* well above many protists with mitochondria and 'typical' eukaryotic endomembrane systems – as if its mitochondria had been secondarily lost. If this were so, Clark and Roger (1995) reasoned, some genes of mitochondrial origin might still be in the nuclear genome. They found (by PCR and sequencing) both a gene for the inner mitochondrial membrane protein pyridine nucleotide transhydrogenase and a cpn60 gene (cpn60 is a chaperonin which refolds proteins after their transport into organelles) in *Entamoeba* nuclear DNA. They took these genes as remnants of a vanished mitochondrial genome (genes now possibly serving in the construction of reduced mitochondrial relic particles lacking DNA).

Emboldened by this, Clark and Roger (1995) and Keeling (personal communication) have begun to look for mitochondrial relic genes in the nuclear genomes of more deeply diverging protist lineages, lineages about whose primitive amitochondriate condition we thought ourselves more certain. Roger (personal communication) has found cpn60 in the parabasalian *Trichomonas vaginalis*. Soltys and Gupta (1994) have already presented immunological evidence for its presence in *Giardia*. Additionally, these two protists seem to contain Fe-SODs (superoxide dismutases) most like those found in proteobacteria and cyanobacteria.

Keeling (personal communication), motivated by the possibility that Archaea lack triose-phosphate isomerase (Danson, 1993), reasoned that this key catalyst in the Embden–Meyerhof pathway might be of endosymbiotic origin in eukaryotes – even though not a 'mitochondrial enzyme'. After adding new sequences to the eubacterial dataset, he can conclude with reasonable certainty that this eukaryotic cytosolic enzyme derives from the same eubacterial 'phylum' as does the mitochondrion. Moreover, among these eukaryotic cytosolic enzymes are two *Giardia* sequences. A similar conclusion was drawn by Markos, Miretsky and Muller (1993) and Henze *et al.* (1995) for another key cytosolic enzyme of glycolysis, glyceraldehyde phosphate dehydrogenase. Although the exact eubacterial source(s) will be very difficult to untangle for this complex dataset, again the eukaryotic sequences include enzymes from *Giardia* and *Trichomonas*.

Assume that the implications of these preliminary observations hold up, and thus that important eukaryotic cytosolic enzymes are of eubacterial origin (perhaps even specifically α-proteobacterial origin), even when these enzymes have nothing directly to do with mitochondrial functions, and even in 'archezoan' protists thought never to have harboured the α-proteobacterial endosymbiont which became the mitochondrion. What can be done to make sense of this?

Imagine that all surviving eukaryotes once had premitochondrial endosymbionts, but that the deepest branching lineages have lost them, secondarily (and perversely, given how misleading this has been to cell evolutionists), i.e. what are now called 'archezoa' are not that, even though there could well have once existed true primitively symbiont-free eukaryotes. The rationalization of the extinction of these 'real' archezoa can be made by assuming that the acquisition of endosymbionts was such a spur to eukaryote evolution that it swept all before it (Margulis, 1970).

Furthermore, one might imagine that the endosymbiont contributed many genes to the nuclear genome during the first few thousands or millions of years of eukaryote history, to produce a 'fusion' between invading eubacterial (mitochondrial) and resident archaebacterial ('host') genomes. Loss of an independently replicating endosymbiont DNA and full respiratory function would then have been the usual course – only one lineage of the four or more primary eukaryotic branchings retained both and was thus 'preadapted' to live in an oxygen-rich world, although others may have retained DNA-less organellar derivatives of the original endosymbiont (for example, hydrogenosomes; Finlay & Fenchel, 1989; Muller, 1993). There would be no way, from data available now, to distinguish between such 'genomic fusion' through endosymbiosis followed by gene transfer, and some singular event involving the coming together of one archaebacterial and one eubacterial cell. However, the former formulation might be preferred because there are many models now of endosymbiotic relationship involving anaerobic eukaryotic hosts (Fenchel, this volume).

The mutual advantages that symbiont and host confer on each other need to be rethought. The molecular data supporting the α-proteobacterial origin of mitochondria do not demand that symbionts spread because they brought the gift of aerobic respiration to proto-eukaryotic anaerobes which were beginning to lose ground in an oxygen-intoxicated world. Woese (1987), for instance, suggested that the symbiont might have been a photosynthetic *Rhodobacter*-like cell, which later evolved into a respiratory organelle, just as certain free-living respiring eubacteria are thought to have evolved (Broda, 1975, 'conversion hypothesis') from photosynthetic α-proteobacterial ancestors. Fenchel and Bernard (1993) have recently described 'purple protists' of just the sort that Woese had imagined.

Searcy (1992) has elaborated still another endosymbiont story, focusing on sulphur metabolism: the symbiont was an α-proteobacterium making a living by oxidizing H_2S provided by the host cell, which used sulphur to oxidize exogenous organic compounds. This host might have been quite like some modern archaebacteria. Searcy nominated *Thermoplasma acidophilum* for the honour, because of its sulphur metabolic activities but even more because it lacks a cell wall, and appears to have some form of cytoskeleton. (Note, however, that *Thermoplasma* is a euryarchaeote, not the crenarchaeote that most formulations of a specific origin of the proto-eukaryote from within the Archaea would demand.)

THE PROMISE AND THREAT OF GENOMICS

Peanuts are often just snacks before supper. Such speculations based on limited quantifiable data whet our appetites for the multi-course meal of genome sequences about to be served, but by the time we get to the dessert of comparative genome analysis, we will have forgotten many of them. The courses we cell evolutionists will remember best will have been whole genome sequences (or even extensive random sequencing surveys) for eubacteria, archaebacteria and small-genomed archezoa.

Microsporidia, archezoa by most analyses, have small genomes (some less than 10 Mbp; Biderre *et al.*, 1994). Several are pathogens of humans or of animals humans eat, so work on them may be supported even at the current nadir in public enthusiasm for knowledge unconnected to profit. If some sort of 'genome fusion', or extensive gene transfer from endosymbionts were really involved at the root of the eukaryotes, then microsporidian genes should fall into two classes; specifically archaebacteria-like and specifically eubacteria-like. There should be strong statistical support for this division of genes into two classes: at the moment, the possibility that many of the incongruencies between trees used as the springboard for chimaeric theories reflect inadequate sampling of taxa or limitations in tree construction algorithms should not be ruled out. It should be possible to find out whether genes that are of one ancestry rather than the other belong to certain functional types or are often linked in eubacterial and archaebacterial genomes. It should also be possible to find out whether these smallest and most divergent (from our own) eukaryotic genomes retain vestiges of the operon-type organization of genes which characterized their prokaryotic ancestors, of whatever type. Surely one of the biggest mysteries of genome evolution is the abandonment by eukaryotes of this form of genetic order.

The bad thing about the genomics dinner is that we may not all get seats at the table. Economics of scale (often overstated) and the rhetoric of genomics funding (frequently overblown), together with the current drive to tie research more directly to wealth generation, will concentrate much sequencing activity in a few centres with strong commercial ties. Although we might imagine a free flow of sequence data from the sequencing mills to the personal computers of molecular evolutionists around the world, and indeed the first two prokaryotic genomes have been made available with remarkable speed and generosity, there are serious reasons to worry about the concentration of sequencing power. The better publicly funded centres will naturally want to do more than 'just sequence', and will increasingly seek to add value to their data by extensive annotation, phylogenetic and functional analyses – all inexpensive compared to the sequencing itself, and thus minor additions to the budget of a major centre. Genome sequencing projects in privately funded centres may go altogether unpublished, but governments will not likely fund parallel public efforts on the same organisms. The rest of us, molecular evolutionists doing 'little science' in universities, could be left with peanuts.

REFERENCES

Ashlock, P. D. (1971). Monophyly and associated terms. *Systematic Zoology*, **20**, 63–9.

Biderre, C., Pages, M., Metenier, G., David, D., Bata, J., Prensier, G. & Vivares, C. P. (1994). On small genomes in eukaryotic organisms: molecular karyotypes of two microsporidian species (Protozoa), parasites of vertebrates. *Compte Rendu Academie Science III*, **317**, 399–404.

Bork, P., Sander, C. & Valencia, A. (1992). An ATPase domain common to prokaryotic cell cycle proteins, sugar kinases, actin, and hsp70 heat shock proteins. *Proceedings of the National Academy of Sciences, USA*, **89**, 7290–4.

Broda, E. (1975). *The Evolution of the Bioenergetic Processes*. Pergamon Press, Oxford, 220 pp.

Brown, J. R. & Doolittle W. F. (1994). Root of the universal tree of life based on ancient aminoacyl-tRNA synthetase gene duplications. *Proceedings of the National Academy of Sciences, USA*, **92**, 2441–5.

Cavalier-Smith, T. (1983). A 6-kingdom classification and a unified phylogeny. In *Endocytobiology II,* Schwemmler, W. & Schenk, H. E. A., eds., pp. 1027–34. De Gruyter, Berlin.

Cavalier-Smith, T. (1987). The origin of cells: a symbiosis between genes, catalysts and membranes. *Cold Spring Harbor Symposium in Quantitative Biology*, **52**, 805–24.

Cavalier-Smith, T. (1993). Kingdom Protozoa and its 18 Phyla. *Microbiological Reviews*, **57**, 953–94.

Clark, C. G. & Roger, A. J. (1995). Direct evidence for secondary loss of mitochondria in *Entamaoeba histolytica. Proceedings of the National Academy of Sciences, USA*, **92**, 6518–21.

Danson, M. J. (1993). Central metabolism of the archaea. In *The Biochemistry of Archaea (Archaeobacteria)*, Kates, M., Kushner, D. J. & Matheson, A. T., eds., pp. 1–24. Elsevier, London.

Darnell, J. E., Jr (1978). Implications of RNA splicing in evolution of eukaryotic cells. *Science*, **202**, 1257–60.

Doolittle, W. F. (1978). Gene-in-pieces, were they ever together? *Nature*, **272**, 581–2.

Doolittle, W. F. & Brown, J. R. (1994). Tempo, mode, the progenote and the universal root. *Proceedings of the National Academy of Sciences, USA*, **91**, 6721–8.

Erickson, H. P. (1995). FtsZ, a prokaryotic homolog of tubulin? *Cell*, **80**, 367–70.

Fenchel, T. & Bernard, C. (1993). A purple protist. *Nature*, **362**, 300.

Finlay, B. J. & Fenchel, T. (1989). Hydrogenosomes in some anaerobic protozoa resemble mitochondria. *FEMS Microbiology Letters*, **65**, 311–14.

Galtier, N. & Guoy, M. (1995). Inferring phylogenies from DNA sequences of unequal base composition. *Proceedings of the National Academy of Sciences, USA*, **92**, 11317–21.

Gogarten, J. P. & Kibak, H. (1992). The bioenergetics of the last common ancestor and the origin of the eukaryotic endomembrane systems. In *The Origin and Evolution of the Cell,* Hartman, H. & Matsuno, K., eds., pp. 131–54. World Scientific, Singapore.

Gogarten, J. P., Kibak, H., Dittrich, P., Taiz, L., Bowman, E. J., Bowman, B. J., Manolson, M. F., Poole, R. J., Date, T., Oshima, T., Konishi, J., Denda, K. & Yoshida, M. (1989). Evolution of the vacuolar H+-ATPase: implications for the origin of eukaryotes. *Proceedings of the National Academy of Sciences, USA*, **86**, 6661–5.

Golding, G. B. & Gupta, R. S. (1995). Protein-based phylogenies support a chimeric origin for the eukaryotic genome. *Molecular Biology of Evolution*, **12**, 1–6.

Gray, M. W. & Doolittle, W. F. (1982). Has the endosymbiont hypothesis been proven? *Microbiological Reviews*, **46**, 1–42.

Henze, K., Badr, A., Wettern, M., Cerff, R. & Martin, W. (1995). A nuclear gene of eubacterial origin in *Euglena gracilis* reflects cryptic endosymbioses during protist evolution. *Proceedings of the National Academy of Sciences, USA*, **92**, 9122–6.

Iwabe, N., Kuma, K. I., Hasegawa, M., Osawa, S. & Miyata, T. (1989). Evolutionary relationship of archaebacteria, eubacteria and eukaryotes inferred from phylogenetic trees of duplicated genes. *Proceedings of the National Academy of Sciences, USA*, **86**, 9355–9.

Iwabe, N., Kuma, K. I., Kishino, H., Hasegawa, M. & Miyata, T. (1991). Evolution of RNA polymerases and branching patterns of the three major groups of archaebacteria. *Journal of Molecular Evolution*, **32**, 70–8.

Kates, M., Kushner, D. J. & Matheson, A. T. (1993). *The Biochemistry of Archaea (Archaeobacteria)*. 582 pp. Elsevier, London.

Keeling, P. J. & Doolittle, W. F. (1995). Archaea: narrowing the gap between prokaryotes and eukaryotes. *Proceedings of the National Academy of Sciences, USA*, **92**, 5761–4.

Keeling, P. J., Charlebois, R. L. & Doolittle, W. F. (1994). Archaebacterial genomes: eubacterial form and eukaryotic content. *Current Opinions in Genetic Development*, **4**, 816–22.

Klenk, H.-P. & Doolittle, W. F. (1994). Archae and eukaryotes versus bacteria? *Current Biology*, **4**, 920–2.

Klenk, H.-P. & Zillig, W. (1994). DNA-dependent RNA polymerase subunit B as a tool for phylogenetic reconstruction: branching topology of the archaeal domain. *Journal of Molecular Evolution*, **38**, 420–32.

Lake, J. A. (1988). Origin of the eukaryotic nucleus by rate-invariant analysis of rRNA sequences. *Nature*, **331**, 184–6.

Lake, J. A., Henderson, E., Oakes, M. & Clark, M. W. (1984). Eocytes: a new ribosome structure indicates a kingdom with a close relationship to eukaryotes. *Proceedings of the National Academy of Sciences, USA*, **81**, 3786–90.

Margulis, L. (1970). *Origin of Eukaryotic Cells*. 349 pp. Yale University Press, New Haven.

Margulis, L. (1992). Biodiversity: molecular biological domains, symbiosis and kingdom origins. *BioSystems*, **27**, 39–51.

Markos, A., Miretsky, A. & Muller, M. (1993). A glyceraldehyde-3-phosphate dehydrogenase with eubacterial features in the amitochondriate eukaryote, *Trichomonas vaginalis*. *Journal of Molecular Evolution*, **37**, 631–43.

Muller, M. (1993). The hydrogenosome. *Journal of General Microbiology*, **139**, 2879–89.

Nagel, G. M. & Doolittle, R. F. (1991). Evolution and relatedness in two amino-acyl tRNA synthetase families. *Proceedings of the National Academy of Sciences, USA*, **88**, 8121–5.

Patterson, D. J. & Sogin, M. L. (1992). Eukaryotic origins and protistan diversity. In *The Origin and Evolution of the Cell*, Hartman, H. & Matsuno, K., eds., pp. 13–46. World Scientific, Singapore.

Rivera, M. C. & Lake, J. A. (1992). Evidence that eukaryotes and eocyte prokaryotes are immediate relatives. *Science*, **257**, 74–6.

Sagan, L. (1967). On the origin of mitosing cells. *Journal of Theoretical Biology*, **14**, 225–75.

Schwartz, R. M. & Dayhoff, M. O. (1978). Origins of prokaryotes, eukaryotes, mitochondria and chloroplasts. *Science*, **199**, 395–403.

Searcy, D. G. (1992). Origins of mitochondria and chloroplasts from sulfur-based symbioses. In *The Origin and Evolution of the Cell*, Hartman, H. & Matsuno, K., eds., pp. 47–78. World Scientific, Singapore.

Sogin, M. L. (1991). Early evolution and the origin of eukaryotes. *Current Opinions in Genetic Development*, **1**, 457–63.

Sogin, M. L., Gunderson, J. H., Elwood, H. J., Alonso, R. A. & Peattie, D. A. (1989). Phylogenetic significance of the kingdom concept: an unusual eukaryotic 16S-like ribosomal RNA from *Giardia lamblia*. *Science*, **243**, 75–7.

Soltys, B. J. & Gupta, R. S. (1994). Presence and cellular distribution of a 60-kDa protein related to mitochondrial hsp60 in *Giardia lamblia*. *Journal of Parasitology*, **80**, 580–90.

Stanier, R. Y. (1970). Some aspects of the biology of cells and their possible evolutionary significance. In *Organization and Control in Prokaryotic and Eukaryotic Cells: 20th Symposium of the Society for General Microbiology*, Charles, H. P. and Knight, B. C. J. G., eds., pp. 1–38. Cambridge University Press, Cambridge.

Stanier, R. Y. & van Niel, C. B. (1962). The concept of a bacterium. *Archives in Microbiology*, **42**, 17–35.

Vossbrinck, C. R., Maddox, J. V., Friedman, B. A., Debrunner-Vossbrinck & Woese, C. R. (1989). Ribosomal RNA sequence suggests microsporidia are extremely ancient eukaryotes. *Nature*, **362**, 411–14.

Woese, C. R. (1987). Bacterial evolution. *Microbiological Reviews*, **51**, 221–71.

Woese, C. R. (1993). The archaea: their history and significance. In *The Biochemistry of Archaea (Achaeobacteria)*, M. Kates, D. J. Kushner and A. T. Matheson, eds., pp. vii–xxv. Elsevier Science Publishers, Amsterdam.

Woese, C. R. & Fox, G. E. (1977*a*). Phylogenetic structure of the prokaryotic domain: the primary kingdoms. *Proceedings of the National Academy of Sciences, USA*, **74**, 5088–90.

Woese, C. R. & Fox, G. E. (1977*b*). The concept of cellular evolution. *Journal of Molecular Evolution*, **10**, 1–6.

Woese, C. R., Kandler, O. & Wheelis, M. L. (1990). Towards a natural system of organisms: proposal for domains Archaea, Bacteria and Eukarya. *Proceedings of the National Academy of Sciences, USA*, **87**, 4576–9.

Zillig, W., Palm, P. & Klenk, H.-P. (1992). A model of the early evolution of organisms: the arisal of the three domains of life from the common ancestor. In *The Origin and Evolution of the Cell*, Hartman, H. and Matsuno, K., eds., pp. 47–78. World Scientific, Singapore.

ARE THE OLDEST FOSSILS CYANOBACTERIA?

J. WILLIAM SCHOPF

IGPP Center for the Study of Evolution and the Origin of Life, Department of Earth and Space Sciences, and Molecular Biology Institute, University of California, Los Angeles 90095, USA

INTRODUCTION

The recent discovery of cellularly preserved filamentous microorganisms in the nearly 3500 Ma-old (million year-old) Apex chert of northwestern Western Australia (Schopf, 1992*a*, 1993) provides significant new evidence of the antiquity of biological organization. Because the 11 species of minute fossil filaments recognized in this deposit constitute the oldest unquestionable evidence of life now known, their biological relationships are of special interest. With what groups might they be allied? Although firm assignment is difficult, all of the Apex microorganisms appear to be prokaryotes, and nearly two-thirds of the taxa seem comparable to oscillatoriacean cyanobacteria, a suggested affinity consistent with mineralogical, paleoecological, and carbon isotopic data and supported by the cellular morphology and size range of the Apex microorganisms, their inferred mechanism of cell division, and their mode of preservation and sequence of cell degradation (Schopf, 1992*a*, 1993, 1994*a*, 1995*a*).

This suggested cyanobacterial affinity, however, is at odds with expectations. Cyanobacteria are morphologically and physiologically *advanced* prokaryotes: rRNA phylogenies consistently place this group near or at the apex of the eubacterial lineage (Woese, 1987), and all cyanobacteria are capable both of oxygenic photosynthesis and aerobic respiration, multicomponent metabolic processes that are complicated and highly evolved (Broda, 1975; Schopf, 1995*b*). It hardly seems plausible that evolution could have proceeded so far, so fast, so early. Moreover, for this affinity to be correct the cyanobacterial lineage would have to have survived for literally *billions* ($>10^9$) of years, a possibility that does not mesh well with other theories of the evolutionary process. The usual rules of evolution (exemplified by the familiar progression from photosynthetic protists to flowering plants, from primitive invertebrates to higher mammals) are firmly known: speciation, specialization, extinction (Schopf, 1994*b*,*c*). Groups rise to dominance, thrive, are supplanted by more successful stocks, and ultimately become extinct. To suggest otherwise, for cyanobacteria, seems a flight of wild imagination. Thus, at best, interpretation of the extremely ancient

Apex microbes as fossil cyanobacteria seems open to question. Just how strong *is* the fossil evidence? How can it be tested?

If the majority of the Apex fossils *are* of cyanobacteria, they must be part of an evolutionary continuum that extends from ~3500 Ma ago to the present, a notion that is testable in at least three ways:

(i) Cyanobacteria should comprise an unbroken biological lineage that can be expected to have left its mark in the intervening fossil record.

(ii) Not only should cyanobacteria be present in deposits of intermediate geological age, but this record should also include fossil oscillatoriaceans, members of the specific cyanobacterial family inferred to be represented among the Apex fossils.

(iii) And the suggested evolutionary stasis of the group (the lack of change over geological time in organismal and cellular morphology and, evidently, in habitat and physiology as well) should similarly be evidenced by the intervening fossil assemblages.

In order to apply these tests, it must be shown that fossil cyanobacteria can be distinguished from other comparably minute microbial remnants preserved in very ancient sediments, especially from other fossil prokaryotes. In the following discussion, this issue is therefore addressed first, followed by a consideration of the paleobiology of the Apex biota, a discussion focusing particularly on the inferred affinities of the cellularly preserved Apex microorganisms.

IDENTIFICATION OF CYANOBACTERIA IN THE EARLY FOSSIL RECORD

Divisions of geological time

Geological time is divided into two major segments of unequal duration: the Precambrian Eon, extending from the formation of the planet, ~4500 Ma ago, to the earliest appearance of widespread shelled invertebrates ~550 Ma ago; and the Phanerozoic Eon, extending from the end of the Precambrian (and the beginning of the Cambrian Period of Earth history) to the present. The Precambrian, in turn, is composed of two subsegments, the earlier Archean Era (extending from ~4500 to 2500 Ma) and the later Proterozoic Era (2500 to ~550 Ma in age). And the younger of these eras, the Proterozoic, is divided into three geochronological units: the Paleoproterozoic (2500 to 1600 Ma ago), Mesoproterozoic (1600 to 900 Ma ago), and Neoproterozoic (900 to ~ 550 Ma ago). Thus, microorganisms of the ~3500 Ma-old Apex chert are of Archean age, and fossils required to test the hypothesized cyanobacterial evolutionary continuum between the Apex and modern biotas should be expected to be preserved in intermediate-age units of the Paleo-, Meso-, and Neoproterozoic.

Evidence of plausibility

The presence of cyanobacteria in the ancient fossil record was suggested early in the development of the field by a seminal (but now commonly overlooked) report published some 40 years ago (Tyler & Barghoorn, 1954) and was strongly supported a decade later by three pivotal publications reporting discovery of diverse Proterozoic cyanobacterium-like fossils (Barghoorn & Tyler, 1965; Cloud, 1965; Barghoorn & Schopf, 1965). Since that time, numerous workers worldwide have noted, and regarded as significant, detailed similarities in cellular morphology between preserved Proterozoic microorganisms and extant cyanobacteria. Indeed, it has

Table 1. *Examples of Precambrian generic namesakes coined by various authors to suggest similarity to modern cyanobacterial genera*

FAMILY	Modern Genus	Precambrian Genus and authors	Country of authors
CHROOCOCCACEAE	*Anacystis*	*Palaeoanacystis* Schopf	USA
	Aphanocapsa	*Aphanocapsaopsis* Maithy & Shukla	India
	Aphanocapsa	*Eoaphanocapsa* Nyberg & Schopf	USA
	Aphanothece	*Eoaphanothece* Xu	China
	Eucapsis	*Eucapsamorpha* Golovenoc & Belova	Russia
	Gloeocapsa	*Eogloeocapsa* Golovenoc & Belova	Russia
	Microcystis	*Eomicrocystis* Maithy	India
	Microcystis	*Microcystopsis* Xu	China
	Microcystis	*Palaeomicrocystis* Maithy	India
	Synechococcus	*Eosynechococcus* Hofmann	Canada
ENTOPHYSALIDACEAE	*Entophysalis*	*Eoenentophysalis* Hofmann	Canada
PLEUROCAPSACEAE	*Hyella*	*Eohyella* Zhang & Golubic	China, USA
	Pleurocapsa	*Eopleurocapsa* Liu	China
	Pleurocapsa	Palaeopleurocapsa Knoll, Barghoorn & Golubic	USA
OSCILLATORIACEAE	*Lyngbya*	*Palaeolyngbya* Schopf	USA
	Microcoleus	*Eomicrocoleus* Horodyski & Donaldson	USA, Canada
	Oscillatoria	*Archaeoscillatoriopsis* Schopf	USA
	Oscillatoria	*Oscillatoriopsis* Schopf	USA
	Oscillatoria	*Oscillatorites* Schepeleva	Russia
	Phormidium	*Eophormidium* Xu	China
	Schizothrix	*Schizothropsis* Xu	China
	Spirulina	*Palaeospirulina* Edhorn	Canada
	Spirulina	*Spirillinema* Shimron & Horowitz	Israel

become common practice for Precambrian paleobiologists to coin generic names intended to denote similarity or inferred identity between ancient cyanobacterium-like fossils and their modern morphological counterparts by adding appropriate prefixes (palaeo-, eo-) or suffixes (-opsis, -ites) to the names of living cyanobacterial genera (Table 1). The validity of such comparisons is variable, but the ubiquity of this practice represents broad consensus of the plausibility of the implied relationships. Similarly, convincing evidence is provided by the detailed similarities evident in paired comparisons of Proterozoic stromatolite-forming microorganisms (Fig. 1B, D, F, H) and the cyanobacteria of modern stromatolitic biocoenoses (Fig. 1A, C, E, G).

Nevertheless, questions remain. Fossil generic namesakes and selected fossil-modern comparisons establish plausibility but they do not prove the point generally. What are needed are detailed morphometric comparisons of large representative samples of fossil and living taxa and the means to establish convincingly that the fossils are actually cyanobacteria rather than 'microbial mimics', members of non-cyanobacterial prokaryotic groups of similar size, shape, and cellular organization. Such analyses are discussed below, and although they substantiate the inferred cyanobacterial affinity of diverse types of abundant widespread Proterozoic microfossils, they are not without their own set of assumptions and problems.

Problems of analysis

More than 4000 taxonomic occurrences of prokaryotic microfossils (the presence of a recognized fossil species in a formally named geologic unit) are now known from nearly 500 Proterozoic units worldwide (Schopf, 1992*b*, and more recent reports). Evaluation of the affinities of these fossils is complicated by the following three problems.

Fig. 1. Pairs of optical photomicrographs (transmitted light) showing living cyanobacteria (A, C, E, G), from mat-building stromatolitic communities of Baja, Mexico, for comparison with their Precambrian morphological counterparts (B, D, F, H). A. *Lyngbya aestuarii* Leibm. ex Gomont, Oscillatoriaceae, encompassed by a cylindrical mucilagenous sheath (at arrow); B. *Palaeolyngbya helva* German, a similarly ensheathed (at arrows) oscillatoriacean, shown in an acid-resistant residue of carbonaceous siltstone from the ~950 Ma-old Lakhanda Formation of the Khabarovsk region of Siberia, Russia. C. *Spirulina subsalsa* Oerst. ex Gomont, Oscillatoriaceae; D. *Heliconema turukhania* German (see also *H. funiculum* Schopf & Blacic, Fig. 10B), a *Spirulina*-like oscillatoriacean shown in an acid-resistant residue of carbonaceous siltstone from the ~850 Ma-old Miroedikha Formation of the Turukhansk region of Siberia, Russia. E. *Gloeocapsa* cf. *repestris* Kütz., Chroococcaceae, a four-celled colony having a thick distinct encompassing sheath (at arrow); F. *Gloeodiniopsis uralicus* Krylov & Sergeev, a similarly sheath-enclosed (at arrow) four-celled colonial chroococcacean shown in a petrographic thin section of bedded chert from the ~1500 Ma-old Satka Formation of southern Bashkiria, Russia. G. *Entophysalis* cf. *granulosa* Kütz., Entophysalidaceae; H. *Eoentophysalis belcherensis* Hofmann, an *Entophysalis*-like colonial entophysalidacean from stromatolitic chert of the ~2150 Ma-old Belcher Group of Northwest Territories, Canada.

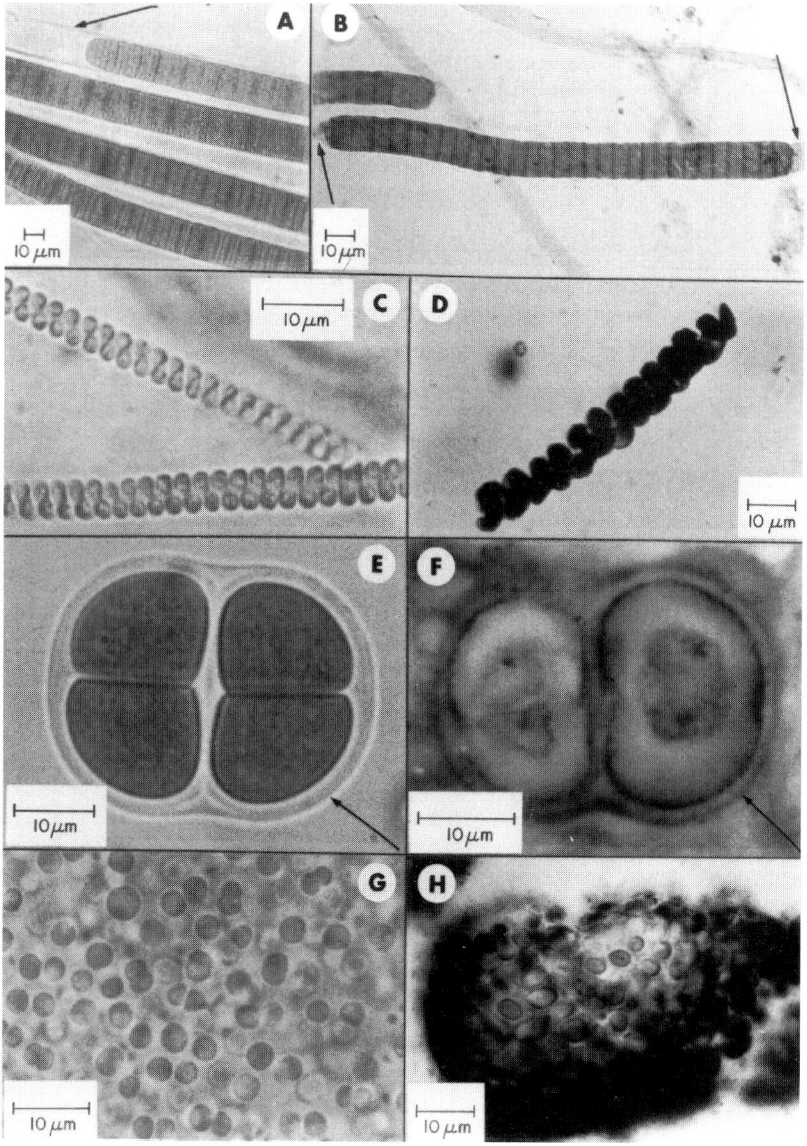

Fig. 1.

Taxonomy

Because of the relative youth of Precambrian paleobiological studies, carried out actively for little more than a quarter-century (Schopf, 1992*c*), the taxonomy of Proterozoic microfossils is confused and unsettled. Different workers have applied different binomials to virtually identical morphotypes detected at differing locales, and some workers have adopted arbitrary, non-biological systems of classification that make comparison of fossil and living taxa virtually impossible. To avoid this confusion, relevant size data for ~1450 species and varieties of living prokaryotes and eukaryotic microalgae have been used to construct a biologically based system of classification that groups morphologically comparable Proterozoic microfossils, regardless of the binomials originally applied, into well-defined species-level categories (Schopf, 1992*d*).

Preservational compression in shales

Historically, micropaleontological studies of Proterozoic strata have advanced along two parallel but largely independent pathways: (i) investigations of members of plankton-dominated assemblages preserved as two-dimensional carbonaceous compressions and studied in acid-resistant residues of shales/siltstones (Fig. 2); and (ii) studies of organic-walled cellular microorganisms preserved three-dimensionally by permineralization (petrifaction) in benthos-dominated stromatolitic communities and studied in petrographic thin sections of carbonates/cherts. Because of the differing types of preservation occurring in these two lithologies, and the differing expertise required for preparation and study of the preserved microfossils, few workers (and even fewer publications) have dealt with microfossils occurring both in Proterozoic shales/siltstones and carbonates/cherts. Only rarely have taxa described from one rock-type been compared with coeval

Fig. 2. Optical photomicrographs (A_1, B_1, E_1, transmitted light; C, D, F_1, G, interference contrast) and interpretive drawings (A_2, B_2, E_2, F_2) showing the effects of preservational compression on originally spheroidal (A, B, E, F), cylindrical–tubular (C), and cylindrical–septate microfossils (D, G) detected in acid-resistant residues of Proterozoic shales/siltstones from the ~950 Ma-old Lakhanda (A, B) and ~850 Ma-old Miroedikha Formations (C, G) of Siberia, and the ~580 Ma-old Redkino (D, F) and ~800 Ma-old Akberdin Formations (E) of Bashkiria, Russia. A, B. Compressed specimens of *Turuchanica ternata* (Timofeev) Jankauskas exhibiting deeply incised radial cracks. C. *Palaeolyngbya sphaerocephala* German, the flattened originally cylindrical–tubular sheath of a filamentous cyanobacterium; note the presence within the compressed ribbon-like sheath of numerous disarticulated discoidal cells. D. *Striatella coriacea* Assejeva, the compressed remnants of an originally cylindrical–septate cyanobacterial trichome; note the occurrence of relatively well-preserved regularly spaced transverse septa. E. A highly compressed specimen of the thick-walled spheroidal microfossil *Kildinella* (= *Leiosphaeridia*) *lophostriata* Jankauskas exhibiting numerous central folds and prominent tangential pleats. F. A highly compressed specimen of the thin-walled coccoidal taxon *Leiosphaeridia incrassatula* Jankauskas exhibiting peripheral tangential accordian-like pleats and flexures. G. *Arctacellularia doliiformis* German, the compressed strap-like remnants of an originally cylindrical-septate filament; note the more or less regular spacing of flattened transverse septa.

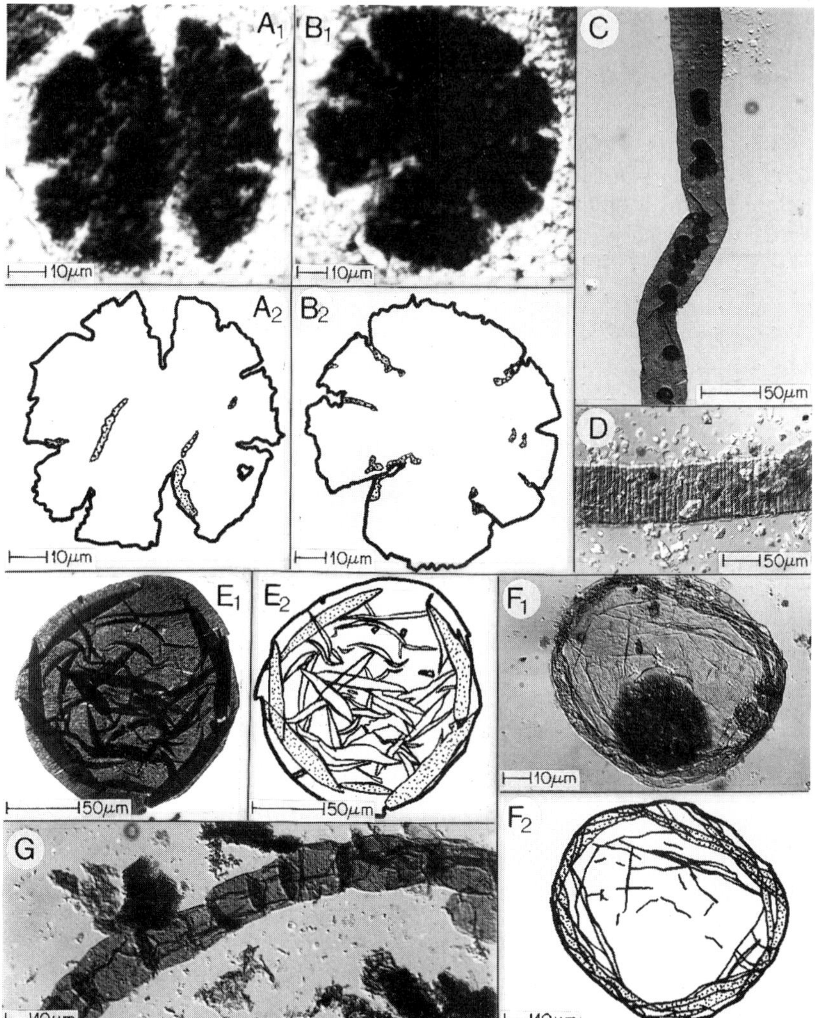
A1
B1
C
10μm
10μm
50μm
A2
B2
D
10μm
10μm
50μm
E1
E2
F1
50μm
50μm
10μm
F2
G
10μm
10μm

Fig. 2.

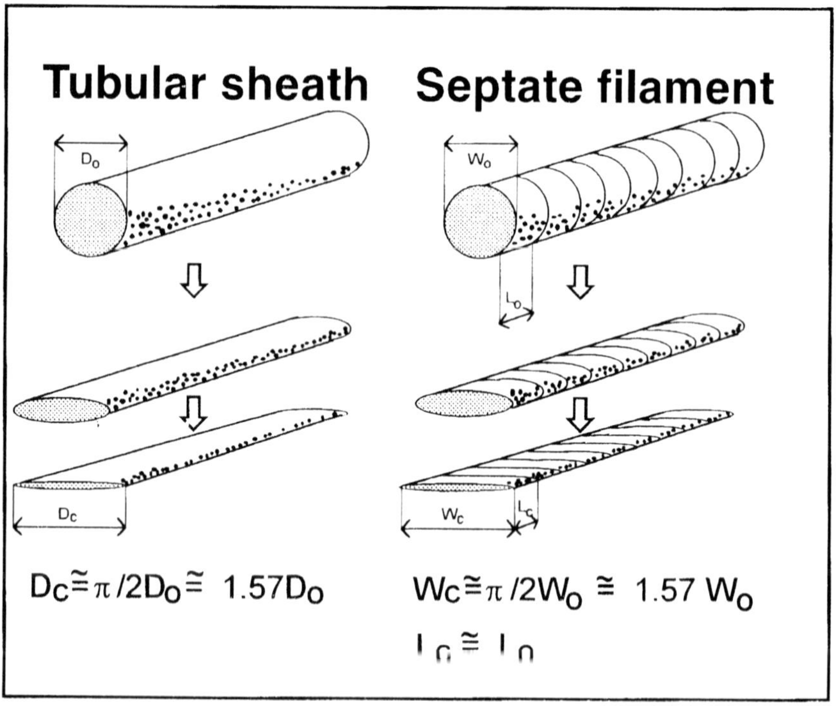

Fig. 3. Correction factors used for morphometric comparisons of uncompressed modern or fossil cylindrical–tubular prokaryotic sheaths or cylindrical–septate filaments with fossil analogues preserved by compression in shales/siltstones. The compressed diameter (D_c) of initially hollow tubular sheaths preserved as flattened carbonaceous compressions in fine-grained clastic sediments is ~1.57 times larger than the original uncompressed diameter (D_o). Similarly, the compressed width (W_c) of the cells of flattened septate filaments preserved in shales/siltstones is ~1.57 times larger than their original three-dimensional width (W_o) but the length of such cells (L_c) is commonly more or less unaltered ($L_c \simeq L_o$) during compression.

taxa reported from the other, chiefly because of the lack of a basis for realistic morphometrical comparison of flattened two-dimensional microfossils with uncompressed three-dimensional specimens of their fossil and modern counterparts.

This difficulty has now been overcome. Means have been devised for converting size measurements of compressed fossils so that they can be compared with three-dimensional fossil and modern analogues (Fig. 3), and detailed tests of this approach (e.g. Fig. 4) have established its validity (Schopf, 1992*b*).

Assessment of affinities

Differentiation of cyanobacteria from other prokaryotic fossils is fraught with difficulty. Because Proterozoic microfossils convincingly identified as cyanobacteria were reported early in the history of the field (Barghoorn &

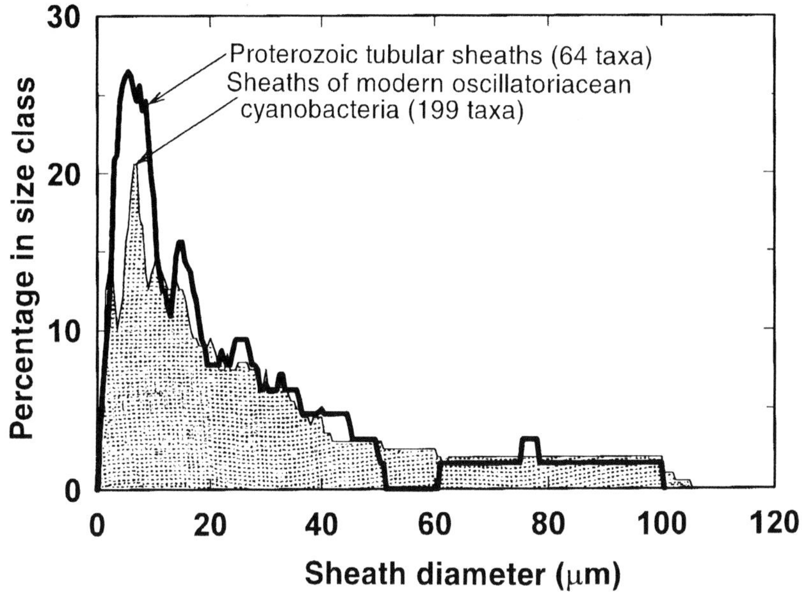

Fig. 4. Comparisons of the size ranges and patterns of size distribution of the diameters of the cylindrical–tubular sheaths of 199 taxa of modern oscillatoriacean cyanobacteria (Desikachary, 1959) with those of the 64 taxa of Proterozoic tubular sheaths recognized by Schopf (1992*d*), with the diameters of fossil specimens preserved in shales/siltstones corrected for the effects of preservational compression as shown in Fig. 3.

Schopf, 1965; Schopf, 1968), the possible relationship of detected fossils to other prokaryotic groups has been ignored by many workers. Indeed, even such a seemingly straightforward problem as determining whether a particular microfossil is of prokaryotic or eukaryotic affinity can be difficult to resolve. For example, in comparison with extant microorganisms, small Proterozoic fossil unicells (i.e. spheroidal microfossils $\leq 5\ \mu m$ in diameter) are morphologically similar to relatively large coccoidal non-cyanobacterial prokaryotes, 'normal-sized' spheroidal prokaryotic cyanobacteria, and small-celled eukaryotic microalgae, but because crucial diagnostic characteristics (flagella and intracellular membranes, nuclei, and other organelles) are almost never preserved in the fossil record, the affinity of such ancient unicells is difficult to assess. Moreover, various biochemical characteristics used to demonstrate the affinities of extant microorganisms (chlorophylls, accessory pigments, enzymes, nucleotide ratios, amino acid sequences) are inapplicable to Proterozoic fossils, having been leached away or obliterated long ago by normal geological processes.

Thus, assessment of the biological relationships of ancient microorganisms is constrained severely by the available fossil evidence. The approach used here is therefore to focus on those characters that are unquestionably preservable in the fossil record and to infer 'probable

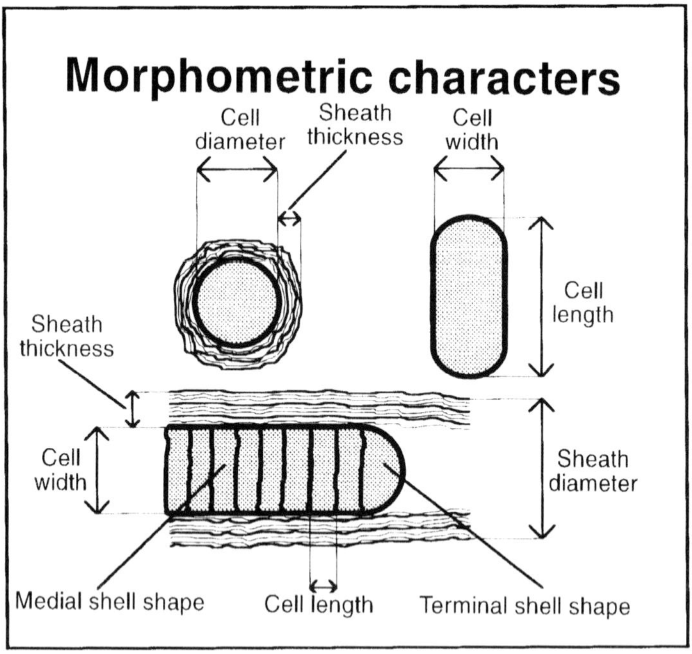

Fig. 5. Ten characters used for morphometric comparisons of modern prokaryotes and comparable Proterozoic microfossils.

affinity' by comparing the limited and necessarily morphologically based suite of such characters with the same characters in extant microbes. Ten such characters, relevant to the great majority of prokaryotes, whether fossil or modern, are illustrated in Fig. 5. Although less reliable than other techniques that can be applied to living microorganisms, this morphologically based approach is valid. Among modern microbes, morphology plays a prominent role in differentiation of many bacterial (Buchanan & Gibbons, 1974) and virtually all cyanobacterial taxa (Desikachary, 1959).

Inferences of affinity based on morphometric analyses

A principal technique used here for inferring the 'probable affinity' of a Proterozoic taxon is that of comparing the size ranges (of such characters as cell diameters, cells lengths and widths, diameters and thicknesses of encompassing sheaths) of morphologically similar fossil and modern taxa. Embedded in this strategy is the assumption that most Proterozoic microorganisms are members of still extant lineages (an assumption consistent with all relevant biological and paleobiological evidence), and that their morphometrical characteristics can therefore be expected to be comparable to those of modern analogues. Fossil–modern comparisons of this type also

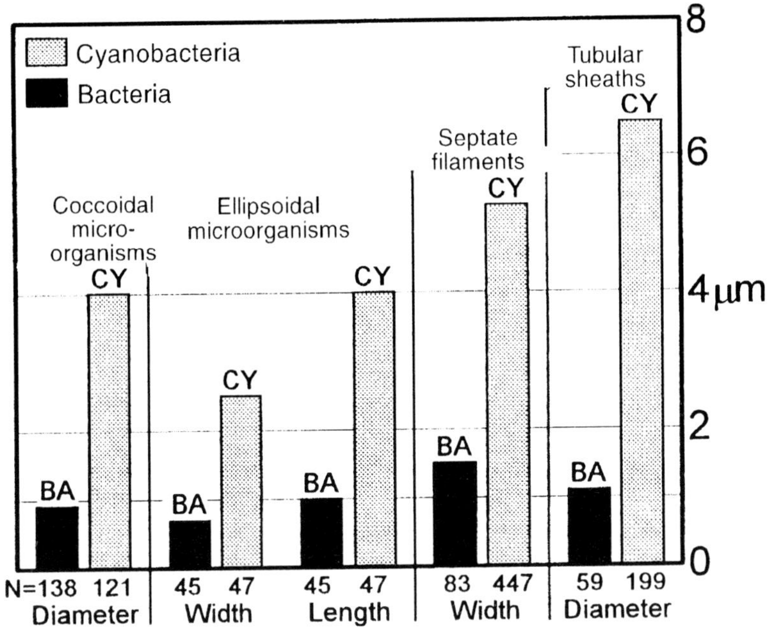

Fig. 6. Comparisons of the median dimensions of various morphological categories of modern cyanobacteria with those of other modern free-living prokaryotes ('bacteria'); N = number of species and varieties considered. (Data from Schopf, 1992*b*.)

suggest means for distinguishing fossil cyanobacteria from other prokaryotic microfossils. For example, the median dimensions (Fig. 6) and patterns of size distribution (Fig. 7) of modern cyanobacteria typically differ substantially from those of other extant free-living prokaryotes ('bacteria'), as do the size ranges of their cells and of other preservable morphological characteristics.

In simple cases, those in which the attributes of compared fossil and modern morphometric characters are essentially identical and in which there is no significant overlap of these with those of similar characters in other extant biological groups, interpretation of fossil affinity is relatively straightforward. Unfortunately, however, the occurrence among modern groups of non-overlapping size ranges (for virtually all of the characters assessed) is the exception rather than the rule. A typical situation is illustrated in Fig. 8, which shows that the pattern of cell size distribution exhibited by modern free-living coccoidal non-cyanobacterial prokaryotes overlaps markedly with that of living coccoidal cyanobacteria for taxa 1.5 to 2.5 μm in diameter. In order to specify the uncertainty inherent in assessment of the affinities of coccoidal microfossils of similar size, fossils $<1.5\,\mu$m in diameter have therefore been interpreted as 'probable bacteria'; those 1.5 to 2.5 μm in diameter as undifferentiated 'prokaryotes'; and those $>2.5\,\mu$m as 'probable cyanobacteria' (Schopf, 1992*b*).

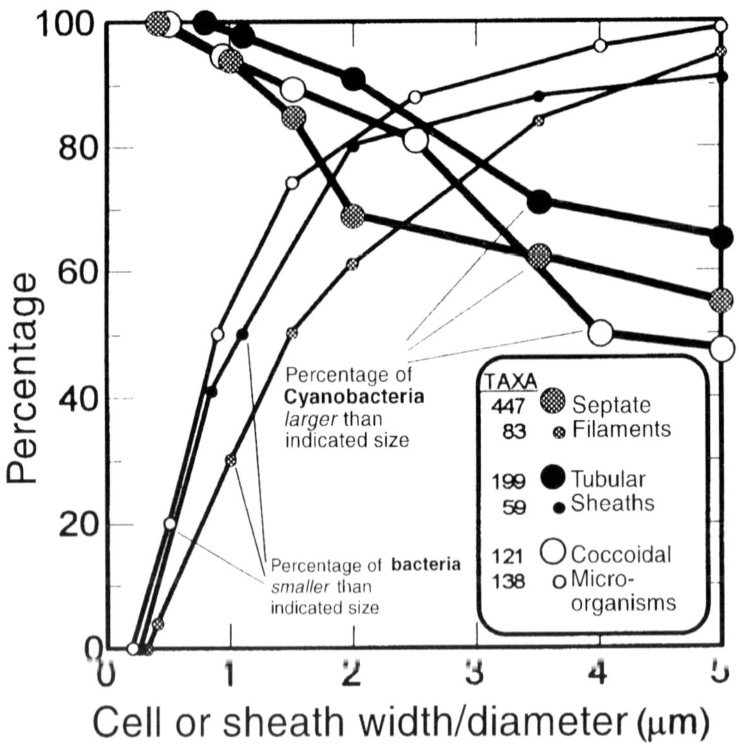

Fig. 7. Prokaryotic size distributions showing for modern cyanobacteria the percentage of septate filaments, cylindrical–tubular sheaths, and coccoidal cells *larger* than the width/diameter indicated, and for other modern free-living prokaryotes ('bacteria') the percentage of taxa exhibiting these characteristics that are *smaller* than the width/diameter indicated. (Data from Schopf, 1992*b*.)

THE PROTEROZOIC PROKARYOTIC FOSSIL RECORD

Among the most commonly occurring Proterozoic prokaryotes are solitary and colonial coccoidal and ellipsoidal microfossils (Figs. 1 and 9), morphotypes that tend to be particularly abundant in cherty carbonate stromatolites (referred to here as 'nearshore carbonates/cherts'). Both sheath-enclosed and naked cellular trichomes (Figs. 1 and 10) are similarly of common occurrence in Proterozoic stromatolitic assemblages, but they also occur in 'offshore' shales/siltstones (many of which actually represent shallow coastal marine facies such as sabkhas, lagoons, and mud flats) as do tubular prokaryotic sheaths (Fig. 11), the presence of which reflects either the preferential preservation of sheaths in comparison with trichomes or the vacating of sheaths by trichomes capable of photo- and chemotactic gliding mobility.

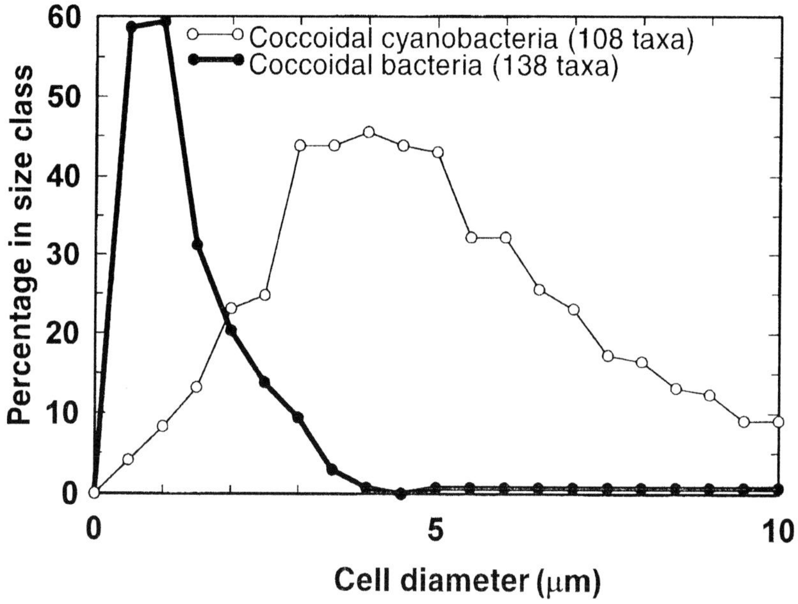

Fig. 8. Comparison of the size distributions of modern coccoidal cyanobacteria ≤10 μm in diameter and other modern small-celled coccoidal free-living prokaryotes ('bacteria') used to infer the probable affinities of fossil analogues. (Data from Desikachary, 1959; Buchanan & Gibbons, 1974; Laskin & Lechevalier, 1977; Starr *et al.*, 1981.)

Coccoidal microorganisms and Proterozoic microfossils

Because the largest modern coccoidal cyanobacteria (*Chroococcus giganteus*, *C. macrococcus*, and *C. turgidus* var. *maximum*) are 50 to 60 μm in diameter (Desikachary, 1959, pp. 99–102) and the largest extant coccoidal noncyanobacterial prokaryote (*Thiovulum majus*) is about half the size (Buchanan & Gibbons, 1974, p. 463), the following discussion is limited to consideration of coccoidal unicells ≤60 μm in diameter. Data regarding larger coccoidal unicells (modern eukaryotic microalgae and fossil sphaeromorph acritarchs >60 μm in maximum diameters) are presented elsewhere (Schopf, 1992*b*).

Nearly 50% of taxa of modern coccoidal cyanobacteria are larger than 5 μm in diameter whereas ~99% of those of other extant free-living spheroidal prokaryotes are smaller than this size (Fig. 7). Similarly, the median cell size of modern coccoidal cyanobacteria is ~4.0 μm, more than four times larger than that of other extant coccoidal prokaryotes (Fig. 6). Indeed, ~75% of such modern non-cyanobacterial prokaryotes are less than 1.5 μm in diameter whereas more than 80% of comparable cyanobacteria are larger than 2.5 μm in size (Figs. 8 and 12). Thus, as explained above, coccoidal Proterozoic microfossils less than 1.5 μm in diameter have been

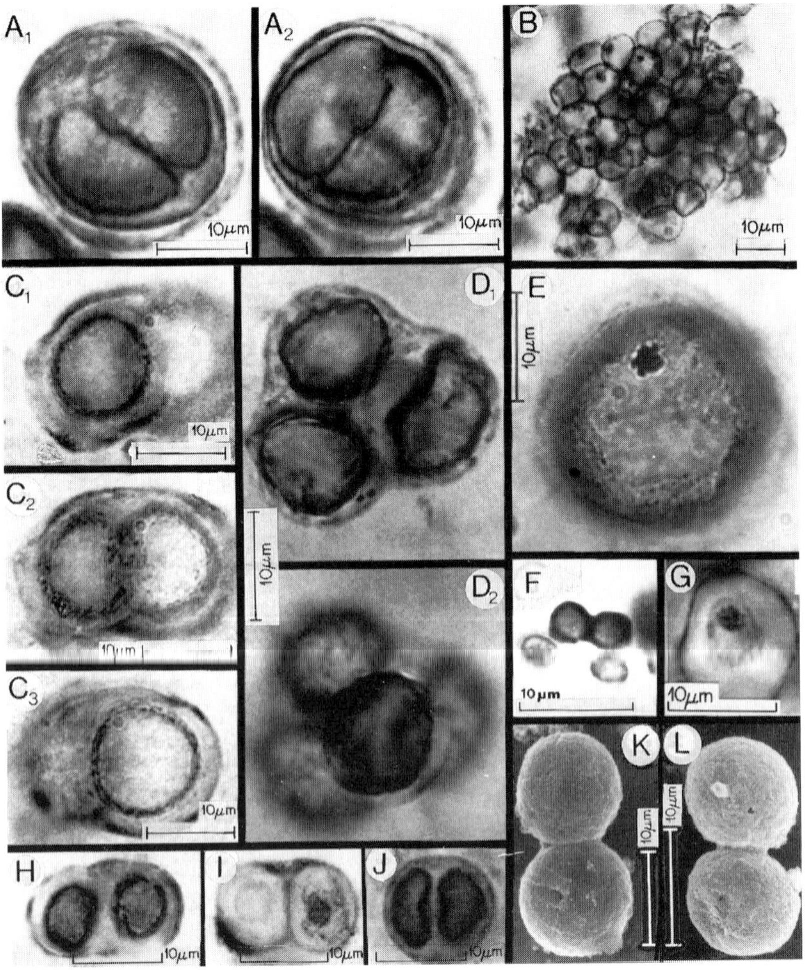

Fig. 9. Optical photomicrographs (A–J, transmitted light) and scanning electron micrographs (K and L) showing Proterozoic cyanobacterial (chroococcacean) coccoidal cells and colonies in petrographic thin sections (A, C–J) and acid-resistant residues (B, K, and L) of finely laminated stromatolitic chert from the ~850 Ma-old Bitter Springs Formation of Northern Territory, Australia. (A_1–A_2), (C_1–C_3), and (D_1–D_2) show single specimens at differing focal depths. A. *Bigeminococcus lamellosus* Schopf & Blacic. B. *Glenobotrydion aenigmatis* Schopf. C. *Eozygion grande* Schopf & Blacic. D. *Eotetrahedrion princeps* Schopf & Blacic. E. *Globophycus rugosum* Schopf. F. *Sphaerophycus parvum* Schopf. G. *Caryosphaeroides pristina* Schopf. H and J. *Eozygion minutum* Schopf & Blacic. I. *Caryospheroides tetras* Schopf. K and L. Unnamed paired coccoid chroococcaceans (Schopf, 1972).

interpreted as 'probable bacteria'; those 1.5 to 2.5 μm in diameter as undifferentiated 'prokaryotes'; and those greater than 2.5 μm as 'probable cyanobacteria' (Schopf, 1992*b*).

Most extant coccoidal cyanobacteria larger than 10 μm in diameter are

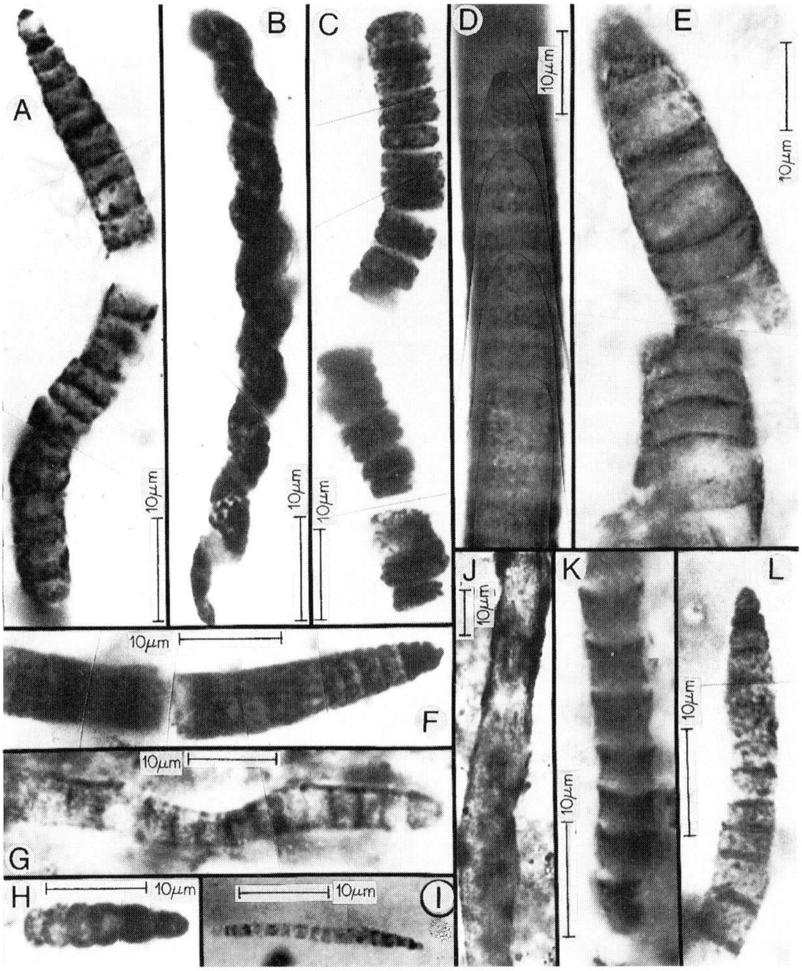

Fig. 10. Optical photomicrographs (transmitted light) showing Proterozoic cyanobacterial (oscillatoriacean) cellular trichomes (A–I, K, and L) and a cylindrical sheath (J) in petrographic thin sections of finely laminated stromatolitic chert from the ~850 Ma-old Bitter Springs Formation of Northern Territory, Australia. Because of the sinuous, three-dimensional preservation of these permineralized (petrified) carbonaceous microfossils, all except the specimen in (J) are shown in composite photomicrographs. (A, F, and L) *Ceophalophytarion laticellulosum* Schopf & Blacic. B. *Heliconema funiculm* Schopf & Blacic. C. *Oscillatoriopsis breviconvexa* Schopf & Blacic. D. Unnamed *Oscillatoria*-like trichome (Schopf, 1974). E. *Obconicophycus amadeus* Schopf & Blacic. G. *Oscillatoriopsis obtusa* Schopf. H. *Filiconstrictosus diminutus* Schopf & Blacic. I. *Cephalophytarion minutum* Schopf. J. *Siphonophycus kestron* Schopf. K. *Halythrix nodosa Schopf.*

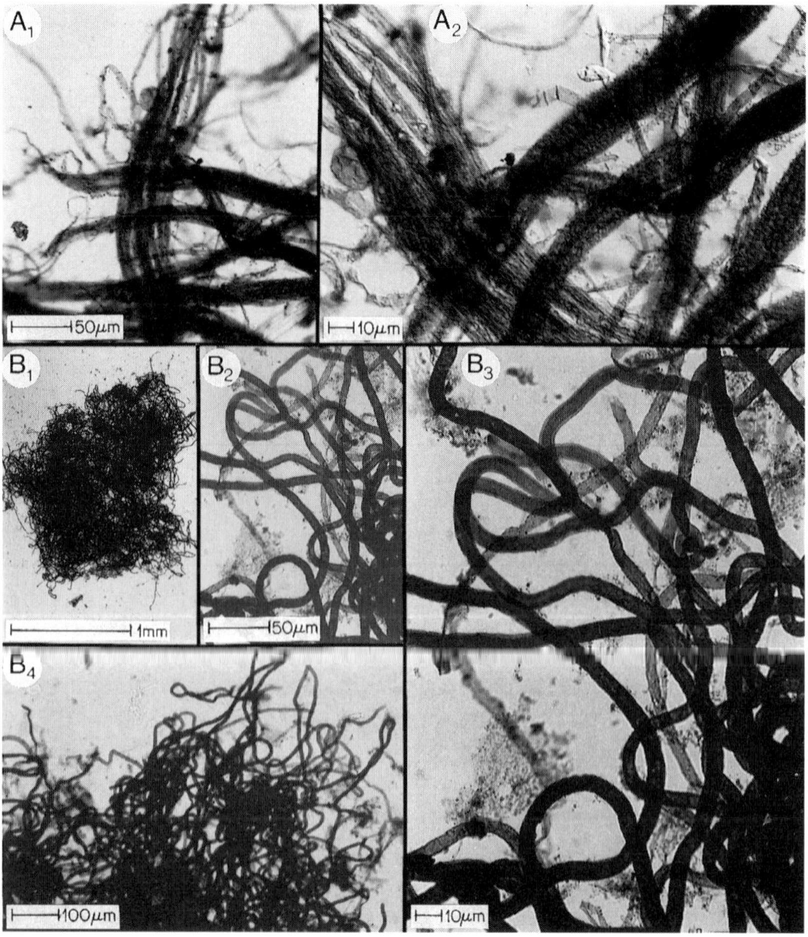

Fig. 11. Optical photomicrographs (A_1 and B_1–B_4, transmitted light; A_2, interference contrast) showing Precambrian cyanobacterial (oscillatoriacean) cylindrical–tubular sheaths in acid-resistant residues of carbonaceous siltstone from the ~850 Ma-old Miroedikha Formation of the Turukhansk region of Siberia, Russia. (A_1–A_2) and (B_1–B_4) show single groups of specimens at differing magnifications. A. *Polythrichoides lineatus* German. B. *Leiothrichoides typicus* German.

morphologically distinctive chroococcaceans (i.e. *Chroococcus* spp. and *Gloeocapsa* spp.) that are encompassed by thick, commonly multilamellated, prominent sheaths. Microfossils of this type, similarly encompassed by thick lamellated sheaths, are particularly distinctive components of the Proterozoic biota (Figs. 1F and 9A). A total of 17 such taxa, having the size distribution summarized in Fig. 13, have been recognized in the Proterozoic fossil record (Schopf, 1992*d*). The ranges of cell size and sheath thickness exhibited by these fossil taxa are notably similar to those of living species of

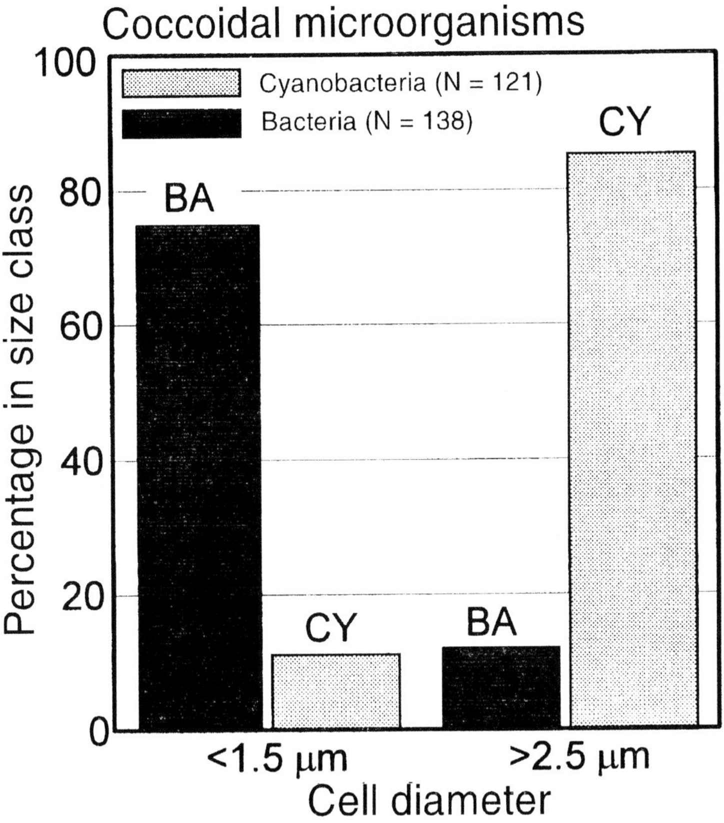

Fig. 12. Size distributions of modern coccoidal cyanobacteria and other modern coccoidal free-living prokaryotes ('bacteria') used to infer the probable affinities of fossil analogues; N = number of species and varieties considered. (Data from Desikachary, 1959; Buchanan & Gibbons, 1974; Laskin & Lechevalier, 1977; Starr *et al.*, 1981.)

Chroococcus and *Gloeocapsa* and, like those of their modern morphological analogues, the cell diameters of the fossil taxa are rather strongly and positively correlated with the thickness of their encompassing sheaths (Fig. 13).

Figure 14 summarizes the size distributions for reported occurrences of non-*Chroococcus*/*Gloeocapsa*-like coccoidal microfossils ≤60 μm in diameter known from nearshore and offshore Paleoproterozoic deposits 1875 to 1600 Ma in age, compared with size data for modern cyanobacteria and eukaryotic microalgae. Similar data are shown in Fig. 15 for Paleo-, Meso-, and Neoproterozoic nearshore carbonates/cherts and offshore shales/siltstones. The model cell size class for such fossils reported from the nearshore (chiefly stromatolitic) carbonates/cherts consistently falls towards the smaller (cyanobacterial) end of the spectrum, whereas that

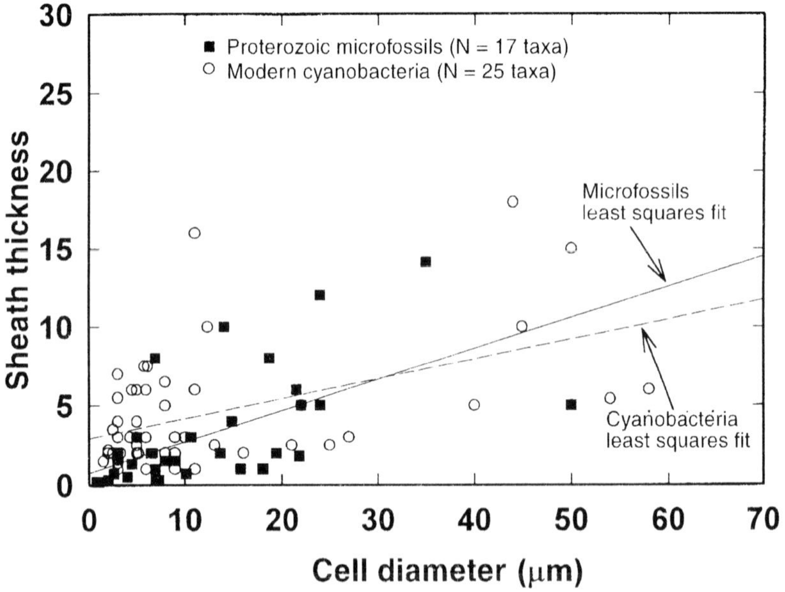

Fig. 13. Minimum and maximum cell diameters versus the sheath thicknesses of modern species of the chroococcacean cyanobacteria *Chroococcus* and *Gloeocapsa* (Desikachary, 1959) compared with those of morphologically similar Proterozoic taxa (Schopf, 1992*a*).

for fossils reported from the offshore units is decidedly larger, broader, and more similar to the pattern of size distribution of extant microalgal eukaryotes.

Inferred affinities and evolutionary trends

In modern nearshore shallow water environments similar to those samples in the Proterozoic, chroococcalean cyanobacteria (e.g. *Chroococcus, Aphanocapsa, Aphanothece* and *Synechococcus*) are common (Golubic, 1976*a*, *b*; Pierson *et al.*, 1992). It is thus not surprising that although coccoidal noncyanobacterial prokaryotes have also been recognized in the Proterozoic fossil record (Schopf, 1968; Mendelson & Schopf, 1992; Fairchild *et al.*, 1995), the vast majority of small Proterozoic coccoids have been inferred to be of chroococcalean (predominantly chroococcacean) affinity (Schopf, 1992*b*, 1994*b*). As shown in Fig. 15, over most of the Proterozoic the patterns of size distribution exhibited by coccoidal microfossils ≤60 μm in diameter remained essentially unchanged, and although the relative abundance of cells >25 μm in size increased in nearshore deposits between the Meso- and Neoproterozoic, this change was almost certainly due to the introduction of relatively large-celled eukaryotic microalgae into this environment rather than to changes in the prokaryotic biota (Schopf, 1992*e*, *f*). Similarly, the known Proterozoic record of diverse types of irregular or

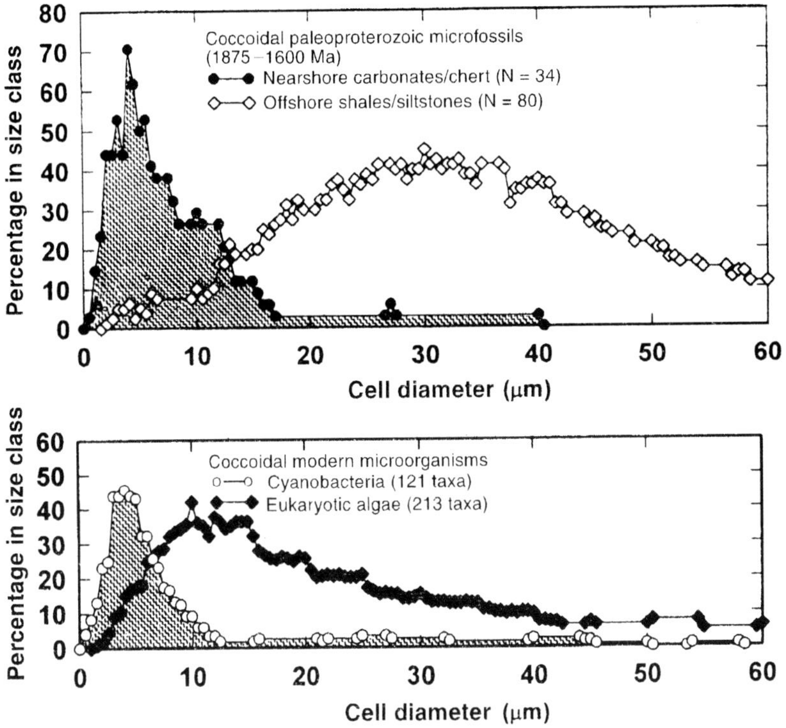

Fig. 14. *Above*: Comparisons of the size distribution of late Paleoproterozoic (1875 to 1600 Ma-old) non-*Chroococcus*/*Gloeocapsa*-like coccoidal microfossils ≤60 μm in diameter reported from petrographic thin sections of nearshore carbonates/cherts (34 occurrences in 7 geological units) with that of similar taxa from acid-resistant macerations of offshore shales/siltstones (80 occurrences in 12 units). (Data from Schopf, 1992*b*; Mendelson & Schopf, 1992.). *Below*: Comparison of the size distributions of modern coccoidal chroococcacean cyanobacteria (Desikachary, 1959) and modern eukaryotic (chlorophycean and rhodophycean) microalgae ≤60 μm in diameter. (Data from West & Fritsch, 1927; Prescott, 1954, 1962; Taylor, 1957; Smith & Bold, 1966; Bourrelly, 1970; La Rivers, 1978.)

ordered colonies of cyanobacterium-like coccoids exhibits no discernable time-related trends (other than a gradual increase over time in the number of taxa reported, an increase demonstrably correlated with the number of geological units sampled: Schopf, 1992*b*). Thus, coccoidal cyanobacterium-like microfossils ≤60 μm in diameter, whether solitary or colonial, appear to have exhibited no major evolutionary changes over Proterozoic time.

Ellipsoidal microorganisms and Proterozoic microfossils

Like those of the modern coccoidal prokaryotes discussed above, the median dimensions (both of widths and lengths) of extant ellipsoidal non-cyanobacterial prokaryotes are about one-fourth as large as those of

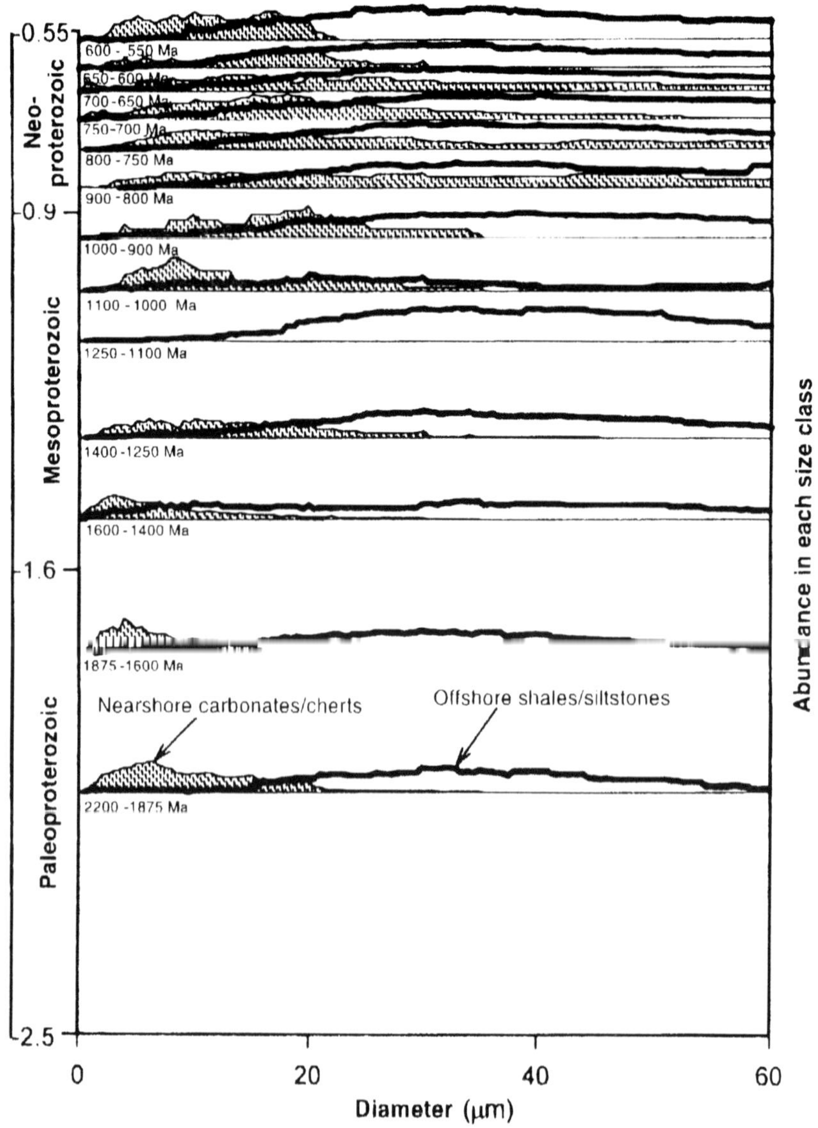

Fig. 15. Summary in various age categories of the size distributions of 1051 Proterozoic occurrences in 259 geological units of the 24 taxa of coccoidal microfossils ≤60 μm in diameter recognized by Schopf (1992*d*) reported from petrographic thin sections of nearshore carbonates/cherts and acid-resistant residues of offshore shales/siltstones. (Data from Schopf, 1992*b*; Mendelson & Schopf, 1992.)

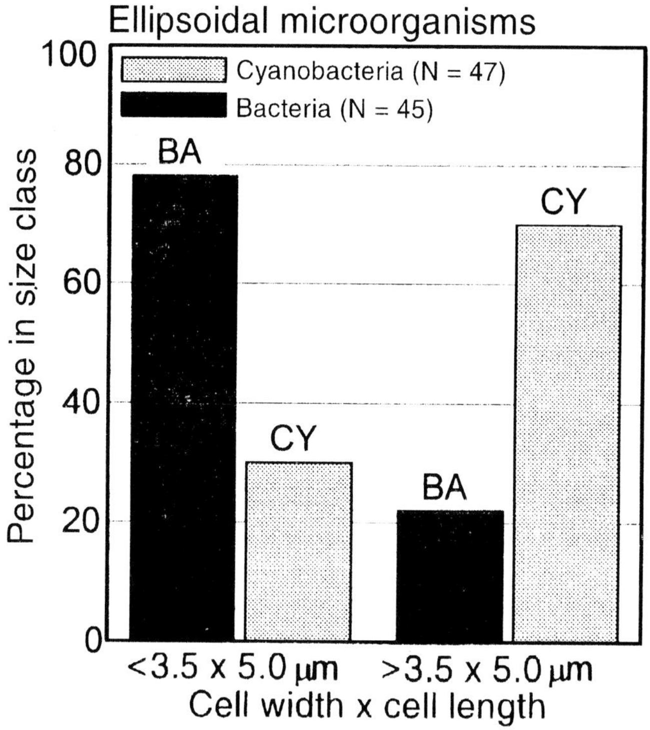

Fig. 16. Comparison of the size distributions of modern ellipsoidal cyanobacteria and other modern small-celled ellipsoidal free-living prokaryotes ('bacteria') used to infer the probable affinities of fossil analogues; N = number of species and varieties considered. (Data from Desikachary, 1959; Buchanan & Gibbons, 1974; Starr *et al.*, 1981.)

comparable cyanobacteria (Fig. 6). Similarly, as shown in Fig. 16, the great majority (about 78% of species and varieties) of modern ellipsoidal bacteria are smaller than $3.5 \times 5.0\ \mu m$, whereas most such cyanobacteria (about 70%) have cells larger than this size. Solitary or colonial ellipsoidal Proterozoic microfossils smaller than $3.5 \times 5.0\ \mu m$ have therefore been regarded as 'probable bacteria' and those $\geqslant 3.5 \times 5.0\ \mu m$ as 'probable cyanobacteria' (Schopf, 1992*b*).

Inferred affinities and evolutionary trends

As shown in Fig. 17, both 'probable bacteria' and 'probable cyanobacteria' (the latter interpreted chiefly as entophysalidaceans) are well represented in the Proterozoic record of ellipsoidal microfossils, and the average cell size of such microfossils increased (resulting in increased occurrences of 'probable cyanobacteria') between the Paleo- and Mesoproterozoic. However, the

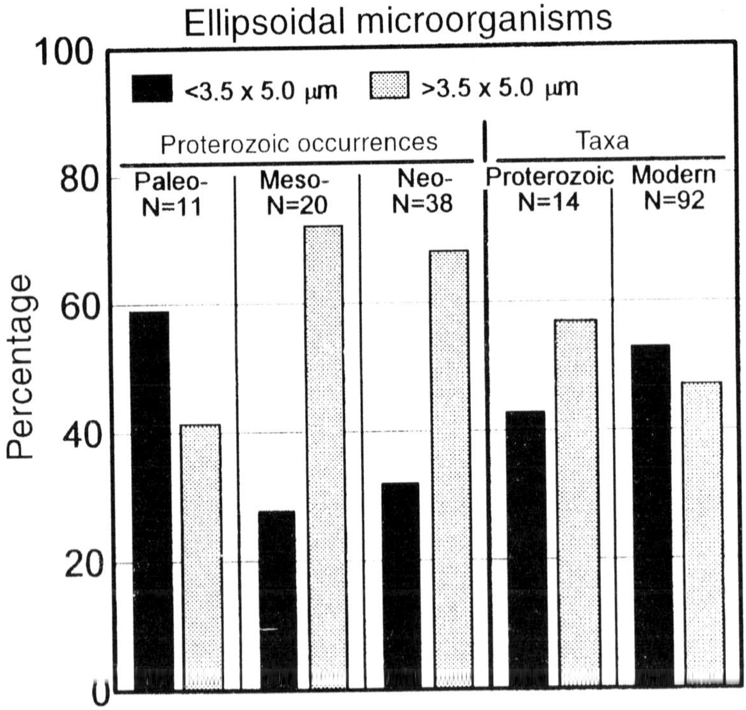

Fig. 17. Comparisons of the relative abundance of ellipsoidal microfossils less than, and greater than, 3.5 × 5.0 µm reported to occur in Paleo-, Meso-, or Neoproterozoic strata, and of the size distributions of Proterozoic ellipsoidal taxa recognized by Schopf (1992*d*) with those of modern prokaryotic ellipsoidal taxa. (Data from Desikachary, 1959; Buchanan & Gibbons, 1974; Starr *et al.*, 1981.)

percentages of 'bacterial' and 'cyanobacterial' taxa of ellipsoidal microfossils recognized in the Proterozoic are essentially the same as those known in the modern biota (Fig. 17), and the apparent paucity of 'probable cyanobacteria' in Paleoproterozoic assemblages may simply reflect inadequate sampling of units of this age (Mendelson & Schopf, 1992). Available data are too few to provide firm evidence of significant evolutionary trends.

Fossil entophysalidaceans and pleurocapsaceans

In addition to the morphometric analyses of coccoidal and ellipsoidal fossil and modern prokaryotes summarized above, detailed studies have been carried out on microfossils referred to two particular cyanobacterial families (the Entophysalidaceae and the Pleurocapsaceae) members of which are decidedly more distinctive morphologically than are those of the predominantly coccoidal Chroococcaceae. Golubic and Hofmann (1976) compared ~2150 Ma-old *Eoentophysalis belcherensis* (Fig. 1H) with two modern entophysalidaceans (*Entophysalis major* and *E. granulosa*). They showed that not only are the fossil and modern species morphologically comparable (in cell shape, and in form and arrangement of originally mucilaginous cellular envelopes) and that they exhibit similar frequency distributions of dividing cells and essentially identical patterns of cellular development (resulting from cell division in three perpendicular planes), but also that both taxa form microtexturally similar stromatolitic structures in comparable intertidal to shallow marine environmental settings, that they undergo similar postmortem degradation sequences, and that they occur in microbial communities that are comparable in both species composition and biological diversity. In a subsequent detailed study, Knoll and Golubic (1979) compared the morphology, cell division patterns, ecology, and postmortem degradation sequences of a second Proterozoic entophysalidacean (~850 Ma-old *Eoentophysalis cumulus*) with those of modern *Entophysalis granulosa* and concluded that the fossil 'microorganism is identical in all its salient characteristics to members of the extant [cyanobacterial] genus' (Knoll & Golubic, 1979, p. 125).

Several species of fossil and living pleurocapsaceans have also been compared in detail. *Polybessurus bipartitus*, first reported from ~770 Ma-old stromatolites of South Australia (Fairchild, 1975; Schopf, 1977), is a morphologically distinctive, gregarious, cylindrical fossil pleurocapsacean composed of stacked cup-shaped envelopes often extended into long tubes orientated perpendicular to the substrate. Specimens of this taxon occurring in rocks of about the same age in East Greenland were interpreted by Green *et al.* to be 'a close morphological reproductive, and behavioral counterpart' to populations of a species of the pleurocapsacean *Cyanostylon* present 'in Bahamian environments similar to those in which the Proterozoic fossils occur' (Green *et al.*, 1987, p. 928). A second fossil pleurocapsacean described from the ~770 Ma-old Australian deposit (*Palaeopleurocapsa wopfnerii*) has been compared by Knoll *et al.* with its living morphological and ecological analogue (*Pleurocapsa fuliginosa*) and regarded as 'further evidence of the evolutionary conservation of [cyanobacteria]' (Knoll, Barghoorn & Golubic, 1975, p. 2492). Two other species of morphologically distinctive fossil pleurocapsaceans (the endolithic taxa *Eohyella dichotoma* and *E. rectroclada*), cited as 'compelling examples of the close resemblance between Proterozoic prokaryotes and their modern counterparts' (Knoll *et al.*, 1986, p. 857), have been described by Green *et al.* from the East Greenland geological sequence as being 'morphologically, developmentally, and behaviorally indistinguishable' from living *Hyella* species of the Bahama Banks (Green, Knoll & Swett, 1988, pp. 837–838).

In concert with the extensive morphometric analyses of coccoidal and ellipsoidal microorganisms discussed above, these in-depth studies of entophysalidaceans and pleurocapsaceans (involving comparisons of environment, taphonomy, development, and behavior, in addition to cellular morphology) provide convincing evidence of species-specific similarities between Proterozoic and modern cyanobacteria.

Septate filamentous microorganisms and Proterozoic microfossils

Like the other categories of extant microorganisms considered here, cellular filamentous types of cyanobacteria tend to be decidedly larger than other such prokaryotes; in median dimensions, modern cyanobacterial trichomes are about three times broader than their non-cyanobacterial analogues (Fig. 6) and, whereas more than 50% of extant cyanobacterial septate filaments are more than 5.0 μm in diameter, only about 5% of other similarly filamentous free-living prokaryotes are of this size (Fig. 7). Moreover, as shown in Fig. 18, there is a significant difference in the abundance among

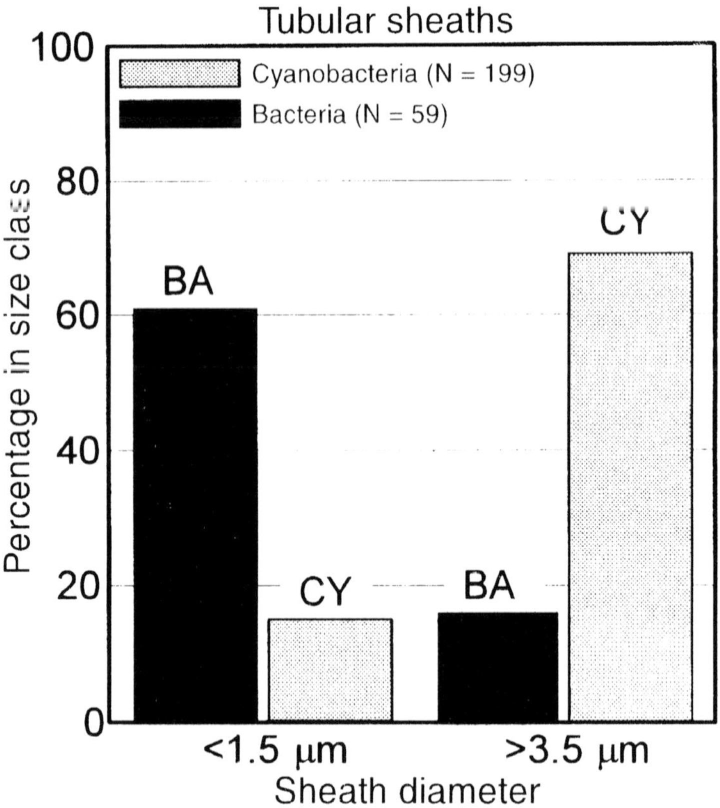

Fig. 18. Comparison of the size distributions of modern cellular cyanobacterial trichomes and other modern septate free-living filamentous prokaryotes ('bacteria') used to infer the probable affinities of fossil analogues; N = number of species and varieties considered. (Data from Desikachary, 1959; Buchanan & Gibbons, 1974; Laskin & Lechevalier, 1977; Clayton & Sistrom, 1978; Starr *et al.*, 1981; Jannasch, 1984.)

such extant taxa of filaments less than 1.5 μm and greater than 3.5 μm in width, a difference leading to interpretation of Proterozoic septate microfossils <1.5 μm in diameter as 'probable bacteria'; those 1.5 to 3.5 μm in diameter as undifferentiated 'prokaryotes'; and those >3.5 μm in diameter as 'probable cyanobacteria' (Schopf, 1992*b*).

Inferred affinities and evolutionary trends

In modern settings, the prokaryotic photoautotrophic biotas of nearshore marine environments (whether of sabkas, lagoons, or mud flats (Knoll & Golubic, 1992) or of flat-laminated to mound-shaped stromatolites (Logan *et al.*, 1974*a*, *b*; Golubic, 1976*a*; Pierson *et al.*, 1992)) are dominated by oscillatoriacean cyanobacteria (predominantly *Oscillatoria*, *Lyngbya*, *Phormidium*, *Spirulina*, *Microcoleus*, and *Schizothrix*). Such dominance, well evidenced in the fossil record by diverse cellularly preserved trichomes (Fig. 10), also evidently occurred during the Proterozoic. Of the 93 taxa of septate filamentous microfossils recognized in Proterozoic strata, >90% have been referred to the Oscillatoriaceae, and nearly 40% appear to be essentially indistinguishable in morphology from living oscillatoriacean species (Schopf, 1992*d*). Although the Proterozoic fossil record of such filaments is incompletely documented (composed of only 10 to 15 taxonomic occurrences per 100 Ma), no obvious evolutionary trends can be discerned; the size distribution of such fossils in Paleoproterozoic units does not differ significantly from those of Meso- or Neoproterozoic sediments (Fig. 19). Thus, like the other categories of fossil prokaryotes discussed here, the majority of Proterozoic septate filaments appear to be of cyanobacterial (oscillatoriacean) affinity with the group having exhibited no evident evolutionary change over Proterozoic time.

Non-septate filaments, prokaryotic sheaths, and Proterozoic microfossils

Figure 20 summarizes the size distribution of the tubular sheaths of modern monotrichomic filamentous bacteria (33 taxa) and modern non-septate non-helical 'thread cell' bacteria with maximum lengths greater than 10 μm and maximum widths greater than 0.5 μm (26 taxa), compared with the diameters of the tubular sheaths of 199 taxa of extant monotrichomic oscillatoriacean cyanobacteria. The median diameter of the bacterial filamentous sheaths and taxa is about six times smaller than that of the cyanobacterial sheaths (Fig. 6), and whereas more than 90% of the bacterial filaments are less than 5.0 μm in diameter, about two-thirds of the oscillatoriacean sheaths are larger than this size (Fig. 7). On the basis of the markedly differing size distributions summarized in Fig. 20, unbranched tubular Proterozoic microfossils less than 2.0 μm in diameter have been interpreted as 'probably bacterial'; those 2.0 to 3.5 μm in diameter as 'prokaryotic'; and those greater than 3.5 μm in diameter as 'probably cyanobacterial' (Schopf, 1992*b*).

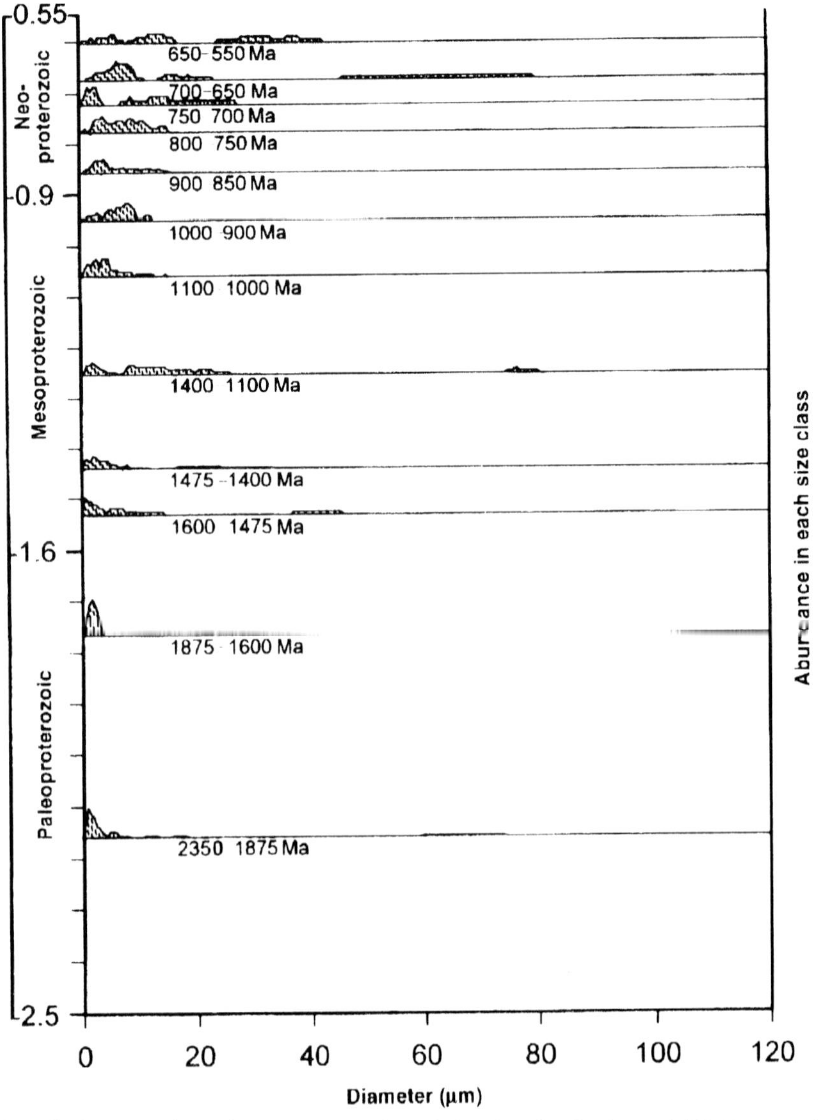

Fig. 19. Summary in various age categories of the size distributions of 251 Proterozoic occurrences in 69 geological units of the 93 taxa of septate filamentous microfossils recognized by Schopf (1992*d*) reported from petrographic thin sections of nearshore carbonates/cherts and acid-resistant residues of offshore shales/siltstones. (Data from Schopf, 1992*b*; Mendelson & Schopf, 1992.)

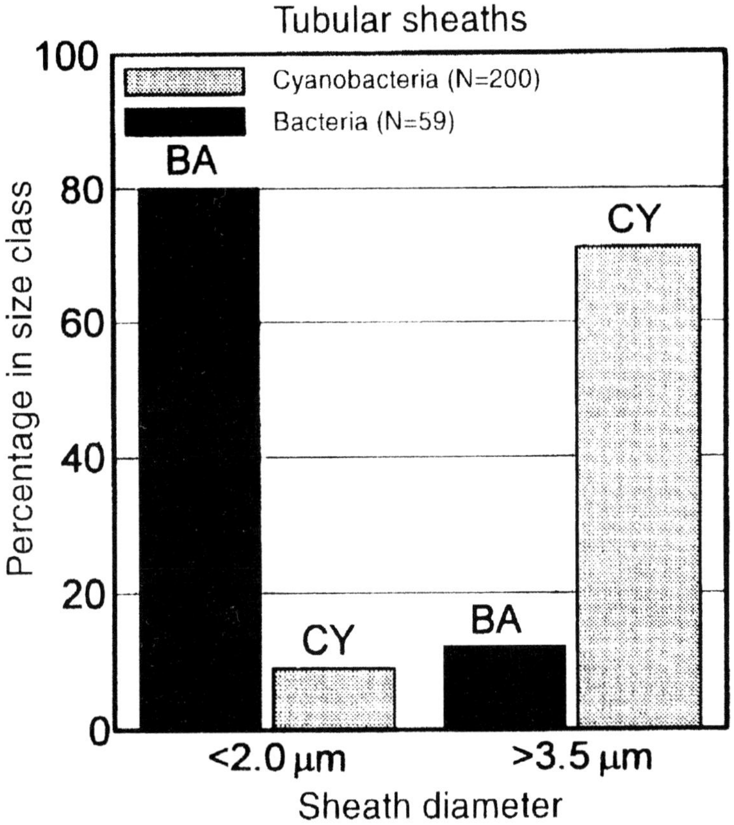

Fig. 20. Comparison of the size distributions of cylindrical–tubular unbranched monotrichomic sheaths of modern cyanobacteria with those of other modern free-living prokaryotes and non-septate 'thread cells' ≥0.5 μm in maximum diameter and ≥10.0 μm in maximum length ('bacteria') used to infer the probable affinities of fossil analogues; N = number of species and varieties considered. (Data from Desikachary, 1959; Buchanan & Gibbons, 1974; Laskin & Lechevalier, 1977; Clayton & Sistrom, 1978; Starr *et al.*, 1981; Jannasch, 1984.)

Inferred affinities and evolutionary trends

Figure 21 summarizes the size distributions of reported occurrences of unbranched tubular microfossils known from units of Paleo-, Meso-, and Neoproterozoic age. As shown in Fig. 4, there is a striking resemblance in both the size range and pattern of size distribution between the sheaths of modern oscillatoriacean cyanobacteria and those of the 64 taxa of tubular sheaths recognized in Proterozoic sediments (Schopf, 1992*d*). Thus, and although small-diameter non-cyanobacterial prokaryotic tubular sheaths have also been recognized in the Proterozoic (Schopf, 1968, 1992*d*; Mendelson & Schopf, 1992; Fairchild *et al.*, 1995), the vast majority of such known Proterozoic taxa, like those shown in Fig. 11, have been referred to the

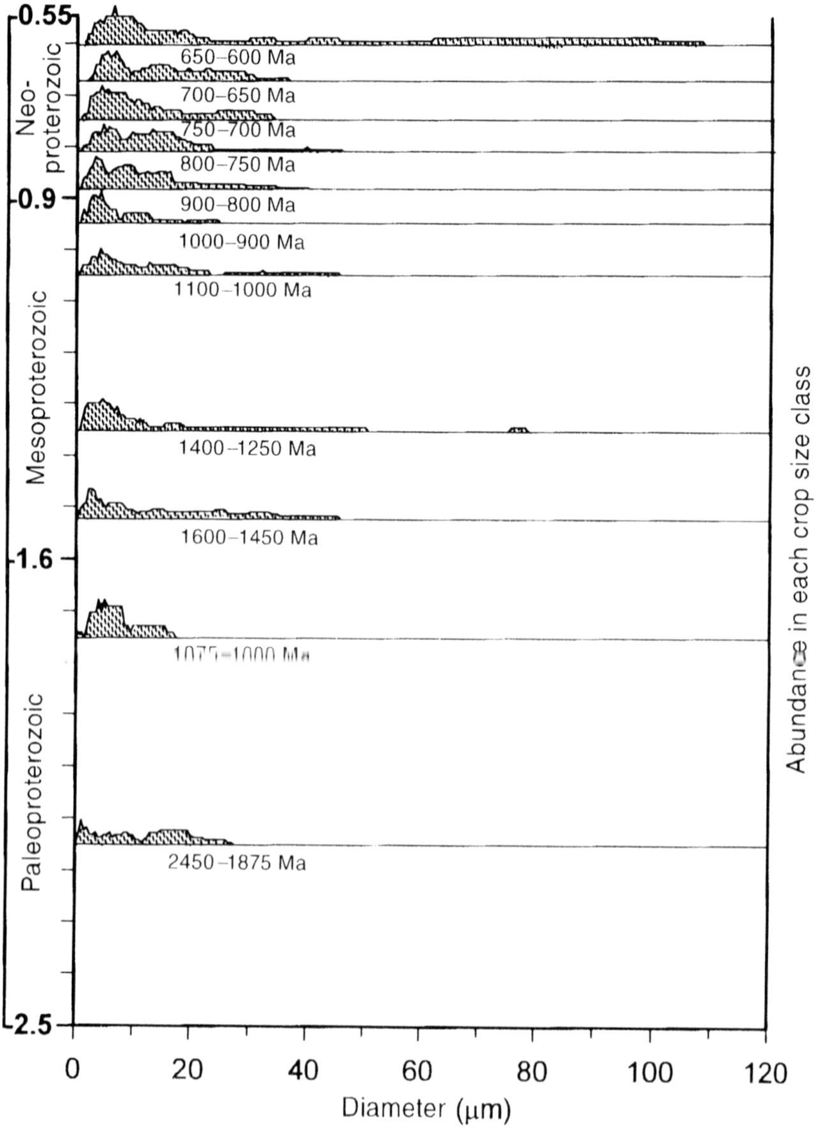

Fig. 21. Summary in various age categories of the size distributions of 447 Proterozoic occurrences in 104 geological units of the 64 taxa of cylindrical–tubular microfossils recognized by Schopf (1992*d*) reported from petrographic thin sections of nearshore carbonates/cherts and acid-resistant residues of offshore shales/siltstones. (Data from Schopf, 1992*b*; Mendelson & Schopf, 1992.)

Oscillatoriaceae (Schopf, 1992*d*; Mendelson & Schopf, 1992). The reported number of each such taxa gradually increases over the Proterozoic, but this increase is strongly coupled to, and is evidently a function of, the number of fossil assemblages sampled (Schopf, 1992*b*). Thus, not surprisingly (and like the Proterozoic septate fossil filaments discussed above (many of which were originally encompassed by extracellular tubular sheaths like those considered here)) no obvious evolutionary trends are reflected in the known Proterozoic record of non-septate filamentous microfossils.

Summary of morphometric analyses

Based on the analyses presented above, the vast majority of known Proterozoic prokaryotes can be inferred to be of a cyanobacterial affinity. Among solitary and colonial coccoidal and ellipsoidal fossils, most are evidently chroococcaleans (predominantly chroococcaceans, with subsidiary entophysalidaceans and pleurocapsaceans), and the great preponderance of known filamentous Proterozoic prokaryotes are comparable to modern nostocalean (i.e. oscillatoriacean) cyanobacteria. The three tests of the linkage between the putatively cyanobacterial fossils of the 3500 Ma-old Apex chert and microorganisms of the modern biota have been met: (i) The unbroken cyanobacterial lineage *has* left its mark in the intervening fossil record; (ii) members of Oscillatoriaceae, the specific cyanobacterial family inferred to be represented among the Apex fossils, *are* well represented in deposits of intermediate geological age; and (iii) the postulated evolutionary stasis of such microorganisms *is* well evidenced in intermediate-age assemblages, indicating that the astonishingly slow, hypobradytelic rate of evolution (Schopf, 1987) exhibited by oscillatoriaceans is probably characteristic of cyanobacteria in general (Schopf, 1994*b*, *c*).

The question remains, however, how strong is the evidence for the suggested cyanobacterial affinity of the Apex fossils? Can these extremely ancient microorganisms be demonstrated to be part of the established evolutionary continuum?

MICROORGANISMS OF THE ~3500 MA-OLD APEX CHERT

The recent discovery of unquestionable microfossils in the Archean (~3500 Ma-old) Apex chert of Western Australia (Schopf, 1992*a*, 1993) represents a major advance toward documenting the earliest records of life. Although questions have been raised regarding some earlier reports of putative Archean microfossils (Schopf, Hayes & Walter, 1983), none of these uncertainties applies to this recently discovered assemblage.

(i) The geographical and stratigraphic provenance of the fossiliferous chert is known with certainty (fossil-bearing samples having been collected from outcrops at the fossiliferous locality by different workers on repeated occasions).

(ii) The early Archean age (i.e. >3458 ± 1.9 Ma; <3471 ± 5 Ma) of the fossiliferous bedded chert is well established, based on U–Pb zircon ages of immediately overlying and stratigraphically underlying units of the rock sequence (Blake & McNaughton, 1984; Thorpe *et al.*, 1992).

(iii) The fossils are demonstrably indigenous to the unit, as demonstrated by their occurrence in petrographic thin sections (Fig. 22).

(iv) The fossils are preserved in transported and redeposited rounded lithic clasts (Schopf, 1992*a*, Fig. 1.5.4A; 1993, Fig. 3A, B) that are demonstrably syngenetic with deposition of the sedimentary conglomerate in which they occur (with the fossils themselves therefore predating deposition of this silicified bedded unit).

(v) Carbon isotopic data (Schopf, 1994*a*) and the morphological complexity and carbonaceous composition of the fossils establish that they are unquestionably biogenic (11 filamentous species have been identified, ranging from 0.5 to 19.5 μm in diameter (Fig. 23) and exhibiting rounded to conical terminal cells; quadrate, disc-shaped, or barrel-shaped medial cells; taxon-specific degrees of trichomic attenuation; and evidence of cell division similar to that occurring in modern filamentous prokaryotes).

Paleoecology and geologic setting

The Apex microfossils occur in a conglomerate chert bed of the Salgash Subgroup of the 3000 to 3500 Ma-old Pilbara Supergroup, about 12 km west of Marble Bar, northwestern Western Australia, in a geological terrain referred to as the Pilbara Craton. The geology of this region has been mapped in detail (Hickman, 1983); its geochemistry is exceptionally well defined (Thorpe *et al.*, 1992); and the paleoenvironmental setting of the ~30 km-thick well exposed flat-lying to gently tilted and relatively little metamorphosed (prehnite-pumpellyite and lower greenschist facies) geological sequences has been thoroughly documented (Barley *et al.*, 1979; Groves, Dunlop & Buick, 1981).

The Pilbara Craton is composed of large (20 to 50 km-diameter) low relief ovoid granite domes (batholiths), each surrounded by interbedded sequences of volcanic and sedimentary rock ('greenstone belts'). Within the greenstone belts there is ample evidence of shallow water deposition (e.g. the presence of evaporitic sulphates, scour-and-fill structures, storm-deposited edgewise conglomerates, cross-bedded silts and sands, pillow lavas). Evidently, the scene was dominated by extensive shallow seas, into which basaltic pillow lavas were erupted, with scattered (felsic) volcanic islands fringed by sedimentary debris (river gravels and sands), mud flats, and interspersed evaporitic lagoons. The conglomeratic Apex chert occurs

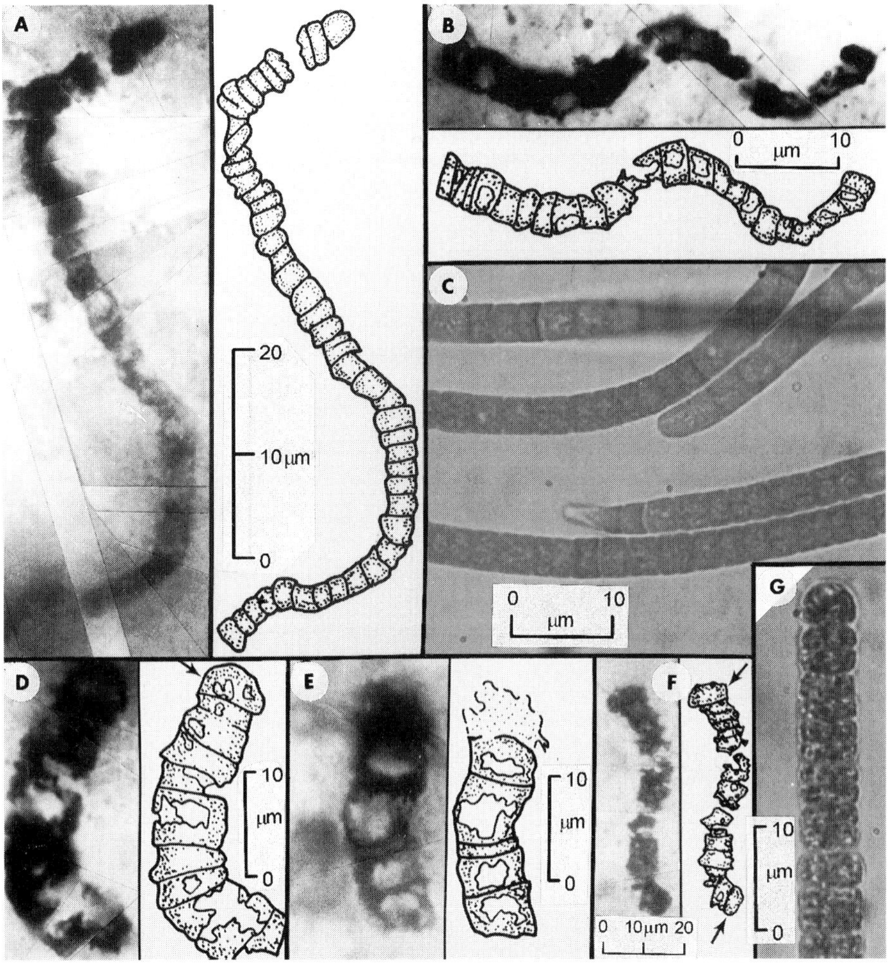

Fig. 22. Optical photomicrographs (transmitted light) showing living oscillatoriacean cyanobacteria (C and G), from mat-building stromatolitic communities of Baja, Mexico, for comparison with Archean microorganisms (A, B, D–F) shown in petrographic thin sections of the ~3500 Ma-old Apex chert of northwestern Western Australia. Because of the sinuous, three-dimensional preservation of these permineralized (petrified) carbonaceous microfossils, specimens in A, B, D, E, and F are shown in composite photomicrographs. A and B. *Primaevifilum amoenum* Schopf, a 2 to 5 μm-broad filamentous prokaryote having hemispheroidal terminal cells and quadrate to discoidal medial cells; C. *Oscillatoria* cf. *brevis* (Kütz.) Gomont, a modern oscillatoriacean having similarly rounded terminal cells and comparably sized medial cells. D, E, and F., *Primaevifilum laticellulosum* Schopf, a 6 to 9 μm-diameter filamentous prokaryote having quadrate to short-cylinder shaped medial cells and 'pillow-shaped' terminal cells (at arrows); G. *Oscillatoria nigroviridis* Thwaites ex Gomont, a modern oscillatoriacean of comparable size and cellular morphology.

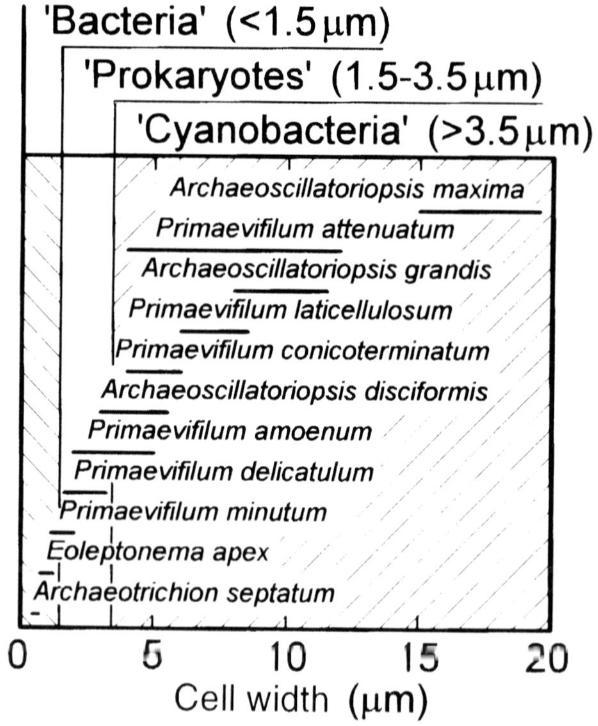

Fig. 23. Size ranges of the 11 species of prokaryotic filaments detected in the ~3500 Ma-old Apex chert of northwestern Western Australia compared with the 'probable affinities' of these taxa suggested by morphometric analyses.

within one such shallow-water sequence, sandwiched between two massive lava flows, on the western flank of the Mount Edgar Batholith.

Although the large scale community structure of the Apex assemblage is difficult to ascertain on the basis of the small (1 to a few mm) rounded clasts in which the microfossils occur (Schopf, 1992*a*, 1993), it evidently differs notably from that typical of Proterozoic stromatolitic assemblages. In particular, the Apex filaments exhibit neither the intertwined subparallel orientation nor the distinctly laminar organization characteristic of most stromatolitic microbiotas. Rather, they occur as irregularly distributed and randomly orientated solitary filaments surrounded by and appearing to 'float' within large wispy to fibrous clumps of flocculent, evidently originally mucilagenous, homogeneous brown to dark-brown finely particulate carbonaceous matter (kerogen). Thus, unlike any previously reported fossil microbial assemblage, the Apex microorganisms appear to have lived embedded within an exceptionally thick mucilagenous matrix. Because this is the only diverse microbial community known from sediments of such antiquity, it can only be speculated whether secretion of copious mucilage was typical of benthic microorganisms of this age. Were this to be established, however, it seems plausible that like other characteristics of modern mat-building prokaryotes (their effective DNA repair mechanisms, synthesis of UV-absorbing scytonemin, phototactic motility, adherence to substrates), secretion of such mucilage may have been an adaptation that enabled colonization of shallow water settings prior to development of an oxygenic atmosphere and an effective UV-absorbing ozone layer (Schopf, 1994*b*).

Affinities of the Apex microorganisms

The Apex assemblage stands out not only because of its great age but also because of its diversity. Indeed, the next oldest comparably diverse microbiotas known (those of the ~2600 Ma-old Campbell Group (Altermann & Schopf, 1995) of South Africa, and the ~2150 Ma-old Belcher Group (Hofmann, 1976) and ~2080 Ma-old Gunflint Formation (Barghoorn & Tyler, 1965) of Canada) are about one billion years younger.

Perhaps even more notable, however, is the fact that nearly two-thirds of the 11 taxa identified in this deposit (comprising ~63% of measured specimens) have been interpreted as probably oscillatoriacean cyanobacteria (Schopf, 1992*a*, 1993). This interpretation is consistent with (i) analyses of the early global ecosystem and cerium and europium concentrations in Archean banded iron formations suggesting that O_2-producing photosynthesis and aerobic respiration, processes characteristic of cyanobacteria, had both evolved in this early stage in Earth history (Towe, 1990, 1991); (ii) the isotopic compositions of organic and carbonate carbon in the Pilbara sediments, evidently indicative of photosynthetic CO_2 fixation like that occurring at the relatively high CO_2 concentrations in extant cyanobacterial populations (Schidlowski, Hayes & Kaplan, 1983; Schopf, 1994*a*); (iii) the occurrence within the Apex filaments of bifurcated cells and cell-pairs that evidently reflect the original presence of partial septations and, thus, of cell division like that occurring in extant oscillatoriacean filaments (Schopf, 1992*a*, Fig. 1.5.6 F, G; 1993, Fig. 5H–J); and (iv) rRNA phylogenies establishing that the Oscillatoriaceae is among the earliest evolved of extant cyanobacterial families (Giovannoni *et al.*, 1988).

The suggested affinity to modern oscillatoriacean cyanobacteria is similarly consistent with the size range, median dimensions, and pattern of size distribution of the Apex filaments, all of which, for the majority of the taxa, are decidedly more like those of extant oscillatoriaceans than of other modern filamentous prokaryotes (Figs. 23 and 24). Moreover, as shown in Fig. 23, several of the Apex taxa, particularly those with broad trichomes (*Primaevivilum laticellulosum*, Fig. 22D–F; *P. attenuatum*; *Archaeoscillatoriopsis grandis*; and *A. maxima*), differ in cell size from almost all

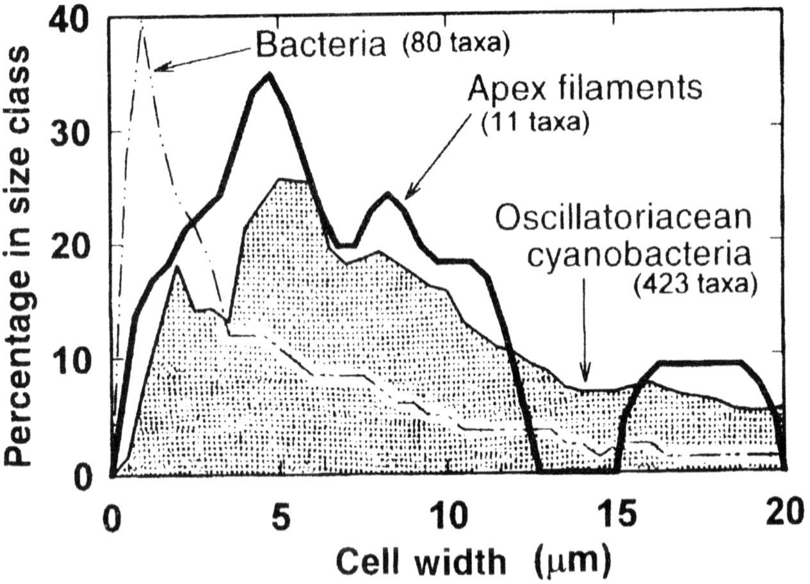

Fig. 24. Cell widths of modern septate oscillatoriacean cyanobacteria and other modern free-living cellular prokaryotic filaments ('bacteria') ≤20 μm in diameter compared with those of taxa detected in the Apex chert.

noncyanobacterial septate prokaryotes (Fig. 7) but are essentially indistinguishable from specific oscillatoriaceans, both Proterozoic (*Oscillatoriopsis* spp.) and modern (*Oscillatoria* spp.).

Taken together these various lines of evidence seem persuasive. There is little doubt that, if the Apex filaments had been discovered in Proterozoic sediments, in which fossil oscillatoriaceans are well known and relatively widespread, or if they had been detected in a modern microbial community and morphology were the only criterion by which to infer biological relationships, the majority would be interpreted as oscillatoriacean cyanobacteria. Their linkage to the modern biota via the intermediate-age Proterozoic fossil record provides powerful evidence of their probable cyanobacterial affinity.

CONCLUSIONS

(i) Over the past quarter-century, detailed genus- and species-level similarities in cellular morphology between described taxa of Precambrian microfossils and extant cyanobacteria have been noted and regarded as biologically and taxonomically significant by numerous workers worldwide.

(ii) On the basis of detailed morphometric analyses of large representative samples of fossil and modern prokaryotes, such similarities have been particularly well documented for members of the Chroococcaceae and Oscillatoriaceae, the two most abundant and widespread Proterozoic cyanobacterial families.

(iii) For members of two additional families, the Entophysalidaceae and Pleurocapsaceae, species-level morphological similarities are supported by in-depth fossil–modern comparisons of environment, taphonomy, development, and behaviour.

(iv) The recent discovery of cellularly preserved oscillatoriacean-like filamentous prokaryotes in the early Archean Apex chert of Western Australia suggests that the cyanobacterial lineage may have been established as early as ~3500 Ma ago, an age about three-quarters that of the Earth.

(v) Three tests to evaluate this suggestion have been met: (i) Cyanobacteria are abundant in the intervening Proterozoic (2500 to ~550 Ma-old) fossil record; (ii) oscillatoriaceans like those inferred to occur in the Apex biota are well represented in these intermediate-age deposits; as is evidence of (iii) widespread evolutionary stasis among cyanobacteria in general.

(vi) Thus, it seems reasonable to conclude that morphologically, and apparently physiologically as well, cyanobacterial 'living fossils' have exhibited an extraordinary slow (hypobradytelic) rate of evolutionary change, a characteristic evidently established early in their evolutionary history as a result of selection of highly successful ecological generalists capable of tolerating a broad range of environmental conditions (Schopf, 1994*b*).

ACKNOWLEDGEMENTS

I thank Carl V. Mendelson of Beloit College for help in compiling morphometric data on Proterozoic microfossils, and Richard Mantonya and J. Shen-Miller of UCLA for assistance in preparation of this manuscript. This study was supported by National Aeronautics and Space Administration Grant NAGW-2147.

REFERENCES

Altermann, W. & Schopf. J. W. (1995). Microfossils from the Neoarchean Campbell Group, Griqualand West Sequence of the Transvaal Supergroup, and their paleoenvironmental and evolutionary implications. *Precambrian Research*, in press.

Barghoorn, E. S. & Schopf, J. W. (1965). Microorganisms from the Late Precambrian of central Australia. *Science*, **150**, 337–9.

Barghoorn, E. S. & Tyler, S. A. (1965). Microorganisms from the Gunflint chert. *Science*, **147**, 563–77.

Barley, M. E., Dunlop, J. S. R., Glover, J. E. & Groves, D. I. (1979). Sedimentary evidence from an Archean shallow-water volcanic-sedimentary facies, eastern Pilbara Block, Western Australia. *Earth Planetary Sciences Letters*, **43**, 74–84.

Blake, T. S. & McNaughton, N. J. (1984). A geochronological framework for the Pilbara region. In Muhling, J. R., Groves, D. K. & Blake, T. S. (eds.), *University of Western Australia Geology Department & University Ext., Publications*, **9**, 1–22.

Bourrelly, P. (1970). *Les Algues d'Eau Douce: Algues Bleues et Rouges*, Vol. 3, 512 pp. Éditions N. Boubée, Paris.

Broda, E. (1975). *The Evolution of Bioenergetic Processes*, 220 pp. Pergamon, New York.

Buchanan, R. E. & Gibbons, M. E. (1974). *Bergey's Manual of Determinative Bacteriology*, 8th edn., 1268 pp. Williams & Wilkins, Baltimore, MD.

Clayton, R. K. & Sistrom, W. R., eds. (1978). *The Photosynthetic Bacteria*, 946 pp. Plenum, New York.

Cloud, P. E., Jr (1965). Significance of the Gunflint (Precambrian) microflora. *Science*, **148**, 27–35.

Desikachary, T. V. (1959). *Cyanophyta*, 686 pp. Indian Council for Agricultural Research, New Delhi.

Fairchild, T. R. (1975). *The Geologic Setting and Paleobiology of a Late Precambrian Stromatolitic Microflora from South Australia.* PhD dissertation, Department of Geology, University of California, Los Angeles, 346 pp.

Fairchild, T. R., Schopf, J. W., Shen-Miller, J., Guimarães, E., Edwards, M. D., Lagstein, A., Li, X., Pabst, M. & Soares de Melo-Felho, L. (1995). Recent discoveries of Proterozoic microfossils in south-central Brazil. *Precambrian Research*, in press.

Giovannoni, S. J., Turner, S., Olsen, G. J., Barns, S., Lane, D. J. & Pace, N. R. (1988). Evolutionary relationships among cyanobacteria and green chloroplasts. *Journal of Bacteriology*, **170**, 3584–92.

Golubic, S. (1976*a*). Organisms that build stromatolites. In *Stromatolites, Developments in Sedimentology 20*, Walter, M. R., ed., pp. 113–126. Elsevier, Amsterdam.

Golubic, S. (1976*b*). Taxonomy of extant stromatolite-building cyanophytes. In *Stromatolites, Developments in Sedimentology 20*, Walter, M. R., ed., pp. 127–140, Elsevier, Amsterdam.

Golubic, S. & Hofmann, H. J. (1976). Comparison of Holocene and mid-Precambrian Entophysalidaceae (Cyanophyta) in stromatolitic mats: cell division and degradation. *Journal of Paleontology*, **50**, 1074–82.

Green, J. W., Knoll, A. H., Golubic, S. & Swett, K. (1987). Paleobiology of distinctive benthic microfossils from the Upper Proterozoic Limestone-Dolomite 'Series', central East Greenland. *American Journal of Botany*, **74**, 928–40.

Green, J. W., Knoll, A. H. & Swett, K. (1988). Microfossils from oolites and pisolites of the Upper Proterozoic Eleonore Bay Group, central East Greenland. *Journal of Paleontology*, **62**, 835–52.

Groves, D. I., Dunlop, J. S. R. & Buick, R. (1981). An early habitat of life. *Scientific American*, **245**, 64–73.

Hickman, A. H. (1983). Geology of the Pilbara Block and its environs. *Western Australia Geological Survey Bulletin*, **127**, 1–268.

Hofmann, H. J. (1976). Precambrian microflora, Belcher Islands, Canada: significance and systematics. *Journal of Paleontology*, **50**, 1040–73.

Jannasch, H. W. (1984). Microbial processes at deep sea hydrothermal vents. In *Hydrothermal Processes at Sea Floor Spreading Centers*, Rona, R. A., ed., pp. 677–709. Plenum, New York.

Knoll, A. H., Barghoorn, E. S. & Golubic, S. (1975). *Paleopleurocapsa wopfnerii gen. et sp. nov*: a late Precambrian alga and its modern counterpart. *Proceedings of the National Academy of Sciences, USA*, **72**, 2488–92.

Knoll, A. H. & Golubic, S. (1979). Anatomy and taphonomy of a Precambrian algal stromatolite. *Precambrian Research*, **10**, 115–51.

Knoll, A. H. & Golubic, S. (1992). Proterozoic and living cyanobacteria. In *Early Organic Evolution*, Schidlowski, M., Golubic, S., Kimberley, M. M., McKirdy, D. M. & Trudinger, P. A., eds., pp. 450–462. Springer, New York.

Knoll, A. H., Golubic, S., Green, J. & Swett, K. (1986). Organically preserved microbial endoliths from the late Proterozoic of East Greenland. *Nature*, **321**, 856–7.

La Rivers, I. (1978). *Algae of the Western Great Basin*, 390 pp. University of Nevada, Las Vegas, NV.

Laskin, A. I. & Lechevalier, H. A., eds. (1977). *CRC Handbook of Microbiology*, 2nd edn., vol. 1, 757 pp. CRC Press, Boca Raton, FL.

Logan, B. W., Hoffman, P. W. & Gebelein, C. D. (1974*a*). Algal mats, cryptalgal fabrics, and structures, Hamelin Pool, Western Australia. *American Society of Petroleum Geologists Memoir*, **22**, 140–95.

Logan, B. W., Read, J. F., Hagen, G. W., Hoffman, P. W., Brown, R. G., Woods, P. J. & Gebelein, C. D. (1974*b*). Evolution and diagenesis of Quaternary carbonate sequences, Shark Bay, Western Australia. *American Association Petroleum Geologists Memoir*, **22**, 1–358.

Mendelson, C. V. & Schopf, J. W. (1992). Proterozoic and selected Early Cambrian microfossils and microfossil-like objects. In *The Proterozoic Biosphere*, Schopf, J. W. & Klein, C., eds., pp. 865–951. Cambridge University Press, New York.

Pierson, B. K., Bauld, J., Castenholz, R. W., D'Amelio, E., Des Marais, D. J., Farmer, J. D., Grotzinger, J. P., Jørgensen, B. B., Nelson, D. C., Palmisano, A. C., Schopf, J. W., Summons, R. E., Walter, M. R. & Ward, D. M. (1992). Modern mat-building microbial communities: a key to the interpretation of Proterozoic stromatolitic communities. In *The Proterozoic Biosphere*, Schopf, J. W. & Klein, C., eds., pp. 245–342. Cambridge University Press, New York.

Prescott, G. W. (1954). *How to Know the Fresh-Water Algae*, 211 pp. W. C. Brown, Dubuque, IA.

Prescott, G. W. (1962). *Algae of the Western Great Lakes Region*, 977 pp. W. C. Brown, Dubuque, IA.

Schidlowski, M., Hayes, J. M. & Kaplan, I. R. (1983). Isotopic inferences of ancient biochemistries: carbon, sulfur, hydrogen, and nitrogen. In *Earth's Earliest Biosphere*, Schopf, J. W., ed., pp. 149–186. Princeton University Press, Princeton, NJ.

Schopf, J. W. (1968). Microflora of the Bitter Springs Formation, Late Precambrian, central Australia. *Journal of Paleontology*, **42**, 651–88.

Schopf, J. W. (1972). Evolutionary significance of the Bitter Springs (Late Precambrian) microflora. *Proceedings, 24th International Geology Congress, Section 1, Precambrian Geology*, pp. 68–77.

Schopf, J. W. (1974). The development and diversification of Precambrian life. *Origins of Life*, **5**, 119–35.

Schopf, J. W. (1977). Biostratigraphic usefulness of stromatolitic Precambrian microbiotas: a preliminary analysis. *Precambrian Research*, **5**, 143–73.

Schopf, J. W. (1987) 'Hypobradytely': comparison of rates of Precambrian and Phanerozoic evolution. *Journal of Vertebrate Paleontology*, **7** (3, Suppl.), 25.

Schopf, J. W. (1992*a*). Paleobiology of the Archean. In *The Proterozoic Biosphere*, Schopf, J. W. & Klein, C., eds., pp. 25–39. Cambridge University Press, New York.

Schopf, J. W. (1992*b*). Proterozoic prokaryotes: affinities, geologic distribution, and evolutionary trends. In *The Proterozoic Biosphere*, Schopf, J. W. & Klein, C., eds., pp. 195–281. Cambridge University Press, New York.

Schopf, J. W. (1992*c*). Historical development of Proterozoic micropaleontology. In *The Proterozoic Biosphere*, Schopf, J. W. & Klein, C., eds., pp. 179–183. Cambridge University Press, New York.

Schopf, J. W. (1992*d*). Informal revised classification of Proterozoic microfossils. In *The Proterozoic Biosphere*, Schopf, J. W. & Klein, C., eds., pp. 1119–1166. Cambridge University Press, New York.

Schopf, J. W. (1992*e*). Tempo and mode of Proterozoic evolution. In *The Proterozoic Biosphere*, Schopf, J. W. & Klein, C., eds., pp. 595–598. Cambridge University Press, New York.

Schopf, J. W. (1992*f*). Times of origin and earliest evidence of major biologic groups. In *The Proterozoic Biosphere*, Schopf, J. W. & Klein, C., eds., pp. 587–593. Cambridge University Press, New York.

Schopf, J. W. (1993). Microfossils of the Early Archean Apex chert: new evidence of the antiquity of life. *Science*, **260**, 640–6.

Schopf, J. W. (1994*a*). The oldest known records of life: stromatolites, microfossils, and organic matter from the Early Archean of South Africa and Western Australia. In *Early Life on Earth*, Bengtson, S., ed., pp. 193–206. Columbia University Press, New York.

Schopf, J. W. (1994*b*). Disparate rates, differing fates: tempo and mode of evolution changed from the Precambrian to the Phanerozoic. *Proceedings of the National Academy of Sciences, USA*, **91**, 6735–42.

Schopf, J. W. (1994*c*). The early evolution of life: solution to Darwin's dilemma. *Trends in Ecology and Evolution*, **9**, 193–206.

Schopf, J. W. (1995*a*). Cyanobacteria: pioneers of the early Earth. *Nova Hedwigia*, in press.

Schopf, J. W. (1995*b*). Metabolic memories of Earth's earliest biosphere. In *Evolution and the Molecular Revolution*, Marshall, C. R. & Schopf, J. W., eds. Jones and Bartlett, Boston, MA, in press.

Schopf, J. W., Hayes, J. M. & Walter, M. R. (1983). Evolution of Earth's earliest ecosystems: Recent progress and unsolved problems. In *Earth's Earliest Biosphere*, Schopf, J. W., ed., pp. 361–384. Princeton University Press, Princeton, NJ.

Smith, R. L. & Bold, H. C. (1966). *Phycological Studies VI. Investigations of the Algal Genera* Eremosphaera *and* Oocystis. *University of Texas Publication No. 6612,* University of Texas, Austin, TX, 121 pp.

Starr, M. P., Stolp, H., Trüper, H. G., Balows, A. & Schlegel, H. G., eds. (1981). *The Prokaryotes* vols. 1 and 2, 2284 pp. Springer, New York.

Taylor, W. R. (1957). *Marine Algae of the Northeastern Coast of North America*, 509 pp. University of Michigan, Ann Arbor, MI.

Thorpe, R. I., Hickman, A. H., Davis, D. W., Mortensen, J. K. & Trendall, A. F. (1992). U–Pb zircon geochronology of Archaean felsic units in the Marble Bar region, Pilbara Craton, Western Australia. *Precambrian Research*, **56**, 169–89.

Towe, K. M. (1990). Aerobic respiration in the Archaean? *Nature*, **348**, 54–6.

Towe, K. M. (1991). Aerobic carbon cycling and cerium oxidation: significance for Archean oxygen levels and banded iron-formation deposition. *Palaeogeography, Palaeoclimatology and Palaeoecology*, **97**, 113–23.

Tyler, S. A. & Barghoorn, E. S. (1954). Occurrence of structurally preserved plants in Pre-Cambrian rocks of the Canadian Shield. *Science*, **119**, 606–8.
West, G. S. & Fritsch, F. E. (1927). *A Treatise on the British Freshwater Algae*, 535 pp. Cambridge University Press, Cambridge, UK.
Woese, C. R. (1987). Bacterial evolution. *Microbiological Reviews*, **51**, 221–71.

RIBOSOMAL RNA AND THE EVOLUTION OF BACTERIAL DIVERSITY

STEPHEN J. GIOVANNONI, MICHAEL S. RAPPÉ, DOUGLAS GORDON, ENA URBACH, MARCELINO SUZUKI AND KATHARINE G. FIELD

Department of Microbiology, 220 Nash Hall, Oregon State University, Corvallis, OR 97331-3804, USA

INTRODUCTION

The development of a satisfactory scheme that explains the diversity of microorganisms has been a challenge to generations of microbiologists. The past two decades have been a period of very rapid progress in this field. Much of this progress resulted from the introduction of molecular phylogenetic methods, particularly the widespread sequencing of ribosomal RNA genes (Olsen & Woese, 1993; Woese, 1987, 1991). The influence of these studies in microbiology has been pervasive: microbial diversity, evolution, diagnostics, and ecology have all been profoundly influenced. A broad outline of microbial evolution has emerged which might best be regarded as a complex theory composed of many parts. It describes the evolution of a single molecule, the ribosomal RNA gene, throughout an extraordinarily broad range of life forms.

While indeed a profoundly useful image of life, the ribosomal RNA phylogenetic gene tree should be regarded as a theory that depends on a particular set of assumptions (Olsen, 1987). Ribosomal RNA phylogenetic trees present the relationships among a single homologous set of molecules. These phylogenetic trees are based on the assumptions that rRNA genes are free from artifacts of convergent evolution or lateral gene transfer between species; both are potentially confounding factors that might obscure the record of cellular descent. Thus, it is with satisfaction that molecular phylogenists have watched the growing congruence between ribosomal RNA phylogenetic trees, phylogenies derived from other molecules, and comparative physiology. It now appears that the resolution of small subunit (SSU) rRNA gene trees, which is determined by factors such as the number of nucleotide residues in the molecule and the distribution of substitution rates among its positions, has been adequate to resolve many of the distant and complex events that occurred in bacterial evolution.

This theory provides the basis for broader explorations of the details of microbial evolution. Vastly improved methods for the acquisition of genetic sequences and their analysis are being used to build on ribosomal RNA phylogenetic models. Nucleic acid sequence information for other molecules, and in some cases entire genomes, is being added to databases, yielding a picture that is more complete in its details, particularly with regard to the evolution of genetic systems, metabolic pathways, and the lateral movement of genetic information between phylogenetically unique cellular lineages. This latter factor, the transfer of genetic information across phylogenetic boundaries, has surely played a key role in the emergence of modern life forms. The full importance of this phenomenon may only be appreciated when comparisons of complete microbial genome sequences become routine.

Here we offer a brief summary of microbial diversity, with descriptions of some of the key problems and promising avenues of research. Particular emphasis is given to environmental studies and some interpretations of Precambrian evolution from a molecular phylogenetic perspective.

MAJOR FEATURES IN THE BACTERIAL PHYLOGENETIC TREE

Figure 1 is a SSU rRNA phylogenetic tree representing the three major lineages of life: the Archaea, Bacteria, and Eukarya (Woese, Kandler & Wheelis, 1990). This tree demonstrates the essentially tripartite nature of cellular life, separating prokaryotes, those cells which lack nuclei, into two domains, the Archaea and Bacteria. Of key interest to many is the position of the node that connects this unrooted, tripartite tree to the earliest cells. By analysing phylogenies of duplicated genes, Gogarten and coworkers and Iwabe and coworkers separately presented convincing evidence that the

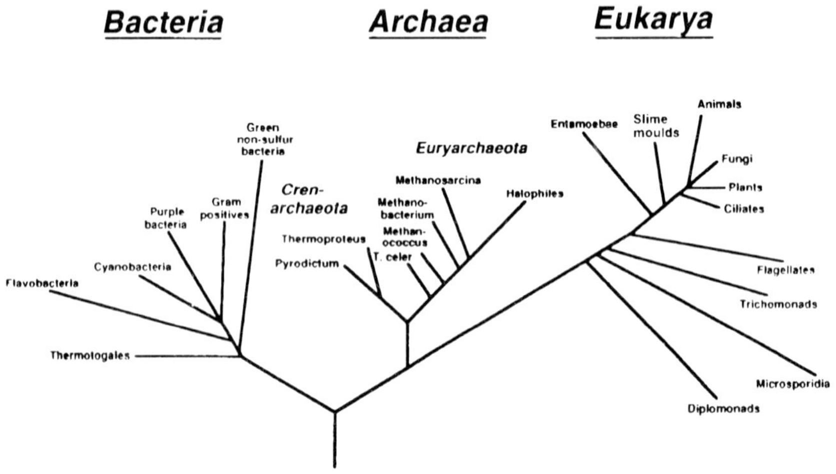

Fig. 1. SSU rRNA phylogenetic tree representing the three major lineages of life: the Archaea, the Bacteria, and Eukarya (from Olsen & Woese, 1993, with permission).

most likely position for this node is on the internal tree segment connecting Bacteria to Archaea and Eukarya (Gogarten *et al.*, 1989; Iwabe *et al.*, 1989). The implication is that Archaea and Eukarya diverged as a common lineage from Bacteria, and that a subsequent divergence created the division between them. The greater similarity between Archaea and Eukarya, first identified in comparisons of ribosomal RNA gene sequences, has found strong support from analyses of other genetic features of cells, and from phylogenetic analyses of protein sequences (Doolittle *et al.*, 1996). The issues surrounding this phylogenetic tree, and the particular relationships among the subdivisions of the Archaea, will not be discussed here.

The evolution of diversity among Bacteria is represented in Fig. 2, a consensus phylogenetic tree constructed from 1232 small ribosomal subunit RNA sequences (Van de Peer *et al.*, 1994). This tree was inferred using an unusual statistical method that controls for the idiosyncratic effects of individual sequences, which may effect the order of divergence among deep branches. The consensus tree corresponds well with other bacterial 16S rRNA phylogenetic trees. Its taxa are separated by distances roughly equivalent to eukaryotic phylum-level distinctions.

In Fig. 2, the earliest branches of the Bacteria are thermophilic organisms, represented here by the Thermotogales, *Thermotoga* and *Fervidobacterium*, and *Aquifex* spp. These organisms possess peptidoglycan cell walls, suggesting that the sacculus was invented early in the diversification of the Bacteria (Kandler, 1994; Koch, 1994). *Fervidobacterium* and *Thermotoga* are heterotrophs, and *Aquifex* is a chemolithotroph that uses hydrogen as an exogenous electron source. Most methods of phylogenetic inference place *Aquifex* unambiguously at the base of the bacterial phylogenetic tree. The fact that *Aquifex* uses oxygen as a terminal electron acceptor is interesting, considering its evolutionary emergence long before the cyanobacterial radiation.

The order in which the remaining bacterial taxa diverged is more difficult to establish, but many publications place the green non-sulphur bacteria as the next clade to emerge after the Thermotogales (Woese, 1987). The green non-sulphur bacteria include heterotrophs as well as species capable of autotrophy by anaerobic photosynthesis. Most of the green non-sulphur bacteria are thermophiles, but a recent study suggests that this may not be a defining characteristic of this phylum (Giovannoni *et al.*, 1996).

The Planctomycetales have also been proposed as a phylogenetically ancient lineage (Fuerst, 1995; Stackebrandt *et al.*, 1984; Van de Peer *et al.*, 1994). This unusual group of budding bacteria does not produce peptidoglycan cell walls, and may prove a sister group to *Chlamydia*, which shares this characteristic (Woese, 1987). The point of divergence of the Planctomycetales from the bacterial lineage has been difficult to establish; its apparent position in the phylogeny changes with different methods of analysis (Fuerst, 1995).

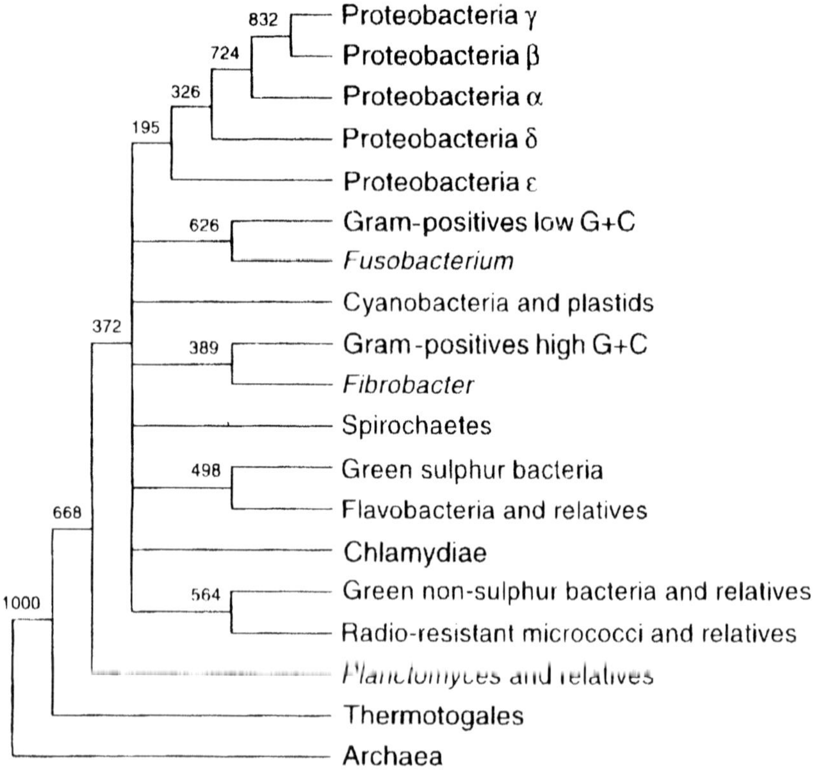

Fig. 2. Consensus phylogenetic tree constructed from 1232 small ribosomal subunit RNA sequences, illustrating the evolution of diversity among the Bacteria. The tree is a majority-rule consensus of 1000 trees, each inferred by the neighbour-joining method from 19 SSU rRNA sequences. Datasets for each of the 1000 trees consisted of one sequence randomly chosen from each of 18 bacterial taxa, with a randomly chosen archaeal sequence designated as the root (from Van de Peer *et al.*, 1994, with permission). The numbers refer to the number of trees (of 1000) which supported the clades.

Chlamydia is a genus of Gram-negative, obligate intracellular parasites that lack peptidoglycan cell walls. The phylogenetic position of this genus is difficult to establish, and the position of *Chlamydia* is left unresolved in Fig. 2.

Deinococcus and *Thermus* are representative genera of the radio-resistant micrococci and their relatives. *Deinococcus* is Gram-positive and highly radiation resistant, while *Thermus* is a Gram-negative thermophile. The 16S rRNA sequence of *Thermus* appears similar to other bacterial thermophiles, but this is thought to be an artefact arising from unusually high G + C content amongst thermophiles (Weisburg, Giovannoni & Woese, 1989).

After the divergence of the green non-sulphur bacteria, the remaining bacterial phyla emerge as a radiation. These include the Proteobacteria, a diverse group including most of the described Gram-negative bacteria. This phylum includes autotrophs, heterotrophs, chemolithotrophs, anaerobes, and aerobes. There are five currently recognized subdivisions of the Proteobacteria: the alpha, beta, gamma, delta and the recently established epsilon. It has been observed that the delta and epsilon divisions are sometimes separated from the alpha, beta and gamma subdivisions in phylogenetic analyses (Van de Peer *et al.*, 1994).

The Gram-positive phylum is divided into high G + C and low G + C phylogenetic groups. While Woese and coworkers consider the Gram-positive bacteria to be monophyletic, Van de Peer and coworkers present evidence that the two Gram-positive clades are sometimes divided by the inclusion of the cyanobacteria (Van de Peer *et al.*, 1994; Woese, 1987). The cyanobacterial cell wall has features, particularly the thickness of the peptidoglycan, that are more similar to Gram-positive than to Gram-negative cell walls.

The heliobacteria appear to be related to the cyanobacteria (Vermaas, 1994). These obligately anaerobic phototrophs contain an unusual chlorophyll, bacteriochlorophyll *g*, which spontaneously oxidizes to chlorophyll *b* in the presence of oxygen.

The cyanobacteria, most of which contain chlorophyll *a*, are defined by the common possession of water-splitting photosystem II (Woese, 1987). In contrast to most bacterial groups, there are a wide variety of cyanobacterial morphological types, but very little variation in physiology.

The spirochaetes are one of the few bacterial clusters that can reliably be classified on the basis of morphology as well as 16S rRNA similarity. A recent study proposed that *Nitrospira* and *Leptospira* constitute a separate bacterial phylum (Ehrich *et al.*, 1995).

The *Bacteroides/Flavobacterium* group includes a mixture of physiological types, from obligate anaerobes to obligate aerobes. This group has been phylogenetically linked with the green sulphur bacteria, a monophyletic group of obligately anaerobic phototrophs requiring exogenous reduced electron sources (Woese *et al.*, 1990). Recent work in our laboratory supports this hypothesis, but also indicates a specific though distant relationship linking the green sulphur bacteria to *Fibrobacter* (Gordon & Giovannoni, 1996).

Advances in phylogenetic methods have enabled researchers to gain a better understanding of bacterial diversity. Maximum likelihood analyses, bootstrapping and the availability of super computers have all combined to give an increasingly clear picture of the domain Bacteria. It is evident, however, that a great deal of ambiguity remains. Early analyses were constrained by the relatively small number of sequences available. As the sophistication of phylogenetic analysis continues to improve and the number and variety of sequences, including both cultured strains and genes cloned directly from the environment, continue to accumulate, an increasingly refined picture of bacterial evolution will inevitably emerge.

The consensus phylogenetic tree of Van de Peer and coworkers shown in Fig. 2 illustrates that the most deeply branching bacterial groups evolved in a radiation with largely unresolved branching orders. A combination of uncertainty in phylogenetic distance estimates and closely spaced internodes connecting bacterial phyla combine to create what is sometimes called a fan or shrub. One of the few groups that clearly emerged early from this radiation are the Thermotogales. Various studies place the Planctomycetales and relatives and the green non-sulphur bacteria as the next phylogenetic groups to emerge, but these conclusions are not well supported by bootstrap methods for estimating confidence. This uncertainty frustrates attempts to reconstruct the order of events in the development of metabolic diversity among microorganisms.

The radiation of bacterial phyla exhibited by ribosomal RNA phylogenies implies that much of the metabolic diversity of microorganisms that we know today arose in an early period of the earth's history. While it is tempting to speculate on the timing of these events, it is perhaps safest to draw the conclusion that evolutionary diversification soon followed the origins of the first bacterial cells. It has been observed that photosynthesis is a recurring theme in the bacterial phylogenetic tree. The green non-sulphur bacteria, which includes phototrophic species, appear to be one of the first branches to emerge following the Thermotogales, although this result is not well supported statistically. Phototrophic members also appear in the Gram-positive bacteria, the cyanobacteria and most subdivisions of the Proteobacteria. It is interesting to note that carbon fixation by the green non-sulphur bacteria occurs by the hydroxypropionate pathway and not by the Calvin cycle.

Achenbach-Richter and coworkers advanced the theory that the earliest life forms were thermophiles (Achenbach-Richter *et al.*, 1992). Support for this conclusion came from ribosomal RNA phylogenetic trees in which the deepest members of both the Archaea and the Bacteria were represented as the thermophilic groups. More recently, the discovery of Crenarchaeota in the cold pelagic ocean has cast doubt on a thermophilic origin for life (DeLong, 1992).

THE IMPLICATIONS OF RIBOSOMAL RNA PHYLOGENIES FOR THE EARLY EVOLUTION OF BACTERIAL LIFE ON EARTH

The primary goal of phylogenetic tree reconstruction is to determine the order of branching among taxa. Attempts to determine the relative timing of phylogenetic 'events' (branchings) are rarely undertaken, though there have been some successes with these endeavours. A fundamental assumption of most phylogenetic analyses is that the pattern of nucleotide substitution in macromolecules displays a 'clock-like' behaviour: that is, substitution is random and a linear function of time. If this assumption were absolutely true, i.e. if substitutions were random and equally likely at all positions within molecules, and if infinitely long macromolecular sequences were available for comparisons, then it would be possible to accurately reconstruct the timing as well as the order of events in the early evolution of bacteria. However, these assumptions represent the ideal case and are met only partially by real molecules. Thus, the best methods for phylogenetic reconstruction are those which are most insensitive to deviations from these assumptions.

Despite these caveats, for some special cases it has been shown that both the order and spacing of taxa in 16S rRNA gene trees are congruent with reliably dated fossil records (Moran *et al.*, 1993). This result suggests (i) that phylogenetic methods provide reasonable models of the evolutionary behaviour of genes, and (ii) that in some cases the timing of evolutionary events might be inferred from such analyses. A question then arises. Can a timeline for early events in microbial evolution be inferred from ribosomal RNA sequence comparisons? Such a timeline might provide an extraordinarily interesting approach to understanding the interaction between the evolution of microbial physiology and the biogeochemistry of the earth during the Archean.

One of the most significant events in the biogeochemical history of earth was the transition from the anoxic atmosphere of the primordial earth to an oxidizing atmosphere. The primary cause of this transition was the appearance of bacteria capable of extracting electrons from water in a light-driven reaction for the reduction of carbon dioxide. One consequence of this metabolism was the production of oxygen as a waste product. Extant organisms that perform this metabolism comprise a holophyletic group in phylogenetic trees based on both ribosomal RNAs (Bhattacharya & Medlin, 1995; Giovannoni *et al.*, 1988; Giovannoni, Wood & Huss, 1993; Helmchen, Bhattacharya & Melkonian, 1995; Turner *et al.*, 1989) and other gene sequences (Delwiche, Kuhsel & Palmer, 1995; Douglas & Murphy, 1994; Morden *et al.*, 1992). The group contains the oxygenic phototrophic bacteria (the cyanobacteria and prochlorophytes) and all photosynthetic plastids of eukaryotes.

It should therefore be possible to use the transition to an oxidizing atmosphere to date the origin of the cyanobacterial radiation. However, evidence used to infer the early geological and chemical history of the earth is subject to different interpretations. According to one interpretation, the presence of banded iron formations in the Archean as early as 3.8 billion years ago (Dymek & Klein, 1988; Klein & Beukes, 1992) argues for the presence of oxygenic photosynthesis by that date (Schopf, 1992*a*); furthermore, contemporaneous Archean microfossils with a cyanobacterial-like morphology are interpreted as remains of cyanobacteria (Schopf, 1992*b*, *d*). In contrast, others argue that abiotic processes are sufficient to explain early banded iron formations (Francois, 1986), and that the cyanobacterial

Table 1. *Average evolutionary distances*[a] *calculated between SSU rRNA sequences of cyanobacteria, Bacteria and Archaea*

Phylogenetic groups	Evolutionary distances[a] Kimura (1980)	Olsen (1987)	Jukes & Cantor (1969)
Cyanobacteria: Cyanobacteria Average of 3 highest distances of 28 distances	12.73	13.93	13.93
Cyanobacteria: other Bacteria Average of 360 distances	21.89	26.01	25.95
Cyanobacteria: other Bacteria Average of 10 highest distances of 360 distances	31.17	31.17	31.08
Cyanobacteria: Archaea Average of 96 distances	32.16	42.40	42.15
Bacteria: Bacteria Average of 1035 distances	21.29	25.26	25.20
Bacteria: Bacteria Average of 10 highest distances of 1035 distances	29.33	37.30	37.20
Bacteria: Archaea Average of 540 distances	33.30	44.48	44.21

Pairwise distances among 45 sequences belonging to the domains Bacteria (Thermotogales (2), green non-sulphur bacteria (2), radio-resistant micrococci and relatives (2), *Planctomyces* and relatives (3), *Chlamydia* (1), Flavobacteria and relatives (2), green sulphur bacteria (1), spirochaetes (3), *Fibrobacter* (2), high G + C Gram-positive (4), *Fusobacteria* and relatives (1), cyanobacteria and plastids (8), low G + C Gram-positive (4), alpha Proteobacteria (2), beta Proteobacteria (2), gamma Proteobacteria (3) and delta Proteobacteria (2)) and 12 sequences belonging to the domain Archaea (Euryarcheota (10) and Crenarcheota (2)) were calculated according to the algorithms proposed by Kimura (1980), Olsen (1987) and Jukes & Cantor (1969). The SSU rRNA alignment used was obtained from the Ribosomal Database Project (RDP, Maidak *et al.*, 1994). The mask was created with the automatic masking module PhyloMask of the alignment editor GDE (Steven Smith). Known hypervariable regions of the SSU rRNA were omitted from the final mask, which contained 784 'homologous' positions.
[a]calculated as percentage substitutions per nucleotide position.

radiation is more likely to have occurred at the time of the sharp rise of oxygen in the atmosphere, at around 2 to 1.8 billion years ago in the Proterozoic (Kasting, Holland & Kump, 1992; Klein & Beukes, 1992; Walker *et al.*, 1983). Can molecular evidence distinguish between these hypotheses?

Ribosomal RNA-based phylogenic trees indicate that sequence diversity among cyanobacteria is low compared to inter-phylum diversity within the bacteria (Table 1; Fig. 3). The largest evolutionary distances among cyanobacterial 16S rRNA genes are 0.13–0.14, whereas the distances between bacterial phyla are 0.29–0.37 (Table 1; Giovannoni *et al.*, 1993). The evolutionary distances separating cyanobacteria and Archaea are even greater: 0.32–0.43. This suggests either that the molecular clock has been

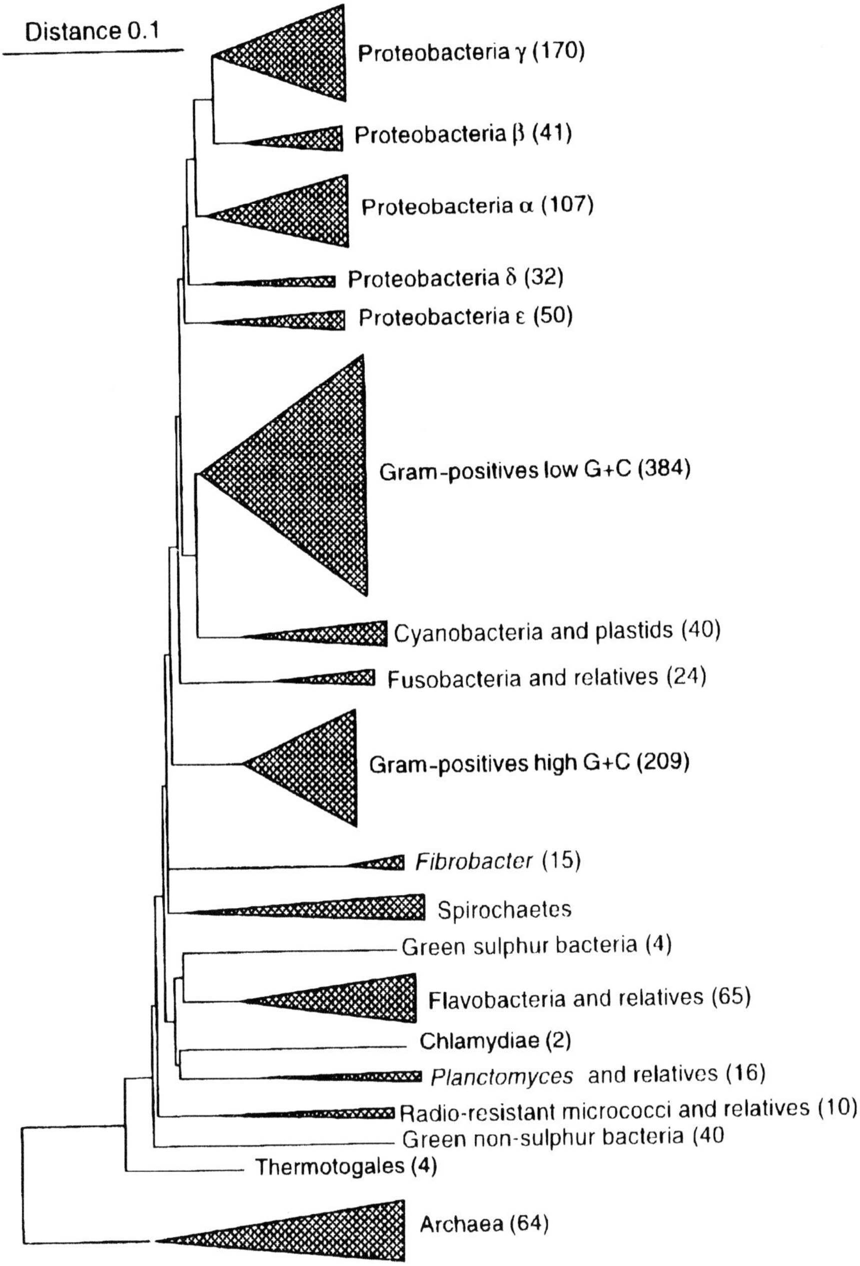

Fig. 3. Bacterial 16S ribosomal RNA phylogenetic tree demonstrating the relative depth of bacterial groups. The tree was constructed from the same data set used in Fig. 2. The numbers indicate the number of gene sequences analysed from each phylum. The bar indicates the number of substitutions per sequence position (from Van de Peer *et al.*, 1994, with permission).

extremely slow in this group, or that cyanobacteria did not evolve early relative to other bacterial groups (Giovannoni *et al.*, 1993).

Doolittle and colleagues have compiled sequence data from a large number of enzymes to attempt to date evolutionary divergence times (Doolittle *et al.*, 1996). They date the split between the Bacteria and Eukarya, which phylogenetic analyses place before the divergence of the oxygenic phototrophs, at approximately 2 billion years ago. The divergence of Gram-negative and Gram-positive bacteria is placed at 1.5 billion years ago. These dates also support the later origin of cyanobacteria.

How can these different lines of evidence be resolved? On one hand, the range of morphological diversity in all prokaryotes is low compared to eukaryotes; it would be most surprising if early prokaryotic microfossils did not resemble living prokaryotes, including living cyanobacteria. Microbiologists are well aware that morphology does not predict physiology or metabolic capacity. Thus, the appearance of microfossils resembling cyanobacteria is not a reliable indicator of the date of origin for this group.

On the other hand, few investigators would believe that cyanobacteria originated after the sharp rise in atmospheric oxygen of 2 to 1.8 billion years ago. How accurate are early divergence times estimated from molecular data? Unfortunately, the earliest branches (nodes) in phylogenetic trees, and the lengths of the segments connecting these nodes, are the most difficult to determine accurately by phylogenetic analysis. This is largely a consequence of the fact that repeated substitutions can occur at the same sites. The most variable regions of ribosomal RNA genes lose their value as molecular chronometers as the distance between the taxa being compared increases, limiting the number of nucleotide positions that are phylogenetically informative about the earliest (most distant) events represented in the trees. The net result of these effects is that the phylogenetic events that are most relevant to Archean evolution are also those which are least well resolved.

It is notable that the metabolic diversity within the oxygenic phototrophic clade is low (restricted to oxygenic photosynthesis) compared to other bacterial lineages, which may comprise extremely diverse metabolic capabilities. This suggests that cyanobacteria, unlike other prokaryote groups, may in fact be 'hypobradytelic', a term originally coined to describe an extremely slow rate of morphological evolution (Schopf, 1992*c*), but extended to imply an equally slow rate of biochemical evolution. Perhaps the transition to the oxygenic phototrophic phenotype was accompanied by extensive genetic and physiological adaptations that limited further evolutionary innovations. The most conservative interpretation of the evidence is that cyanobacteria did not evolve as early as the earliest microfossils, nor as late as extrapolation of molecular clock data backwards might indicate, but somewhere in between. A likely time, of course, would be just before the sharp increase in atmospheric oxygen.

THE DISCOVERY OF NOVEL BACTERIAL DIVERSITY IN CLONES FROM ENVIRONMENTAL DNA

In the early 1980s, Pace and coworkers proposed that microbial diversity might be explored by analyses of nucleic acids isolated from natural samples without resorting to microbial cultivation (Olsen *et al.*, 1986). Early studies focused on the 5S rRNA molecule, but it did not take long for the larger 16S rRNA to become the standard in these studies. The gradual refinement and application of these methods has profoundly changed the fields of microbial ecology and evolution. 16S rDNA sequences have been retrieved from such diverse habitats as soils (Liesack & Stackebrandt, 1992), lake water (Giovannoni, unpublished data), seawater (DeLong, 1992; DeLong, Franks & Alldredge, 1993; Fuhrman, McCallum & Davis, 1992, 1993; Giovannoni *et al.*, 1990; Rappé, Kemp & Giovannoni, 1995; Schmidt, DeLong & Pace, 1991), thermal springs (Barns *et al.*, 1994; Reysenbach, Wickham & Pace, 1994; Ward, Weller & Bateson, 1990), peat bogs (Hales *et al.*, 1996), and the hindguts of termites (Ohkuma & Kudo, 1996; Paster *et al.*, 1996). The results of these studies have often been startling: not only does it appear that the most abundant microbes in natural ecosystems are predominantly undescribed microbial species, but often these microbes cannot confidently be placed within any of the bacterial phyla described previously by Woese (Woese, 1987).

Among the first environments to receive significant attention in these studies were the oceans (Giovannoni *et al.*, 1990; Schmidt *et al.*, 1991). To date, more than 440 prokaryotic 16S rRNA genes cloned from seawater have been phylogenetically characterized. Though the first studies analysed samples from the surface of subtropical regions of the ocean, samples from 80, 100, 250 and 500 metres, the eastern continental shelf of the US, the Oregon coast, and marine snow have also been examined (DeLong, 1992; DeLong *et al.*, 1993; Fuhrman *et al.*, 1992, 1993; Rappé *et al.*, 1995).

Important questions addressed by these studies included: (i) how diverse are bacterial rDNA lineages in a single sample? (ii) do environmental rDNA clones match rDNA sequences of known species? and (iii) are the same phylogenetic lineages found in different oceans?

An unexpected aspect of bacterial diversity uncovered by these studies was the existence of 16S ribosomal RNA 'gene clusters': sympatric, closely related but independent lineages (Giovannoni *et al.*, 1990). The significance of these gene clusters in natural populations is unknown, but many studies of natural microbial populations, from seawater and other habitats, have now observed them (Barns *et al.*, 1994; Britschgi & Giovannoni, 1991; DeLong *et al.*, 1993, 1994; Fuhrman *et al.*, 1993; Hales *et al.*, 1996; Ohkuma & Kudo, 1996; Schmidt *et al.*, 1991).

Analyses of rDNA libraries cloned by different investigators using different methods, as well as supporting hybridization data, have shown that 16S

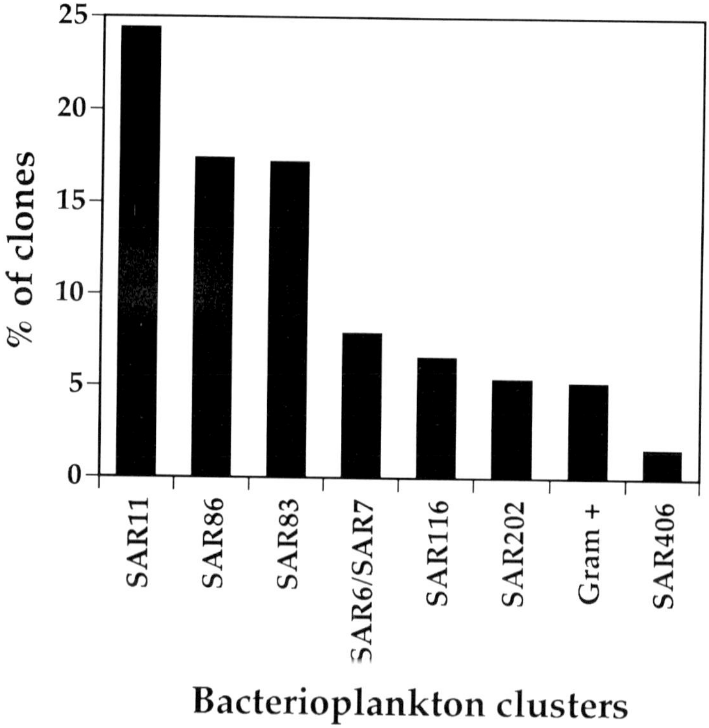

Fig. 4. Frequency of the most common bacterioplankton 16S rRNA gene clusters, calculated from the same data set that was used in Fig. 5. Frequencies were determined by dividing the number of clones of a particular phylotype by the total number of clones in the data set. The data set includes 442 clones. SAR6 and SAR7 are environmental clones from the marine *Synechococcus*/*Prochlorococcus* cluster.

rRNA genes isolated from natural samples of bacterioplankton are diverse, but predominantly fall into ten distinct phylogenetic lineages (Giovannoni, Mullins & Field, 1995). The dominance of a handful of 16S rDNA lineages in planktonic marine 16S rDNA clone libraries is shown graphically in Fig. 4. Of all bacterial 16S rDNAs cloned from marine picoplankton 86% fall into one of eight bacterial phylogenetic groups.

Of the clones, 25% belong to the SAR83 or SAR6/SAR7 (Marine *Synechococcus*/*Prochlorococcus*) cyanobacterial cluster, members of which have been cultured. The remaining six groups are completely composed of uncultured bacterial groups.

Several genes cloned from the Sargasso Sea are nearly identical to genes cloned from Pacific populations (Fig. 5), suggesting that these previously unrecognized bacterial groups are distributed widely in ocean surface waters (Mullins *et al.*, 1995). Further evidence for this conclusion comes from recently analysed 16S rDNA clone libraries from the Oregon coast, the eastern continental shelf of the US, and 80 m and 250 m in the Sargasso Sea, where, along with numerous unique lineages, members of the most common marine bacterioplankton lineages depicted in Fig. 4 were recovered in abundance (Gordon & Giovannoni, 1996; Giovannoni, unpublished data).

The phylogenetic relationships among clusters of prokaryotic 16S rRNA gene clones recovered most frequently from marine environmental clone libraries are shown in Fig. 5. Though the majority of cultured marine microbial species are Proteobacteria, only four of the ten major marine rRNA gene clusters are members of this class: the SAR11, SAR83, and SAR116 clusters of the alpha subclass, and the SAR86 cluster of the gamma subclass. Interestingly, these four clusters are clearly distinct from the major genera of cultivated marine bacteria (mostly of the subclass gamma) for which 16S rDNA sequences are available. The SAR83 cluster is, however, closely related to the aerobic photoheterotroph *Roseobacter denitrificans* (similarity = 0.96). The SAR11 cluster is a rapidly evolving, deeply branching clade of the alpha subclass of Proteobacteria, and often appears as the deepest branch within this group. The marine *Synechococcus*/*Prochlorococcus* cluster is phylogenetically affiliated with the cyanobacteria, in particular to the marine *Synechococcus* group and *Prochlorococcus marinus*. It is not surprising that a cluster of cyanobacteria was recovered in clone libraries from pelagic marine samples since it was previously known that oxygenic phototrophs often constitute a significant proportion of the total bacteria in these environments. Although related to the high G + C division of the Gram-positive bacteria, the Marine Gram-positive cluster shows low similarity to all of the major high G + C Gram-positive lineages, and instead forms a deep, unique branch in this phylum. The phylogenetic positions of the SAR202 and SAR406 clusters, like those of the SAR11 and Marine Gram-positive clusters, are relatively deep with respect to the phyla in which they occur. The SAR202 cluster is the deepest branch of the *Chloroflexus*/*Herpetosiphon* phylum (the green non-sulphur bacteria; Giovannoni *et al.*, 1996). The SAR406 cluster is a distinct relative of the genus *Fibrobacter* and the green sulphur bacteria, which includes *Chlorobium*. It has no definitive phylum affiliation (Gordon & Giovannoni, 1996). The remaining two groups are novel lineages within the Archaea. The Group I Archaea are peripherally related to hyperthermophilic Crenarchaeota, while the Group II Archaea appear to share a common ancestry with the Euryarchaeote *Thermoplasma acidophilum*.

The similarity between genes cloned directly from seawater and their closest relatives among cultured species represented in databases is shown graphically in Fig. 6. Even though approximately 74% of marine bacterial species described in the systematic literature are represented in 16S rDNA sequence databases, the median similarity between marine bacterial genes retrieved from nature and genes from cultured marine microbial species in only *ca.* 0.87 (Fig. 6). The significance of minor sequence variation between environmental gene clones and sequences from cultivated species is not

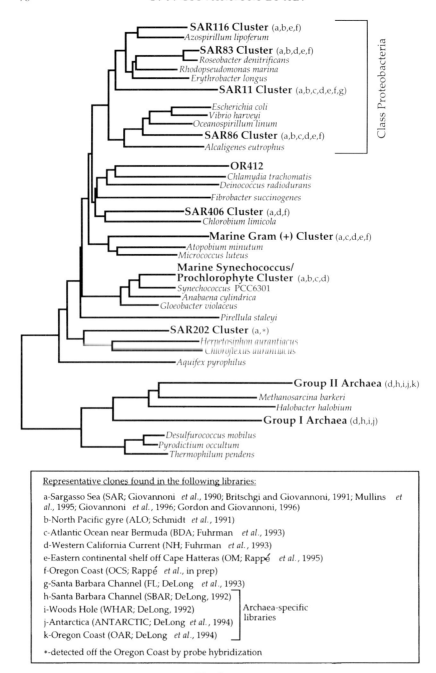

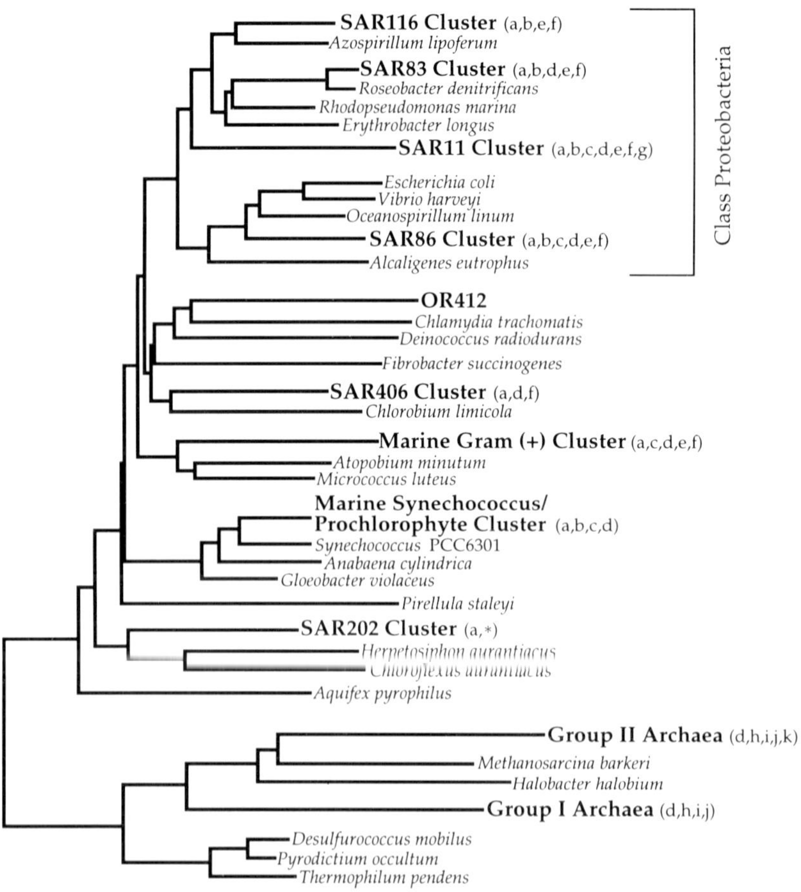

<u>Representative clones found in the following libraries:</u>

a-Sargasso Sea (SAR; Giovannoni *et al.*, 1990; Britschgi and Giovannoni, 1991; Mullins *et al.*, 1995; Giovannoni *et al.*, 1996; Gordon and Giovannoni, 1996)
b-North Pacific gyre (ALO; Schmidt *et al.*, 1991)
c-Atlantic Ocean near Bermuda (BDA; Fuhrman *et al.*, 1993)
d-Western California Current (NH; Fuhrman *et al.*, 1993)
e-Eastern continental shelf off Cape Hatteras (OM; Rappé *et al.*, 1995)
f-Oregon Coast (OCS; Rappé *et al.*, in prep)
g-Santa Barbara Channel (FL; DeLong *et al.*, 1993)
h-Santa Barbara Channel (SBAR; DeLong, 1992)
i-Woods Hole (WHAR; DeLong, 1992)
j-Antarctica (ANTARCTIC; DeLong *et al.*, 1994)
k-Oregon Coast (OAR; DeLong *et al.*, 1994)

(h–k: Archaea-specific libraries)

*-detected off the Oregon Coast by probe hybridization

Fig. 5.

known. However, when an environmental clone is significantly dissimilar to any known gene sequence (similarity < 0.96), it is unlikely to represent a cultured strain. To put these 16S ribosomal RNA sequence similarities in perspective, consider that bacterial 16S rDNAs diverge at an average rate of 1% per 50 million years (Moran *et al.*, 1993). Even allowing for broad variation in rates of evolution, the data still support the conclusion that most marine bacteria do not share recent (i.e. in the Cambrian or later) common ancestors with cultured species.

In summary, the following general conclusions can be drawn from studies of bacterioplankton environmental clone libraries. First, the most abundant rDNA genes recovered in clone libraries do not correspond to cultured species. Secondly, many of the same novel gene lineages are found in clone libraries prepared from different oceans and by different methods. And thirdly, these lineages (phylotypes) are often composed of multiple gene lineages (clusters) more related to each other than to any other gene lineage.

How diverse are bacteria in other natural systems? Liesack and Stackebrandt (1992) analysed the diversity of bacteria in an Australian soil sample by cloning 16S rRNA genes from environmental DNA. Sequence analysis of 30 clones and dot blot hybridization of 83 additional clones with taxon-specific oligonucleotide probes revealed three major bacterial groups. Representing 50% of the total, the most abundant clone type was closely related to nitrogen-fixing members of the α subclass of Proteobacteria, an unsurprising result given the source of the environmental sample. Of the clones, 6% affiliated with the Planctomycetales, though they showed no close similarity to any described species. This was very surprising, since all Planctomycetales described to date come from aqueous habitats. The third major clone type, representing 20% of the total, appeared to form a unique line of descent within the Bacteria, possibly sharing a common ancestry with *Chlamydia* and the Planctomycetales.

A similar pattern of phylogenetic diversity was found in the intestinal microflora of the termite *Reticulitermes speratus* (Ohkuma & Kudo, 1996). Of 55 rDNA clones, 20.0% were members of the Proteobacteria and showed close relationships with cultivated species of enteric and sulphate-reducing bacteria. The remaining clones showed no close relationships to known bacterial taxa, though members of the spirochaetes (18.2%), *Bacteroides/Flavobacterium* group (16.4%), and low G + C division of the

Fig. 5. Composite phylogenetic tree displaying relationships among the most widespread bacterioplankton 16S rRNA gene clusters. The occurrence of these groups in environmental clone libraries from different sources is also shown. This tree was inferred from comparisons of about 1000 nucleotide positions by the neighbour-joining method. Strain OR412 is an undescribed heterotrophic bacterium. Of the eight bacterial clusters and two archaeal groups shown, only the Marine *Synechococcus/Prochlorococcus* cluster and the SAR83 cluster have members that are in culture. Many other lineages found in marine aggregates or in a single clone library are not shown.

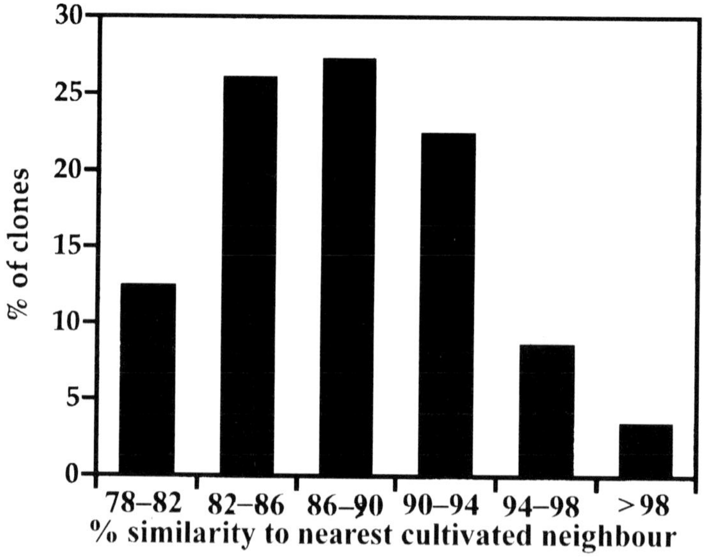

Fig. 6. Chart summarizing the percentage similarity to the nearest phylogenetic neighbour, calculated for a combined data set of 442 environmental 16S rRNA gene clones from pelagic marine samples. This data set is composed of all published pelagic marine 16S rRNA gene clone libraries as well as unpublished data from surface samples from the Oregon coast, the eastern continental shelf of the US, and 80 m and 250 m in the Sargasso Sea.

Gram-positive bacteria (27.3%) were recovered. In addition, 18.2% of the clones, among which 14.5% formed a single cluster, could not be placed within any of the major phyla of the Bacteria, and appeared to form unique lines of descent.

The termite and soil studies reinforced the observation that the majority of microbial diversity found in natural habitats consists of uncultured, uncharacterized taxa. It is also evident that some clone types may be sufficiently divergent to form unique lines of descent with the Bacteria.

Direct cloning of ribosomal RNA genes from the environment has revealed previously unknown diversity among the Archaea as well as the Bacteria. All cultured members of the domain Archaea have 'extremophile' modes of existence: methanogens, extreme halophiles, sulphate reducers and extreme thermophiles. However, PCR amplification and cloning have revealed that DNA from subsurface marine bacterioplankton is up to 4% archaean, despite mild temperatures and oxygenic conditions (DeLong, 1992; DeLong *et al.*, 1994; Fuhrman, McCallum & Davis, 1992). The unexpectedly high proportion of archaeal DNA in marine samples may indicate an ecological importance for anaerobic microenvironments associated with metazoan digestive systems or decaying detrital particles. Alternatively, the archaean sequences may signal the existence of previously unknown, mesophilic archaeal groups. Novel archaeal sequences have also

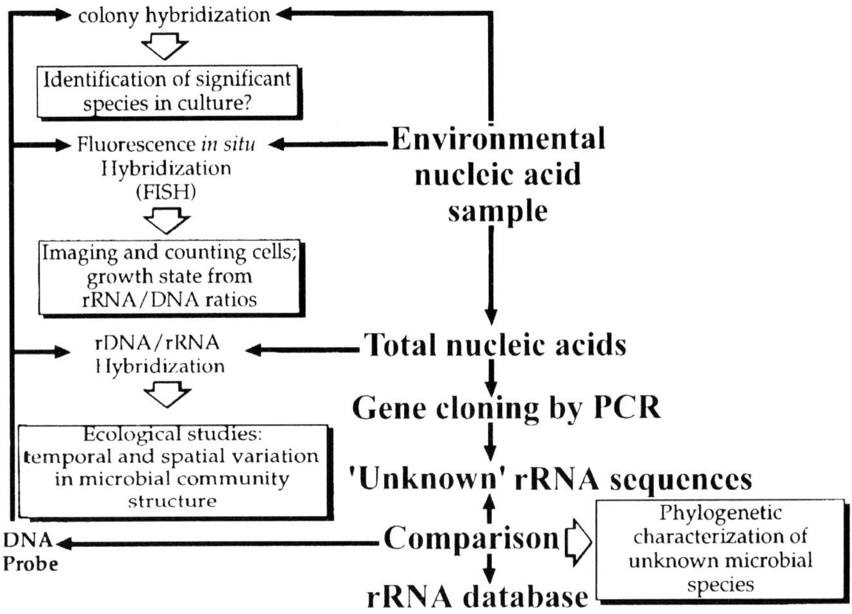

Fig. 7. Schematic outline of the strategies used for genetic investigations of bacterioplankton diversity. The approaches do not rely on cultivation, but instead focus on natural populations of cells obtained by filtration.

been recovered from freshwater hot springs, which apparently support Archaea representing all known hyperthermophilic groups, as well as organisms unrelated to cultured isolates (Barns *et al.*, 1994). A previously uncultured Archaean from the hot springs environment was isolated into clonal culture by a combination of fluorescent *in situ* hybridization using ribosomal RNA probes and manipulation using optical tweezers (Huber *et al.*, 1995).

Environmental rDNA clone libraries are providing more than just a measure of unculturable microbial diversity. rDNA genes cloned from natural ecosystems are also serving as genetic markers for microbes, thus enabling studies which focus on the ecology of uncultured bacteria in natural settings or which seek to isolate them into culture. Some of these applications are illustrated in Fig. 7. Sequences of 16S ribosomal RNAs show a non-random distribution of rapidly evolving and less variable nucleotide positions. By targeting different regions of the molecule, short oligonucleotide probes can be designed to identify phylogenetic groups with varying degrees of relatedness. Rapidly evolving hypervariable regions are used to distinguish very closely related lineages, even to the subspecies level, while more highly conserved regions identify broader groups on the species, genus, phylum or even domain levels (Amann, Ludwig & Schleifer, 1995; Salama, Sandine & Giovannoni, 1993). Phylogenetic group-specific probes have been used to monitor changes in population densities of uncultured marine bacterioplankton groups over space and time, and probes of nested phylogenetic specificity have been used to assess the numbers of sulphate-reducing bacteria in a bioreactor, related by varying degrees to cultured isolates (Gordon & Giovannoni, 1996; Kane, Poulsen & Stahl, 1993). Oligonucleotide probes with attached fluorescent side groups can be hybridized *in situ* to fixed bacterial cells, enabling cell counts for specific phylogenetic groups by fluorescence microscopy or flow cytometry (DeLong *et al.*, 1989; Amann *et al.*, 1990).

CONCLUSION: WHAT THE FUTURE MAY HOLD

The discovery of new microorganisms by means of cultivation and gene cloning from natural ecosystems is leading to an increasingly detailed map of microbial diversity based on ribosomal RNAs. One direction of future work will certainly be the continued refinement of our phylogenetic picture of the Bacteria as it is seen through the inheritance of ribosomal RNA genes. More exciting, though, is the application of the phylogenetic information obtained from ribosomal gene sequencing to other explorations of microbial diversity. Two of the most exciting of these are the sequencing of complete microbial genomic DNAs and the exploration of microbial ecology by molecular methods (Fleischmann *et al.*, 1995; Fraser *et al.*, 1995; Mullins *et al.*, 1995; Stein *et al.*, 1996). In both of these research areas, ribosomal RNA phylogenies are playing a key role. The selection of organisms for genomic sequencing is greatly influenced by perceptions of the uniqueness of species as indicated by their ribosomal RNA gene sequences. Moreover, the genomic sequences themselves will provide vastly improved information to extend phylogenetic comparisons from single gene sequences to much larger sequence arrays. This will greatly reduce the level of statistical error in phylogenetic analyses, and will furthermore permit the reconstruction of important events in the evolution of microbial physiology. In molecular microbial ecology, ribosomal RNA gene sequences are playing an equally pivotal role by serving as genetic markers for microbial species (Giovannoni & Cary, 1993; Gordon & Giovannoni, 1996). Currently, efforts are under way to obtain microbial genomes from natural ecosystems for a variety of uses, including ecological studies and the isolation of industrially useful genes and enzymes (Stein *et al.*, 1996). In these latter investigations, the organisms are known entirely from gene sequences; thus the process of reconstructing a genome begins with a single marker gene to which other genes are linked, either by physical co-occurrence on the same fragment cloned into a cosmid vector, or as part of a series of fragments linked into a contig by hybridization analysis. It seems almost inevitable that ribosomal RNAs will continue to be the marker genes used to begin the process of genetic linkage. These investigations will eventually lead to the reconstruction of genomes and the inference of microbial physiology for organisms that have never been grown in a laboratory. Although the great age of reconstructing microbial evolution by means of ribosomal RNA gene sequencing may have passed, the central role of this molecule in a wide range of disciplines is far from over.

ACKNOWLEDGEMENTS

This work was supported by National Science Foundation grant OCE 9016373 for the study of microbial diversity at the Bermuda Atlantic Time Series Station and Department of Energy grant FG 0693ER61697 from the Ocean Margins program for the study of bacterioplankton population dynamics over continental shelves. We are grateful to the Bermuda Atlantic Time Series group (BATS) for sampling, and to Kevin Vergin and Nanci Adair for processing the nucleic acid samples.

REFERENCES

Achenbach-Richter, L., Gupta, R., Stetter, K. O. & Woese, C. R. (1992). Were the original eubacteria thermophiles? *Systematic and Applied Microbiology*, **9**, 34–9.

Amann, R. A., Ludwig, W. & Schleifer, K. H. (1995). Phylogenetic identification and in situ detection of individual microbial cells without cultivation. *Microbiological Reviews*, **59**, 143–69.

Barns, S. M., Fundyga, R. E., Jeffries, M. W. & Pace, N. R. (1994). Remarkable archaeal diversity detected in a Yellowstone National Park hot spring environment. *Proceedings of the National Academy of Sciences, USA*, **91**, 1609–13.

Bhattacharya, D. & Medlin, L. (1995). The phylogeny of plastids: a review based on comparisons of small-subunit ribosomal RNA coding regions. *Journal of Phycology*, **31**, 489–98.

Britschgi, T. B. & Giovannoni, S. J. (1991). Phylogenetic analysis of a natural marine bacterioplankton population by rRNA gene cloning and sequencing. *Applied and Environmental Microbiology*, **57**, 1313–18.

DeLong, E. F. (1992). Archaea in coastal marine bacterioplankton. *Proceedings of the National Academy of Sciences, USA*, **89**, 5685–9.

DeLong, E. F., Wickham, G. S. & Pace, N. R. (1989). Phylogenetic stains: Ribosomal RNA-based probes for the identification of single cells. *Science*, **243**, 1360–3.

DeLong, E. F., Franks, D. G. & Alldredge, A. L. (1993). Phylogenetic diversity of aggregate-attached vs. free-living marine bacterial assemblages. *Limnology and Oceanography*, **38**, 924–34.

DeLong, E. F., Wu, K. Y., Prezellin, B. B. & Jovine, R. V. M. (1994). High abundance of Archaea in Antarctic marine picoplankton. *Nature*, **371**, 695–7.

Delwiche, C. F., Kuhsel, M. & Palmer, J. D. (1995). Phylogenetic analysis of *tuf*A sequences indicates a cyanobacterial origin of all plastids. *Molecular Phylogeny and Evolution*, **4**, 110–28.

Doolittle, R. F., Feng, D.-F., Tsang, S., Cho, G. & Little, E. (1996). Determining divergence times of the major kingdoms of living organisms with a protein clock. *Science*, **271**, 470–7.

Douglas, S. E. & Murphy, C. A. (1994). Structural, transcriptional, and phylogenetic analyses of the *atp*B gene cluster from the plastid of *Cryptomonas* ϕ (Cryptophyceae). *Journal of Phycology*, **30**, 500–8.

Dymek, R. F. & Klein, C. (1988). Chemistry, petrology and origin of banded iron-formation lithologies from the 3800 MA Isua supercrustal belt, West Greenland. *Precambrian Research*, **39**, 247–302.

Ehrich, S., Behrens, D., Lebedeva, E., Ludwig, W. & Bock, E. (1995). A new obligately chemolithoautotrophic, nitrite-oxidizing bacterium, *Nitrospira moscoviensis* sp. nov. and its phylogenetic relationship. *Archives of Microbiology*, **164**, 16–23.

Fleischmann, R. D., Adams, M. D., White, O., Clayton, R. A., Kirkness, E. F., Kerlavage, A. R., Bult, C. J., Tomb, J.-F., Dougherty, B. A., Merrick, J. M., McKenney, K., Sutton, G., FitzHugh, W., Fields, C., Gocayne, J. D., Scott, J., Shirley, R., Liu, L.-I., Glodek, A., Kelley, J. M., Weidman, J. F., Phillips, C. A., Spriggs, T., Hedblom, E., Cotton, M. D., Utterback, T. R., Hanna, M. C., Nguyen, D. T., Saudek, D. M., Brandon, R. C., Fine, L. D., Fritchman, J. L., Fuhrman, J. L., Geoghagen, N. S. M., Gnehm, C. L., McDonald, L. A., Small, K. V., Fraser, C. M., Smith, H. O. & Venter, J. C. (1995). Whole-genome random sequencing and assembly of *Haemophilus influenzae*. *Science*, **269**, 496–512.

Francois, L. M. (1986). Extensive deposition of banded iron formations was possible without photosynthesis. *Nature*, **320**, 352–4.

Fraser, C. M., Gocayne, J. D., White, O., Adams, M. D., Clayton, R. A., Fleischmann, R. D., Bult, C. J., Kerlavage, A. R., Sutton, G., Kelley, J. M., Fritchman, J. L., Weidman, J. F., Small, K. V., Sandusky, M., Fuhrman, J. L., Nguyen, D., Utterback, T. R., Saudek, D. M., Phillips, C. A., Merrick, J. M., Tomb, J.-F., Dougherty, B. A., Bott, K. F., Hu, P.-C., Lucier, T. S., Peterson, S. N., Smith, H. O., Hutchison, C. A., III & Venter, J. C. (1995). The minimal gene complement of *Mycoplasma genitalium*. *Science*, **270**, 397–403.

Fuerst, J. A. (1995). The planctomycetes: emerging models for microbial ecology, evolution and cell biology. *Microbiology*, **141**, 1493–506.

Fuhrman, J. A., McCallum, K. & Davis, A. A. (1993). Phylogenetic diversity of subsurface marine microbial communities from the Atlantic and Pacific oceans. *Applied and Environmental Microbiology*, **59**, 1294–302.

Fuhrman, J. A., McCallum, K. & Davis, A. A. (1992). Novel major archaebacterial group from marine plankton. *Nature*, **356**, 148–9.

Giovannoni, S., Rappé, M., Vergin, M. & Adair, N. (1996). 16S rRNA genes reveal stratified open ocean bacterioplankton populations related to the *Chloroflexus/Herpetosiphon* phylum. *Proceedings of the National Academy of Sciences, USA*, in press.

Giovannoni, S. J., Britschgi, T. B., Moyer, C. L. & Field, K. G. (1990). Genetic diversity in Sargasso Sea bacterioplankton. *Nature*, **345**, 60–3.

Giovannoni, S. J. & Cary, C. (1993). Probing marine systems with ribosomal RNAs. *Oceanography*, **6**, 95–104.

Giovannoni, S. J., Mullins, T. D. & Field, K. C. (1995). Microbial diversity in oceanic systems: rRNA approaches to the study of unculturable microbes. In *Molecular Ecology of Aquatic Microbes*, Joint, I., ed., vol. 38, pp. 217–248. Springer-Verlag, New York.

Giovannoni, S. J., Turner, S., Olsen, G. J., Barns, S., Lane, D. J. & Pace, N. R. (1988). Evolutionary relationships among cyanobacteria and green chloroplasts. *Journal of Bacteriology*, **170**, 3584–92.

Giovannoni, S. J., Wood, N. & Huss, V. (1993). Molecular phylogeny of oxygenic cells and organelles based on small-subunit ribosomal RNA sequences. In *Origins of Plastids*, Lewin, R. A., ed., pp. 159–170. Chapman and Hall, New York and London.

Gogarten, J. P., Kibak, H., Dittrich, P., Taiz, L., Bowman, E. J., Bowman, B. J., Manolson, M. F., Poole, R. J., Date, T., Oshima, T., Konishi, J., Denda, K. & Yoshida, M. (1989). Evolution of the vacuolar H+-ATPase: implications for the origin of eukaryotes. *Proceedings of the National Academy of Sciences, USA*, **86**, 6661–5.

Gordon, D. & Giovannoni, S. J. (1996). Stratified microbial populations related to *Chlorobium* and *Fibrobacter* detected in the Atlantic and Pacific oceans. *Applied and Environmental Microbiology*, **62**, 1171–7.

Hales, B. A., Edwards, C. R., Ritchie, D. A., Hall, G., Pickup, R. W. & Saunders, J. R. (1996). Isolation and identification of methanogen-specific DNA from blanket bog peat by PCR amplification and sequence analysis. *Applied and Environmental Microbiology*, **62**, 668–75.

Helmchen, T., Bhattacharya, D. & Melkonian, M. (1995). Analysis of ribosomal RNA sequences from glaucocystophyte cyanelles provide new insights into the evolutionary relationships of plastids. *Journal of Molecular Evolution*, **41**, 203–10.

Huber, R., Burggraf, S., Mayer, T., Barns, S. M., Rossnagel, P. & Stetter, K. O. (1995). Isolation of a hyperthermophilic archaeon predicted by *in situ* RNA analysis. *Nature*, **376**, 57–8.

Iwabe, N., Kuma, K., Hasegawa, M., Osawa, S. & Miyata, T. (1989). Evolutionary relationships of archaebacteria, eubacteria and eukaryotes inferred from phylogenetic trees of duplicated genes. *Proceedings of the National Academy of Sciences, USA*, **86**, 9355–9.

Jukes, T. H. & Cantor, C. R. (1969). Evolution of protein molecules. In *Mammalian Protein Metabolism*, Munro, H. N., ed., pp. 21–132. Academic Press, New York.

Kane, M. D., Poulsen, L. K. & Stahl, D. A. (1993). Monitoring the enrichment and isolation of sulfate-reducing bacteria by using oligonucleotide hybridization probes designed from environmentally derived 16S rRNA sequences. *Applied Environmental Microbiology*, **59**, 682–6.

Kandler, O. (1994). Cell wall biochemistry and three-domain concept of life. *Systematic and Applied Microbiology*, **16**, 501–9.

Kasting, J. F., Holland, H. D. & Kump, L. R. (1992). Atmospheric evolution: the rise of oxygen. In *The Proterozoic Biosphere*, Schopf, J. W. and Klein, C., eds., pp. 159–163. Cambridge University Press, Cambridge, New York.

Kimura, M. (1980). A simple method for estimating evolutionary rate of base substitutions through comparative studies of nucleotide sequences. *Journal of Molecular Evolution*, **16**, 111–20.

Klein, C. and Beukes, N. J. (1992). Time distribution, stratigraphy, and sedimentologic setting, and geochemistry of Precambrian iron-formations. In *The Proterozoic Biosphere*, Schopf, J. W. and Klein, C., eds., pp. 139–141. Cambridge University Press, New York.

Koch, A. L. (1994). Development and diversification of the last universal ancestor. *Journal of Theoretical Biology*, **168**, 269–80.

Liesack, W. & Stackebrandt, E. (1992). Occurrence of novel groups of the Domain Bacteria as revealed by analysis of genetic material isolated from an Australian terrestrial environment. *Journal of Bacteriology*, **174**, 5072–8.

Maidak, B. L., Larsen, N., McCaughey, M. J., Overbeek, R., Olsen, G. J., Fogel, K., Blandy, J. & Woese, C. R. (1994). The Ribosomal Database Project. *Nucleic Acid Research*, **22**, 3485–7.

Moran, N. A., Munson, M. A., Baumann, P. & Ishikawa, H. (1993). A molecular clock in endosymbiotic bacteria is calibrated using the insect hosts. *Proceedings of the Royal Society of London B*, **253**, 167–71.

Morden, C. W., Delwiche, C. F., Kuhsel, M. & Palmer, J. D. (1992). Gene phylogenies and the endosymbiotic origin of plastids. *BioSystems*, **28**, 75–90.

Mullins, T. D., Britschgi, T. B., Krest, R. L. & Giovannoni, S. J. (1995). Genetic comparisons reveal the same unknown bacterial lineages in Atlantic and Pacific bacterioplankton communities. *Limnologia Oceanographica*, **40**, 148–58.

Ohkuma, M. & Kudo, T. (1996). Phylogenetic diversity of the intestinal bacterial community in the termite *Reticulitermes speratus*. *Applied and Environmental Microbiology*, **62**, 461–8.

Olsen, G. J. (1987). The earliest phylogenetic branchings: comparing rRNA-based evolutionary trees inferred with various techniques. *Cold Spring Harbor Symposia in Quantitative Biology*, **52**, 825–38.

Olsen, G. J., Lane, D. L., Giovannoni, S. J., Pace, N. R. & Stahl, D. A. (1986). Microbial ecology and evolution: a ribosomal RNA approach. *Annual Review of Microbiology*, **40**, 337–66.

Olsen, G. J. & Woese, C. R. (1993). Ribosomal RNA: a key to phylogeny. *Federation of American Societies for Experimental Biology*, **7**, 113–23.

Paster, B. J., Dewhirst, F. E., Cooke, S. M., Fussing, V., Poulsen, L. K. & Breznak, J. A. (1996). Phylogeny of not-yet cultivated spirochetes from termite guts. *Applied and Environmental Microbiology*, **62**, 347–52.

Rappé, M. S., Kemp, P. F. & Giovannoni, S. J. (1995). Chromophyte plastid 16S ribosomal RNA genes found in a clone library from Atlantic Ocean seawater. *Journal of Phycology*, **31**, 979–88.

Reysenbach, A. N., G. S. Wickham & Pace, N. R. (1994). Phylogenetic analysis of the hyperthermophilic pink filament community in Octopus Spring, Yellowstone National Park. *Applied and Environmental Microbiology*, **60**, 2113–19.

Salama, M. S., Sandine, W. E. & Giovannoni, S. J. (1993). Isolation of *Lactococcus lactis* subsp. *cremoris* from nature by colony hybridization with rRNA probes. *Applied and Environmental Microbiology*, **59**, 3941–5.

Schmidt, T. E., DeLong, E. F. & Pace, N. R. (1991). Analysis of a marine picoplankton community by 16S rRNA gene cloning and sequencing. *Journal of Bacteriology*, **173**, 4371–8.

Schopf, J. W. (1992*a*). *Major Events in the History of Life*. Jones and Bartlett, Boston.

Schopf, J. W. (1992*b*). Paleobiology of the Archean. In *The Proterozoic Biosphere*, Schopf, J. W. and Klein, C., eds., pp. 25–39. Cambridge University Press, New York.

Schopf, J. W. (1992*c*). Tempo and mode of Proterozoic evolution. In *The Proterozoic Biosphere*, Schopf, J. W. and Klein, D., eds., pp. 595–598. Cambridge University Press, New York.

Schopf, J. W. (1992*d*). Times of origin and earliest evidence of major biologic groups. In *The Proterozoic Biosphere*, Schopf, J. W. and Klein, D., eds., pp. 587–593. Cambridge University Press, New York.

Stackebrandt, E., Ludwig, W., Schubert, W., Klink, F., Schlesner, H., Roggentin, T. & Hirsch, P. (1984). Molecular genetic evidence for early evolutionary origin of budding peptidoglycan-less eubacteria. *Nature*, **307**, 735–7.

Stein, J. L., Marsh, T. L., Wu, K. Y., Shizuya, H. & DeLong, E. F. (1996). Characterization of uncultivated prokaryotes: isolation and analysis of a 40-kilobase pair genome fragment from a planktonic marine archaeon. *Journal of Bacteriology*, **178**, 591–9.

Turner, S., Burger-Wiersma, T., Giovannoni, S. J., Mur, L. R. & Pace, N. R.

(1989). The relationship of a prochlorophyte, *Prochlorothrix hollandica*, to green chloroplasts. *Nature*, **337**, 380–2.

Van de Peer, Y., Neefs, J. M., de Rijk, P., de Vos, P. & de Wachter, R. (1994). About the order of divergence of the major bacterial taxa during evolution. *Systematic and Applied Microbiology*, **17**, 32–8.

Vermaas, W. F. J. (1994). Evolution of heliobacteria: implications for photosystem reaction center complexes. *Photosynthesis Research*, **41**, 285–94.

Walker, J. C. G., Klein, C., Schidlowski, M., Schopf, J. W., Stevenson, D. J. & Walter, M. R. (1983). Environmental evolution of the Archean–Early Proterozoic Earth. In *Earth's Earliest Biosphere: Its Origin and Evolution*, Schopf, J. W., ed., pp. 260–290. Princeton University Press, Princeton.

Ward, D. M., Weller, R. & Bateson, M. (1990). 16S rRNA sequences reveal numerous uncultured organisms in a natural community. *Nature*, **345**, 63–5.

Weisburg, W. G., Giovannoni, S. J. & Woese, C. R. (1989). The *Deinococcus-Thermus* phylum and the effect of rRNA composition on phylogenetic tree construction. *Systematic and Applied Microbiology*, **11**, 128–34.

Woese, C. R. (1987). Bacterial Evolution. *Microbiological Reviews*, **51**, 221–71.

Woese, C. R., Mandelco, L., Yang, D., Gherna, R. & Madigan, M. T. (1990). The case for the relationship of the flavobacteria and their relatives to the green sulfur bacteria. *Systematic and Applied Microbiology*, **13**, 258–62.

Woese, C. R. (1991). The use of ribosomal RNA in reconstructing evolutionary relationships among bacteria. In *Evolution at the Molecular Level*, Selander, R. K., Clark, A. G. and Whittam, T. S., eds., pp. 1–24. Sinauer Associates, Sunderland.

Woese, C. R., Kandler, O. & Wheelis, M. L. (1990). Towards a natural system of organisms: proposal for the domains Archaea, Bacteria and Eucarya. *Proceedings of the National Academy of Sciences USA*, **87**, 4576–9.

THE PROKARYOTIC ANCESTRY OF EUKARYOTES

JAMES A. LAKE AND MARIA C. RIVERA

Molecular Biology Institute and MCD Biology, University of California, Los Angeles, Los Angeles, CA 90095, USA

The prokaryotes, unlike their larger eukaryotic relatives such as the multicellular animals, plants, and fungi, have few morphological features that are useful for phylogenetic studies. Within the last decade, with the availability of DNA sequences from phylogenetically diverse organisms, it has become possible to use sequence data to probe the relationships among the most diverse prokaryotic groups and even to investigate the remote prokaryotic origins of eukaryotes. This review describes the current understanding of relationships among the major groups of prokaryotes and eukaryotes, and emphasizes the search for the prokaryotic group of organisms that has contributed the majority of genes to the eukaryotic nucleus.

Understanding the origin of eukaryotes is made more difficult because the nucleus is a chimera of genes from various sources. Many eukaryotic nuclear genes have been imported from eukaryotic organelles, for example, the chloroplast, the mitochondrion, and even the mechanism by which the nucleus was formed may have involved an endosymbiosis between two bacterial types (Gray, 1993; Baldauf & Palmer, 1990; Lake, 1982, also see the concluding remarks in this chapter).

THE TRADITIONAL PROKARYOTIC–EUKARYOTIC CLASSIFICATION

Early attempts to understand the diversity of life on earth emphasized eukaryotic cells, since their large nuclei were readily visible by light microscopy. According to these studies, all life could be divided into either eukaryotes (organisms with a nucleus) or prokaryotes (organisms that lack a nucleus). Although the nucleus is a positive (derived) feature that unites the eukaryotes into a monophyletic group (a group containing the last common ancestor of the group and all of its descendants, see de Queiroz & Gauthier, 1990), the lack of a nucleus in prokaryotes is a negative feature and cannot be used to define a group (Eldridge & Cracraft, 1986). Gradually it has become clear that the prokaryotes are not a valid, monophyletic group. To understand this, imagine that one is classifying all life by placing all organisms into one of two boxes. Into the first, one places cells with nuclei and into the second cells that lack nuclei. When the sorting is completed, one will have a group of related organisms, those containing a nucleus, and a group that may or may not be related, those lacking a nucleus. One cannot tell whether they are related, because they were grouped by a negative feature. A major challenge in microbiology is to ascertain the phylogenetic relationships among the prokaryotes. The central question which arises: 'Which group of prokaryotes is the closest relative of the eukaryotic nucleus?', is addressed in this review.

POTENTIAL PROKARYOTIC RELATIVES OF EUKARYOTES

Before attempting to identify the prokaryotic group that is most closely related to the eukaryotic nucleus (sister taxon) it is useful to survey briefly their phenotypic and phylogenetic diversity. A phylogenetic tree representing the major known bacterial groups is shown in Fig. 1. The relationships shown are generally accepted by the microbiological community except that the origin of the eukaryotic lineage (the subject of this chapter) has been controversial. There are several groups with prokaryotic organization, any one of which might be related to the eukaryotic nucleus. These include: (1) the Bacteria, (2) the Halobacteria, (3) the Methanogens, (4) relatives of the Methanogens, and (5) sulphur-metabolizing, high-temperature organisms known as the eocytes.

The Bacteria are a diverse group that includes all the photosynthetic bacteria (except for the halobacteria) as well as many non-photosynthetic groups. Some of the better known representatives of the Bacteria are the cyanobacteria, the purple bacteria, the gram-positive bacteria, the bacteroides-flavobacteria group, the green nonsulphur bacteria, the green sulphur bacteria, the spirochetes, the planctomyces, the thermotogales, and many others (see especially Giovannoni, this volume). Most Bacteria are mesophiles; however, a few are extreme thermophiles and grow optimally at temperatures above 80°C. In Fig. 1, the Bacteria are represented by *Thermotoga maritima* and *Aquifex pyrophilus*, which can grow up to 90°C and 95°C, respectively, utilizing a currently unknown fermentative process (Huber & Stetter, 1992). The lipids of Bacteria are primarily of the ester type although *Thermotoga, Aquifex*, and their relatives also contain branched ether lipids.

The Halobacteria are extreme halophiles. They are carbon heterotrophs that can use an unusual photosynthesis system, namely a light-driven proton pump based on bacteriorhodopsin. Surprisingly, they have many molecular properties in common with the Bacteria that are not found in other bacterial groups. For example, like Bacteria, they contain the biochemical pathways for the synthesis of C40 and C50 carotenoids (Goodwin, 1980). Halobacteria also have a bacterial life *rrn* anti-termination system, including several

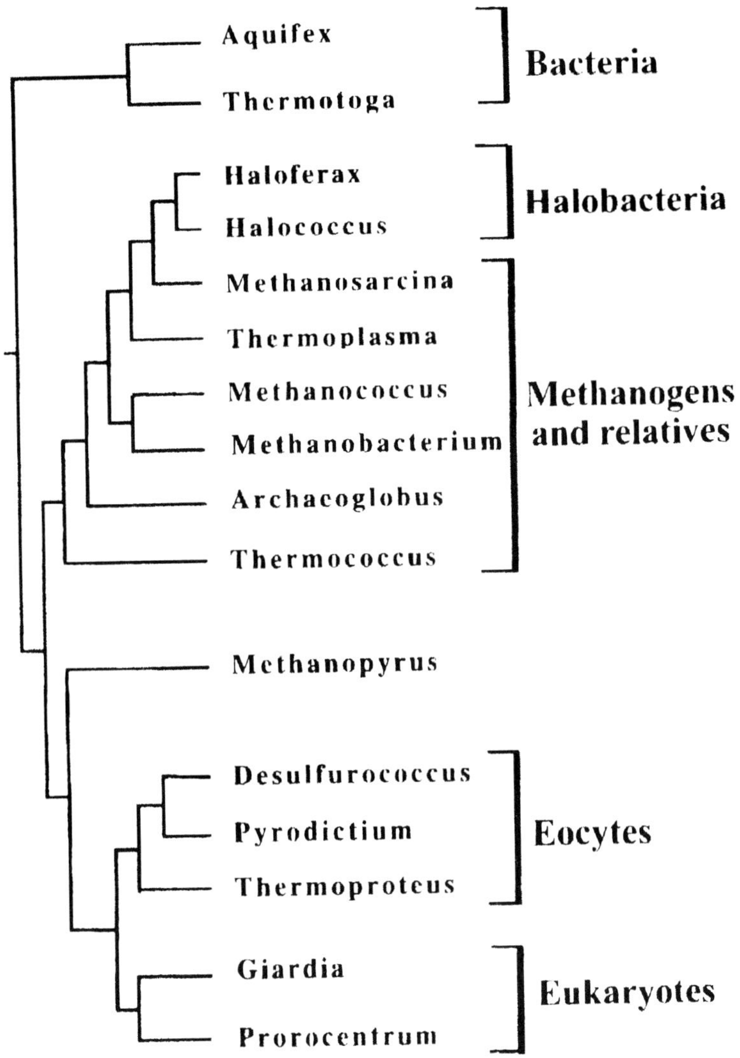

Fig. 1. The rooted phylogenetic tree relating prokaryotic and eukaryotic organisms reconstructed from 16S and 18S ribosomal RNA sequences using paralinear distances (Lockhart *et al.*, 1994; Lake, 1994) and the bootstrappers gambit multi-taxon tree reconstruction algorithm (Lake, 1995). Two hundred bootstrap replicates supported the Halobacteria + Methanogens and Relatives clade at the 99% level and supported the eocytes + eukaryotes clade at the 100% level.

genes, notably those for RNA P, at least two of the *nus* loci, and others. Halobacteria also contain traces of ester lipids and functional fatty acid synthetases, although their lipids are predominantly biphytanol ether lipids (Kamekura & Kates, 1988). It is puzzling that these properties which are uniquely shared with the bacteria are not consistent with the three derived from ribosomal RNA sequences (Fig. 1).

The Methanogens are a phylogenetically diverse group, despite the fact that they share a common phenotype, namely they are strict anaerobes with the ability chemically to reduce carbon compounds to methane to provide energy. According to most rRNA phylogenetic studies some Methanogens, such as *Methanospirillum* and *Methanosarcina*, are more closely related to the halobacteria than they are to other methanogens.

Associated with the Methanogens is a phenotypically diverse group of organisms represented by such organisms as *Thermococcus celer*, and *Archaeoglobus fulgidus*, and, possibly, *Methanopyrus kandleri* (Rivera & Lake, 1996) (in the tree in Fig. 1, *Methanopyrus* is not within the same clade as the methanogens). *Thermococcus celer* is not a methanogen. It lives at high temperatures (up to 97°C) and metabolizes sulphur to yield H_2S. *Archaeoglobus*, can reduce CO_2 to methane, but also can reduce S^o to H_2S in the absence of CO_2. *Archaeoglobus*, like the Methanogens, possesses the characteristic blue-green fluorescence at 420 nm that results from factor 420 (Stetter *et al.*, 1990). Others like *Methanopyrus kandleri*, which is phenotypically a Methanogen, can grow at temperatures up to 112°C, but genotypically is probably not a Methanogen (Burggraf *et al.*, 1991; Rivera & Lake, 1996).

The final prokaryotic group, the eocytes, consists of thermophilic, mostly sulphur-metabolizing organisms, many of which can grow at temperatures in excess of 100°C. The eocytes include *Sulfolobus, Desulforococcus, Thermoproteus, Pyrodictium, Pyrobaculum* etc. *Sulfolobus sulfataricus*, oxidizes sulphur to H_2S. Others, such as *Acidianus infernus*, can oxidize or reduce S^o to H_2SO_4 or to H_2S, respectively. The organisms with the highest maximum growth temperatures (112°C) are *Pyrodictium occultum* and *Pyrodictium abyssum*. The group is metabolically diverse, uniformly thermophilic, and phylogenetically monophyletic.

ORIGIN OF THE EUKARYOTES

Theories for the origin of Eukaryotes

The ability of eukaryotes and prokaryotes to transfer genes laterally is well known. Numerous genes originally contained in the mitochondrial and chloroplast genomes have been transported and are now encoded in the nucleus. Well-documented examples include the incorporation of diverse mitochondrial and chloroplast genes into the nuclear genome (for a thoughtful review, see Smith, Feng & Doolittle, 1992). Thus not all nuclear genes are indicative of the ancestry of the bulk of the nuclear genes. As a result, in this review we concentrate on the genes of the translational and transcriptional machinery. Since these genes are so well integrated into the cellular machinery, they are unlikely to have been imported from outside sources. They are generally accepted to be good markers of the evolutionary history of the nucleus.

The two most extensively studied prokaryotic genes are the 16/18S ribosomal RNA genes and the genes of protein synthesis factor EF-Tu (EF-1α in eukaryotes). Based on their analyses, two theories have been proposed to explain the origin of the eukaryotic nucleus; the archaebacterial (archaea) theory and the eocyte theory (but see also Sogin, this volume). Trees corresponding to both theories, reconstructed from 16/18S rRNA sequences, are shown in Fig. 2. Both are rooted in the branch leading to the Bacteria, as in Fig. 1 (Gogarten *et al.*, 1989; Iwabe *et al.*, 1989). The fundamental difference between these two theories is that in the eocyte tree (at the top of the Figure) the eukaryotic nucleus shares a most recent common ancestor solely with the eocytes, whereas in the Archael theory the eukaryotes are most closely related to an ancestral organism that gave rise to the Halobacteria, the Methanogens and their relatives and to the crenarchaea (= eocytes). These two theories are based on the topology of the phylogenetic trees. Therefore, no re-rooting of the tree can convert one tree into the other. This makes the two theories mutually exclusive and hence eminently testable because, if one theory is correct, the other must be incorrect.

Molecular sequences are our most informative source of data for testing these theories; however, there are significant artefacts associated with their analysis. During the time since eukaryotes and prokaryotes have been separated, frequent nucleotide and amino acid changes have occurred in their sequences. Frequently individual nucleotides have undergone multiple substitutions. For example, a nucleotide that is identical in an eukaryotic and an eocyte sequence may have undergone multiple substitutions, perhaps changing from a C to a U and back to a C. Most algorithms would underestimate the number of these changes in computing the mean number of nucleotide substitutions. These underestimations can have dangerous consequences in aligning sequences and in calculating phylogenetic trees.

When sequences have diverged extensively, three different artefacts can cause long branch attraction. The artefacts plague phylogenetic reconstructions. The observed effects are that rapidly evolving taxa (long branches) are placed with other rapidly evolving taxa, whether or not the taxa are phylogenetically related. These three types of artefacts are relevant to any discussion of the origin of eukaryotes because the archaebacterial tree groups the two longest branches of the tree together. It places the rapidly evolving Bacteria with the rapidly evolving eukaryotes and therefore could be caused by these artefacts. The Archael tree is shown in Fig. 3, and the longer branches leading to the Bacteria and eukaryotes are apparent. One comprehensive paper, unusual for its attention to detail, indicates that the

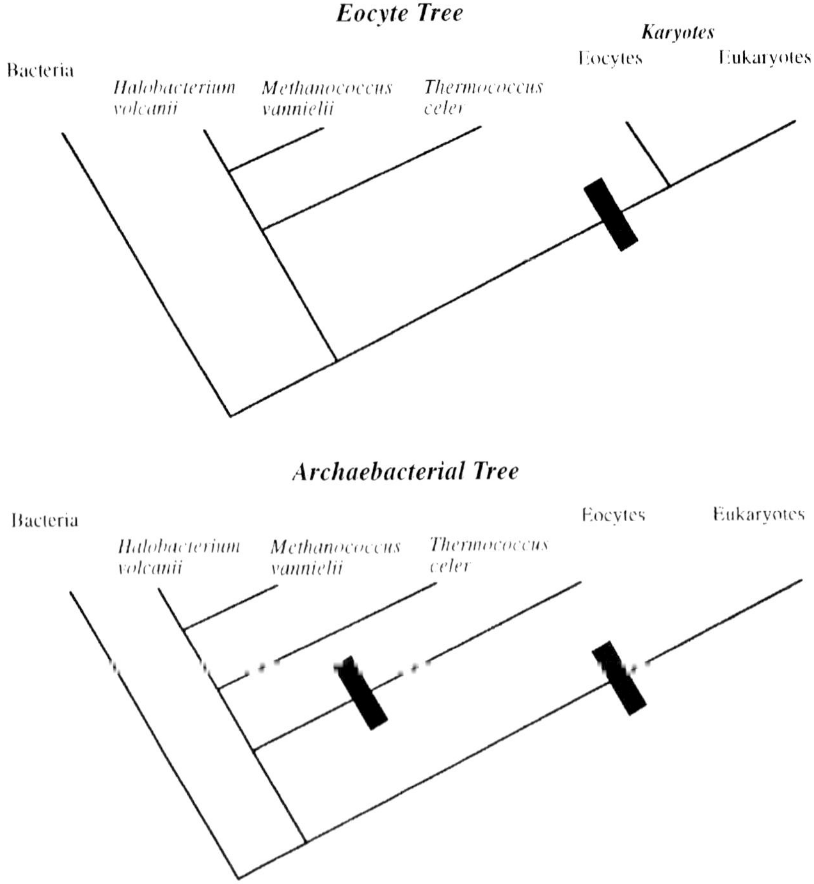

Fig. 2. Rooted trees illustrating the two theories proposed to explain the origin of the eukaryotic nucleus. The trees corresponding to both theories are reconstructed from 16S/18S sequences and rooted in the branch leading to the Bacteria. The solid boxes indicate changes from the 4-amino acid segment to the 11-amino acid form. The eocyte tree is favoured by parsimony since it requires only a single change to the 11-amino acid segment whereas the archaebacterial tree is opposed since it requires two independent changes (the same distribution could also be explained by one appearance of the 11-amino acid form and reappearance of the 4-amino acid form, but still two changes would be required).

Archael topology is probably caused by unequal rate effects (Volters & Erdmann, 1989).

The three artefacts that can cause these effects are: unequal rate effects that result when tree reconstruction algorithms fail to account adequately for multiple substitutions, site-to-site variation that results when the variation of rates of evolution within sequences is not properly accounted for, and alignment artefacts. Although the unequal rate effects have recently been solved in mathematically exact form through the use of paralinear/

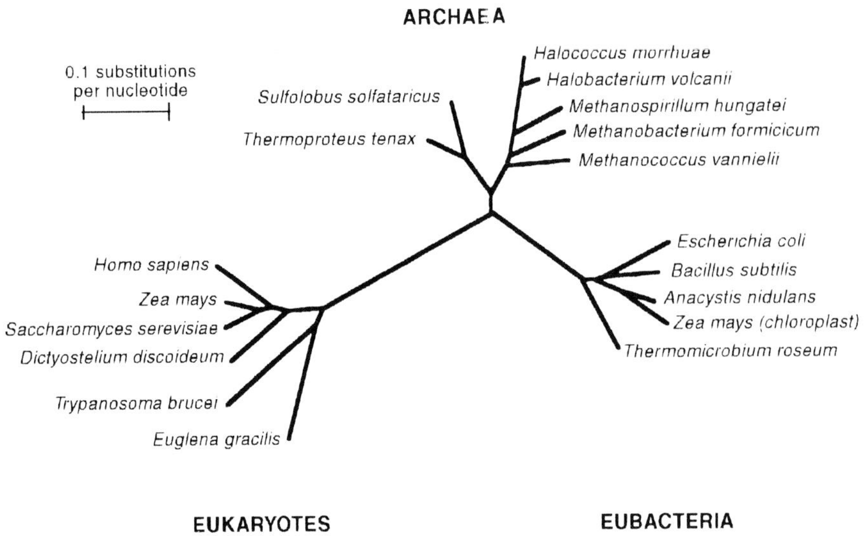

Fig. 3. The Archael unrooted tree relating all known groups of organisms. The tree was reconstructed from 16S and 18S ribosomal RNA sequences using the least-squares, distance matrix method. (Adapted from Olsen & Woese, 1989.)

LogDet distances (Lockhart *et al.*, 1994; Lake, 1994), comprehensive solutions for site-to-site variation and alignment artefacts are currently lacking.

MAPPING THE ORIGIN OF EUKARYOTES WITH ELONGATION FACTOR GENES

As a result of the artefacts inherent in reconstructing deep divergences from sequence data, the interpretation of the 'tree of life' which defines the origin of eukaryotes has been controversial (Penny, 1988; Hoffman, 1992). Since sequence analyses of identical sequences differing only in their alignments can support either the Archael or the eocyte trees, depending upon the alignment (Lake, 1991), we searched for molecular sequences which contained structural features, such as inserted segments, that might evolve much more slowly than individual nucleotides and hence be more easily interpreted.

The molecule we chose to study was protein synthesis elongation factor EF-Tu (EF-1α in eukaryotes). EF-Tu is an ubiquitous protein that transports aminoacyl-tRNAs to the ribosome and participates in their selection by the ribosome. The structure of the guanosine diphosphate (GDP)-binding domain of EF-Tu from *Escherichia coli* has been determined by X-ray diffraction (Jurnak, 1985; la Cour *et al.*, 1985). Within this domain (Fig. 4), amino acid sequence $DCPGH_{84}$ ends a β strand, forms part of the GDP binding site, and initiates a short α helix. The sequence $KNMITG_{94}$,

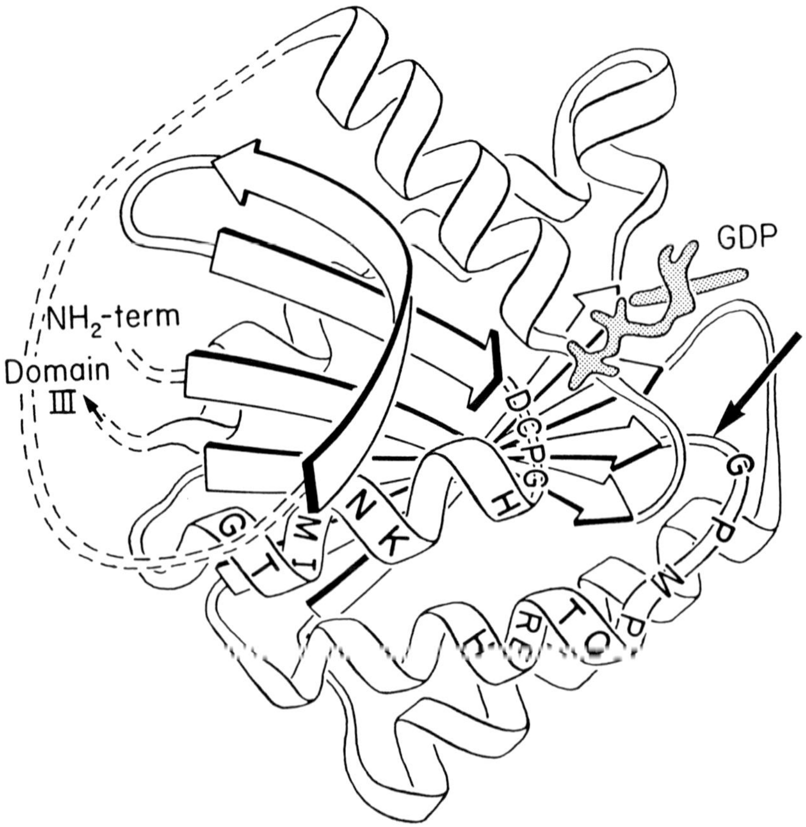

Fig. 4. Stylized schematic of the structure of the GDP binding domain of trypsin-modified elongation factor EF-Tu from *E. coli* (Jurnak, 1985; la Cour, 1985). Arrows represent β-strands, spirals represent helices; and GDP is shown schematically. The dashed line represents sequence removed from this domain by trypsin treatment. The arrow indicates the N-terminus of the four amino acid segment. The DCPGH, KNMITG, and GPMP regions are indicated.

which is conserved in EF-1α and EF-Tu sequences, terminates this helix. The β-strand that follows is terminated by $GPMP_{113}$ at the GDP binding site; $QTREH_{118}$ then starts a 3_{10} helix. The amino acid motifs of the eukaryotic EF-1α are similar, except that the four-amino acid sequence $GPMP_{113}$ is replaced by the 11-amino acid sequence GEFEAGISKDG and its variants (see Table 1).

Since the eukaryotic 11 amino acid insert is so well conserved among eukaryotic sequences, we thought that eocyte sequences might also contain the 11 amino acid insert. Using the polymerase chain reaction and DNA primers designed for use with the KNMITG and QTREH sites, we amplified, cloned, and sequenced the insert region with the results shown in Table 1. The eocyte amino acid sequences, translated from DNA, shared the eukaryotic motif (11 amino acids) rather than that found in Methanogens, Halobacteria, and Bacteria (4 amino acids). The longer 11-amino acid segment, present in eocytes and eukaryotes, shares little obvious similarity with the shorter 4-amino acid segment found in other prokaryotes.

In order to ascertain which form of the segment existed first, we compared the sequences of the related (paralogous) proteins EF-2 (termed EF-G in Bacteria) and IF-2 because both diverged from EF-1α before the last common ancestor of eukaryotes and prokaryotes (Bourne, Sanders & McCormick, 1991; Iwabe *et al.*, 1989; Gogarten *et al.*, 1989). Since all sequences from EF-2 and IF-2 contained the 4-amino acid segment, this indicated that $GPMP_{113}$ (or one of its variants) is the original (plesiomorphic) form of this segment and that the 11-amino acid segment must be the derived (synapomorphic) form.

Our conclusions are shown in Fig. 2. We have mapped the changes on to the trees representative of both theories. Starting from the 4-amino insert at the root of the tree, each solid box indicates a change from the 4-amino acid segment to the 11-amino acid form. The eocyte tree is favoured because it requires only a single change, whereas the Archael tree requires two independent but identical changes. The Archael tree could also be explained by one appearance of the 11-amino acid form and one reappearance of the 4-amino acid form, but even so, two changes would be required. The EF-1α synapomorphy cannot discriminate among the possible branchings below the eocyte–eukaryote node, so that any tree having eocytes and eukaryotes as sister taxa would be consistent with these data. In Fig. 5, these results are compared with the larger set of taxa shown in Fig. 1. Again, the agreement is complete.

Several lines of reasoning buttress the interpretation that eocytes are the closest relatives of the eukaryotes. First, the 11-amino acid segments present in eocytes and eukaryotes are very likely homologous. Eight of eleven amino acids (seven in *Sulfolobus* and *Acidianus*) are identical to the consensus eukaryotic sequence. Amino acid shuffling of the segments produced random alignments that score 6–7 standard deviations lower than those found for the eukaryotic–eocyte alignment, thereby implying homology (Waterman & Eggert, 1987). Secondly, the alignments are well defined. No gaps are needed to align the eukaryotic and eocytic EF-1α sequences, and no gaps are needed to align the Bacteria, Methanogen, and Halobacterial sequences. Thirdly, the sequences encoding EF-1α are not likely to have been laterally transferred between organisms, since EF-1α is present in all cells and, during protein synthesis, interacts with cellular components encoded by genes dispersed throughout the bacterial genome, including aminoacyl-tRNAs, ribosomal proteins, elongation factor EF-Ts, and 16S and 18S ribosomal RNAs (Hill *et al.*, 1990). Thus these results lend strong support to the proposal that the eukaryotes and eocytes are sister taxa within the 'tree of life'.

Table 1. *Comparison of the* Methanopyrus kandleri *EF-1 sequence to the sequences from methanogens, halobacteria, eubacteria, eocytes, and eukaryotes*

Taxon	Organism	11 amino-acid segment	4-amino acid segment		
Methanogens and relatives					
	Mp.kand	KNMITGASQADAAILVVAADD		*---GVMP*	qtreh
	T.celer	KNMITGASQADAAVLVVAVTD		*---GVMP*	QTKEH
	P.woes.	KNMITGASQADAAVLVVAATD		*---GVMP*	QTKEH
	A.fulg.	knmitgASQADAAVLVMDVVE		*---KVQP*	qtreh
	Mc.vann.	KNMITGASQADAAVLVVNVDD		*AKSGIQP*	QTREH
	T.acido.	KNMITGTSQADAAILVISARD		*-GEGVME*	QTREH
Halobacteria					
	H.maris.	KNMITGASQADNAVLVVAADD		*---GVQP*	QTQEH
Eubacteria					
	Th.mar.	KNMITGAAQMDGAILVVAATD		*---GPMP*	QTREH
	D.sal.	KNMITGAAQMDGAIIVCSAAD		*---GPMP*	QTREH
	E.coli.	KNMITGAAQMDGAILVVAATD		*---GPMP*	QTREH
Eocytes					
	Su.acid.	KNMITGASQADAAILVVSAKK	*GEYEAGMSAEG*		QTREH
	Td.mari.	KNMITGASQADAALLVVSARK	*GEFEAGMSAEG*		qtreh
	P.occu.	knmitgASQADAAILVVSARK	*GEFEAGMSAEG*		qtreh
	D.muco.	knmitgASQADAAILVVSARK	*GEFEAGMSAEG*		qtreh
	A.infe.	knmitgASQADAAIIAVSAKK	*GEFEAGMSEEG*		qtreh

Eukaryotes	Giardia	KNMITGESQADVAILVVAAGQ	*GEFEAGISKDG*	QTREH
	Tetrahy.	KNMITGTSQADVAILMIASPQ	*GEFEAGISKDG*	QTREH
	Yeast	KNMITGTSQADCAILIIAGGV	*GEFEAGISKDG*	QTREH
	Tomato	KNMITGESQADCAVLIIDSTT	*GGFEAGISKDG*	QTREH
	Droso.	KNMITGTSQADCAVQIDAAGT	*GEFEAGISKND*	QTREH
	Rat	KNMITGTSQADCAVLIVAAGV	*GEFEAGISKNG*	QTREH
	Human	KNMITGTSQADCAVLIVAAGV	*GEFEAGISKNG*	QTREH

Four amino acid and eleven amino acid segments are italicized. Small letters represent sequences from the PCR primers. The sequences compared are the following: The methanogens and their relatives are Mp. kan., Methanopyrus kandleri, T. celer, *Thermococcus celer*; P. woesei, *Pyrococcus woesei*; A. fulg., *Archaeoglobus fulgidus*; Mc. van., *Methanococcus vannielii*; and T. acido., *Thermoplasma acidophilum*. The halobacteria H. maris. *Halobacterium marismortui*. The eubacterial sequences are the thermophilic eubacteria Tt. mar., *Thermotoga maritima* and the halophilic cyanobacteria D. salina, *Dactylococcopsis salina*. The eocytes are Td. mar., *Thermodiscus maritimus*; P. occu., *Pyrodictium occultum*; A. infe., *Acidianus infernus*; S. acid., *Sulfolobus acidocaldarius*. The eukaryotes are: Giardia, *Giardia lamblia*; Tetrahy., *Tetrahymena pyriformis*; yeast, *Saccharomyces cerevisiae*; tomato, *Lycopersicion esculentum*; Droso., *Drosophila malanogaster*; rat, *Rattus norvegicus*; and human, *Homo sapiens*. Original sources for the sequences are listed in Rivera & Lake (1996).

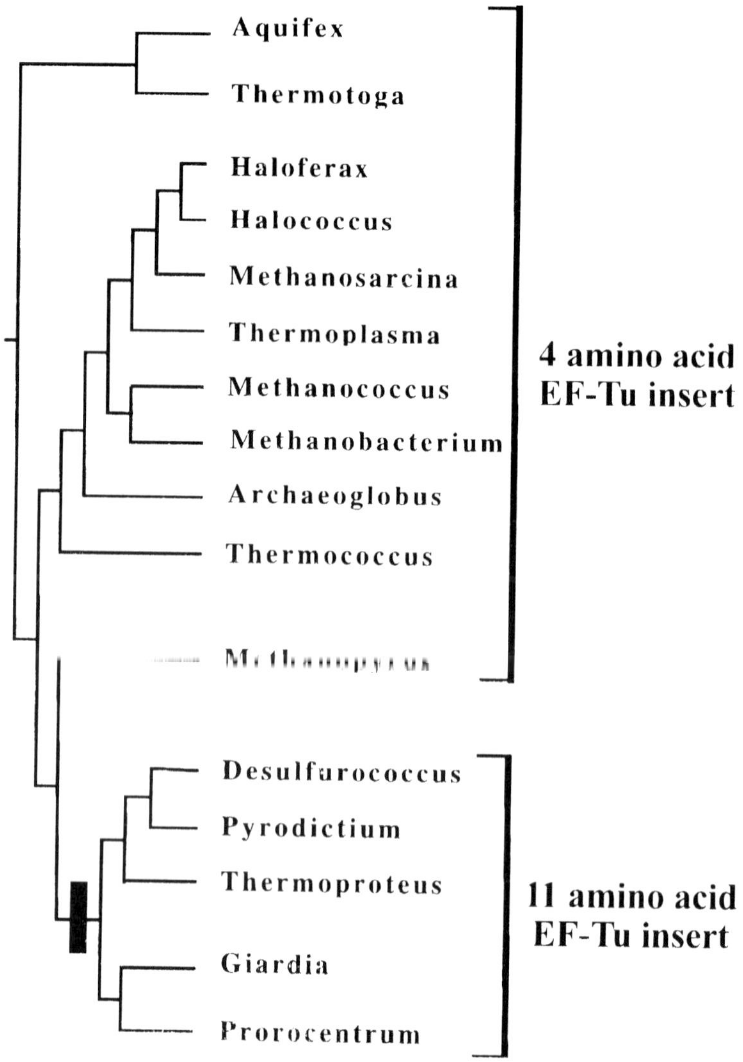

Fig. 5. The phylogenetic tree from Fig. 1 illustrating the distribution of 4- and 11-amino acid EF-Tu inserts. The dark box indicates the approximate location for the change from the 4-amino acid insert to the 11-amino acid insert.

OTHER MOLECULAR PROPERTIES SUPPORTING AN EUKARYOTE–EOCYTE CLADE

Organization of ribosomal operons

A number of fundamental molecular properties have been thought to have an idiosyncratic distribution on the tree of life, principally because they did not fit the Archael tree. Yet, these same molecular properties fit the eocyte theory perfectly well. In the following section two examples, ribosomal rRNA operons and prokaryotic TATA-like boxes, are compared with the eocyte tree.

Because small subunit ribosomal RNA sequences are the standard for defining the phylogenetic positions of organisms, a large database of ribosomal RNAs exists, and one knows far more about the organization of ribosomal operons than about any other operons. Bacteria, Halobacteria, Methanogens, and eocytes contain three rRNAs, 16S, 23S, and 5S, which are homologous to the eukaryotic 18S, 5.8S + 28S, and 5S. (For simplicity we will refer to both the eukaryotic and prokaryotic homologues using the prokaryotic labels.) The number of ribosomal rRNA transcriptional units varies between one and four in the Halobacteria and Methanogens. Ribosomal operons are arranged in the same general pattern in Bacteria, Halobacteria and Methanogens, namely 16S–tRNA–23S–5S. Occasionally, an additional tRNA gene will be found between the 16S and 23S genes or following the 5S gene (for review see Brown, Daniels & Reeve, 1989). *Thermoplasma*, which is phylogenetically related to the Methanogens, is an exception to this general rule and unlike any other prokaryotes, contains unlinked 16S, 23S and 5S genes (Tu & Zillig, 1982). The pattern in eocytes and eukaryotes is different from the Bacteria, Halobacteria, and Methanogens. In the eocytes the 16S–23S genes are linked without a tRNA spacer and there is a variable linkage of 5S rRNA encoding genes to the 16S–23S unit. The non-operon-associated 5S rRNA gene of *D. mobilis* forms its own transcriptional unit (Kjems & Garrett, 1988) but those of many other eocytes contain a 16S–23S–5S unit. The eukaryotic pattern is similar with a 16S–23S (equivalent) transcription unit lacking tRNA spacers and with the 5S either separately transcribed or linked (Gerbi, 1985). An exception to this rule is found among the Cryptomonads where the rRNA genes are unlinked (Gray, 1992).

Although this pattern of rRNA operon organization cannot be easily explained by the Archael theory, it fits the eocyte tree well. In Fig. 6, the tree is labelled with the types of operons. To accommodate this distribution only a single change of operon type is required. Namely, the 16S–tRNA–23S–5S pattern found in Bacteria, Halobacteria, and Methanogens is substituted by the derived 16S–23S type at the position on the tree shown by the box. Depending upon the operon organization in *Methanopyrus* (presently unknown) the site will be either before or after *Methanopyrus* branches. In either case still only a single change will be required.

Organization of transcriptional promoters

In eukaryotes there are three classes of specialized DNA-dependent RNA polymerases, I, II, and III. These polymerases, responsible for transcribing rRNAs, mRNAs, and tRNAs, respectively, employ different sets of

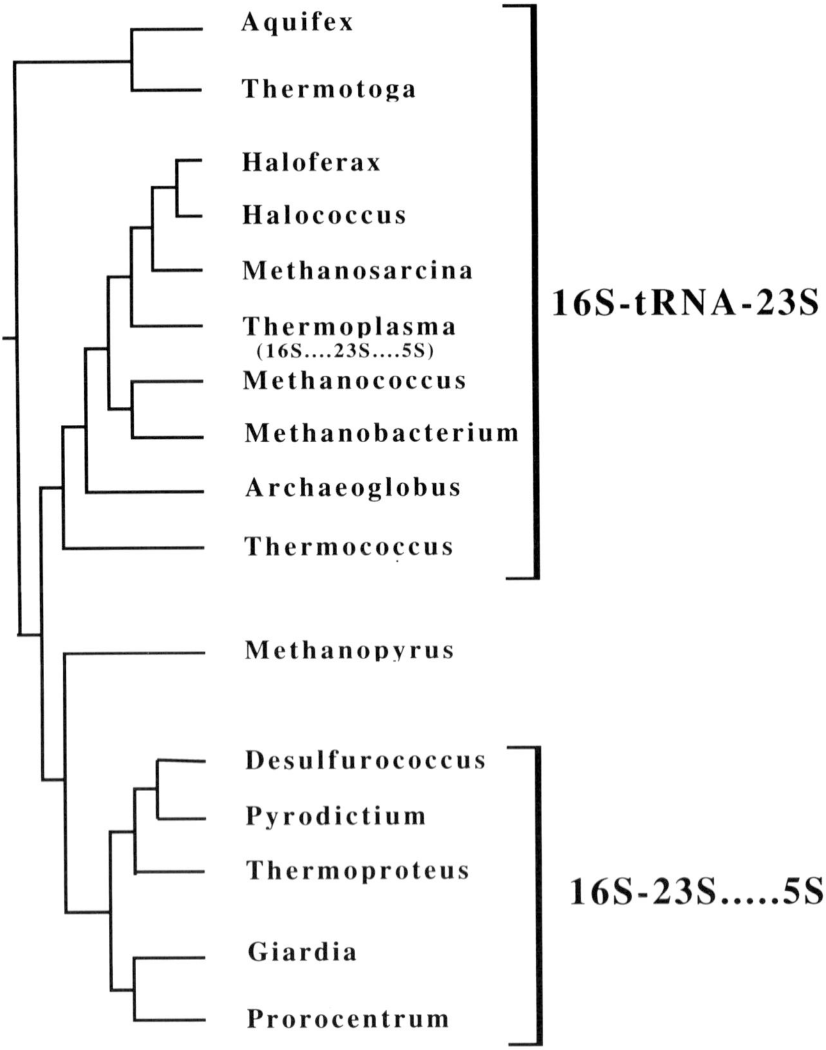

Fig. 6. The phylogenetic tree from Fig. 1 illustrating the distribution of ribosomal operon types. In the 16S–tRNA–23S type the operon contains 16S rRNA + tRNA (possibly two) + 23S rRNA + 5S rRNA. In the 16S–23....5S type the operon contains the 16S rRNA + 23S rRNA with the 5S rRNA frequently transcribed separately.

transcriptional initiation factors. The TATA-binding protein (TBP) is the only one of these factors that is required by all three classes of polymerase, whether or not the recognized promoter contains a TATA box (White & Jackson, 1992). Recently TBP homologues have been identified in *Pyrococcus* (Rowlands, Baumann & Jackson, 1994) and *Thermococcus* (Marsh *et al.*, 1994) suggesting that this protein, which is not found in the Bacteria originated at the time of the common ancestor of Halobacteria, Methanogens, eocytes and eukaryotes. In *Sulfolobus*, a similarity between eocytic and eukaryotic transcription was shown by the site-specific mutational analyses of the ribosomal RNA promoter. This established a role for a TATA-box sequence resembling the polymerase II promoter in transcriptional efficiency and start-site selection in eocytes (Reiter, Hudepohl & Zillig, 1990).

Unfortunately, due to the high visibility of the archael theory, many authors have assumed that what is true for eocytes is also true for Methanogens and Halobacteria, and therefore all Archaea have a TATA-box. That does not seem to be true.

Brown *et al.* (1989) have extensively reviewed promoter sites and specifically characterized the differences in promoter sequence in Halobacteria, Methanogens and eocytes. In *Methanococcus vannielii* initiation of transcription of stable RNA genes occurs at the first G within the conserved TGCAAGT (box B) (Wich *et al.*, 1986; Wich, Sibold & Bock, 1986). Approximately 20 bp upstream of Box B is the highly conserved 'box A' sequence. Sequences similar to box A, particularly to the TTATATA portion of the sequence, have been identified approximately 25 bp upstream of the site of transcription initiation in numerous methanogen genes. Brown *et al.* (1989) list the Methanogen box A consensus sequence AAANNTT-TATATA. In the Halobacteria a somewhat less well-conserved sequence GANGCCYTTAAGTA can be identified in the −25 to −30 region, but this consensus sequence is considerably different from that in Methanogens. Finally, they compile the consensus sequence for eocytes as AAANNTT-TAA. This latter sequence is quite close to the TATA sequence in multicellular eukaryotes, TATAAA. These results are summarized in Fig. 7, with the presumed region of homology indicated by the vertical lines. Within this region of homology, eocytes share five out of six promoter nucleotides with eukaryotes, Methanogens share four out of six, and Halobacteria share three out of six. This last figure is similar to that found for the bacterial −35 sequence, which is normally described as being unrelated to the TATA box. Thus one can appreciate the great power of an incorrect theory (Archaea) to mislead. Because eocytes have a TATA-like promoter, it has been (incorrectly) assumed that all Archaea have TATA-like promoters, even though the promoters of Halobacteria, Methanogens, and eocytes are quite different. Again, the data are best fit by the eocyte tree.

Recently discovered novel eocyte-related prokaryotic taxa

Within the last several years advances in DNA amplification have greatly expedited the characterization of bacterial environments through sampling the 16S rRNA sequences of natural populations. This has allowed the phylogenetic identification of organisms that cannot be cultured in the

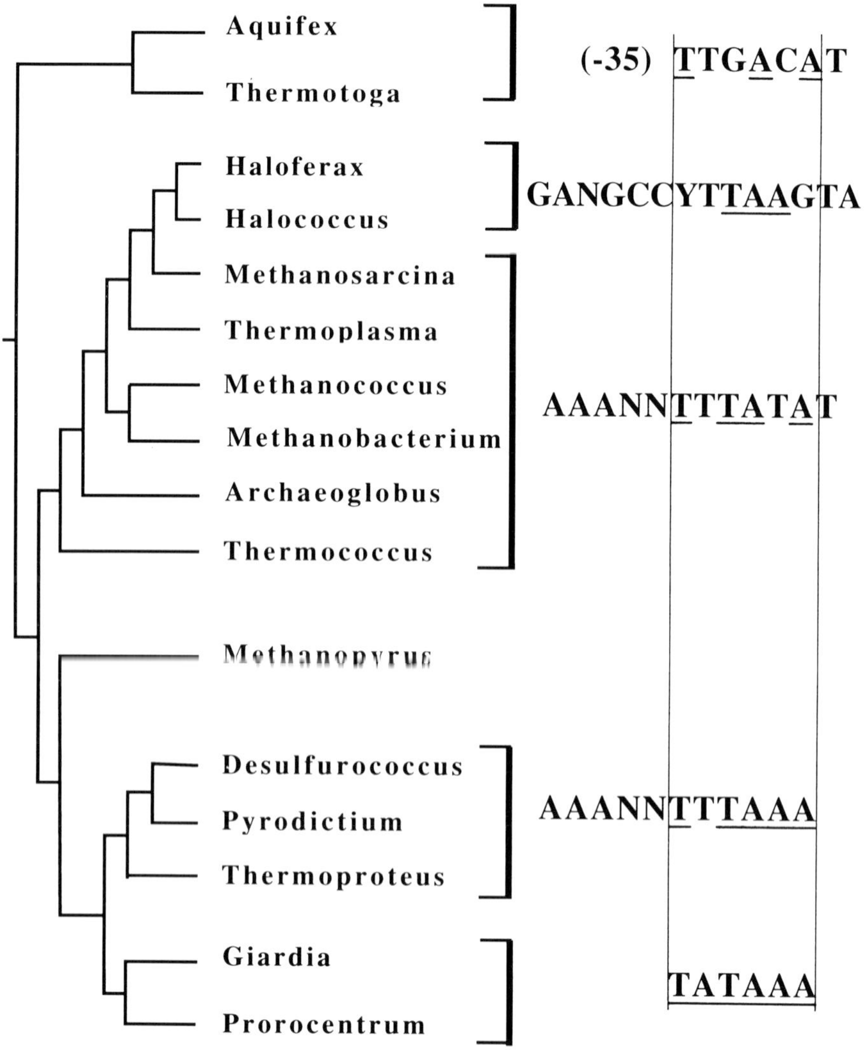

Fig. 7. The phylogenetic tree from Fig. 1 illustrating the distribution of TATA-like promoter sequences. The vertical lines indicate the conserved portions of the TATA-box sequence. The locations of the conserved nucleotides are underlined.

laboratory. One of the most surprising developments is that ribosomal RNA sequences obtained from diverse environments, including marine plankton as well as geothermal springs, have yielded organisms that are phylogenetically related to the eocytes in prokaryotic trees (Giovannoni *et al.*, 1990; Fuhrman, McCallum & Davis, 1992; Barns *et al.*, 1994). This suggested that these organisms might also be sister taxa to the eukaryotes.

With the development of the paralinear/LogDet algorithm which is insensitive to unequal rate effects (Lockhart *et al.*, 1994; Lake, 1994), one of the three major barriers to reconstructing deep trees from sequences has been solved, and it seemed that our chances of reconstructing a reasonable phylogeny were improved. Hence we included two clones from a Yellowstone National Park site that were associated with eocytes (pJ89 and pJ78, Barns *et al.*, 1994), and a marine isolate clone (actually the 5′ and 3′ ends from two closely related clones, Fuhrman *et al.*, 1992). The results are shown in Fig. 8. As suggested by the published prokaryotic trees, the clones are the sister group of the eocytes + eukaryotes.

Several years ago the node that separates the eukaryotes and their eocyte sister taxon from the Halobacteria + methanogen (Archael) clade was identified (Lake, 1988) and that ancestor plus all of its descendants was defined as the Karyotes. The location of that node is shown by the dark circle at the bottom of Fig. 8. Also indicated is the node corresponding to a monophyletic Archae (the dark circle at the top of Fig. 8). Although the data are preliminary, it appears that clones 78 and 89 as well as the marine isolates are also karyotes, as illustrated in Fig. 8. It will be interesting to see if other genes from these organisms (such as EF-Tu) continue to support this grouping.

CONCLUSIONS

Of all the sequences obtained to date, the EF-Tu molecule seems to offer the most reliable indication of early divergences. It is one of, if not the, slowest evolving sequence yet found and is unlikely to be laterally transferred between organisms because it is present in all cells and, during protein synthesis, interacts with cellular components that are dispersed throughout the bacterial genome. Furthermore, the direct analyses of the 11-amino acid insert as well as phylogenetic analyses of EF-Tu sequences by most authors support the eocyte tree. Additional support for the eocyte tree also comes from the finding that eukaryotic ribosomal operons are organized like those of *Sulfolobus, Desulfurococcus*, and *Thermoproteus* and not organized like the tRNA containing rRNA operons of Halobacteria, Methanogens, and Eubacteria.

Finally, it is important to recognize that the eukaryotic cell may itself have been created as a result of an early endosymbiotic event. Several theories exist for the origin of the nucleus. Most workers in the field subscribe to the *Karyogenic Hypothesis*, shown in Fig. 9. In this theory the nucleus and its enclosing membranes were gradually acquired through some (unspecified) segregating process. The competing theory is less well known. The *Endokaryotic Hypothesis* is based on the observation that the nucleus, like the other eukaryotic organelles enclosed in double membranes (the chloroplast and mitochondrion) has been derived through capture by an engulfing bacterium (Lake, 1982). The latter proposal is simple and parsimoniously

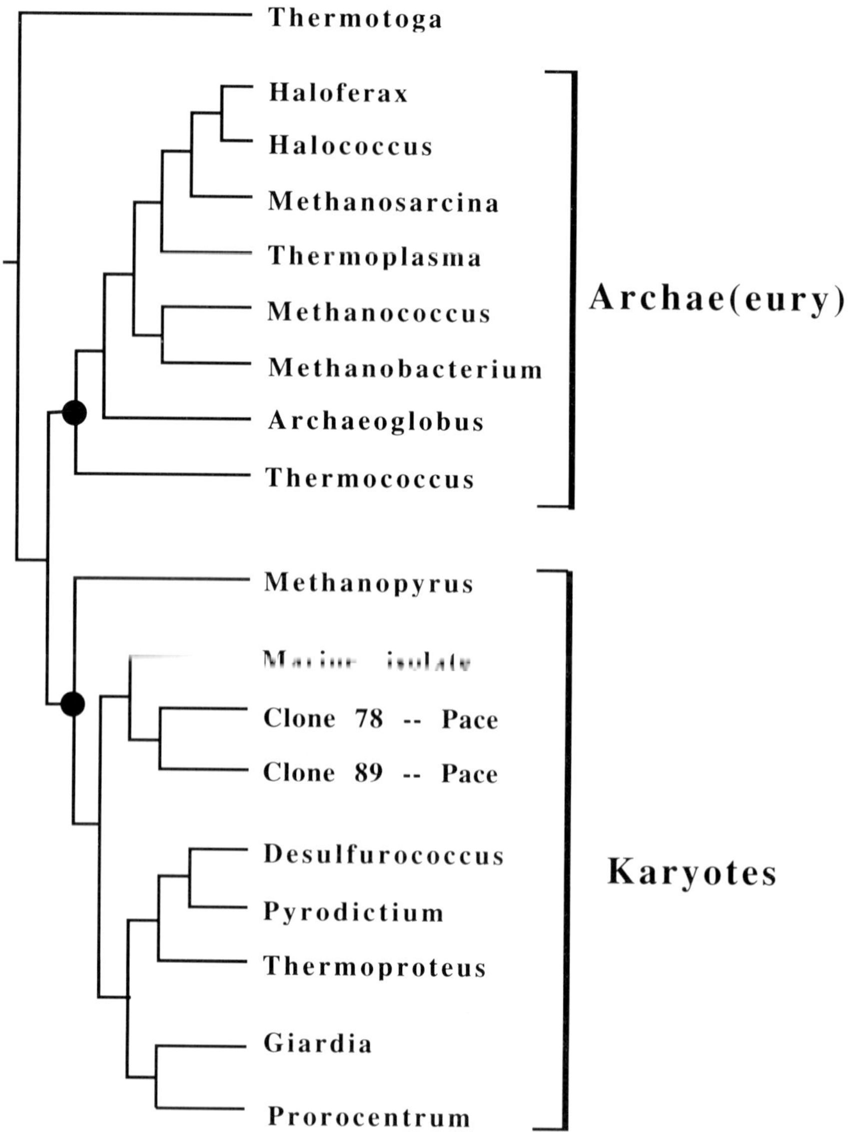

Fig. 8. The rooted phylogenetic tree relating prokaryotic and eukaryotic organisms, including recently discovered novel eocyte-related prokaryotic taxa. The tree was reconstructed from 16S and 18S ribosomal RNA sequences using the same algorithms described for the tree in Fig. 1.

explains the origin of all double-membrane organelles through a single mechanism rather than requiring two different mechanisms (one for the mitochondrion and chloroplast and another for the nucleus). Until recently there were few data to support either theory. Recently, however, Gupta *et*

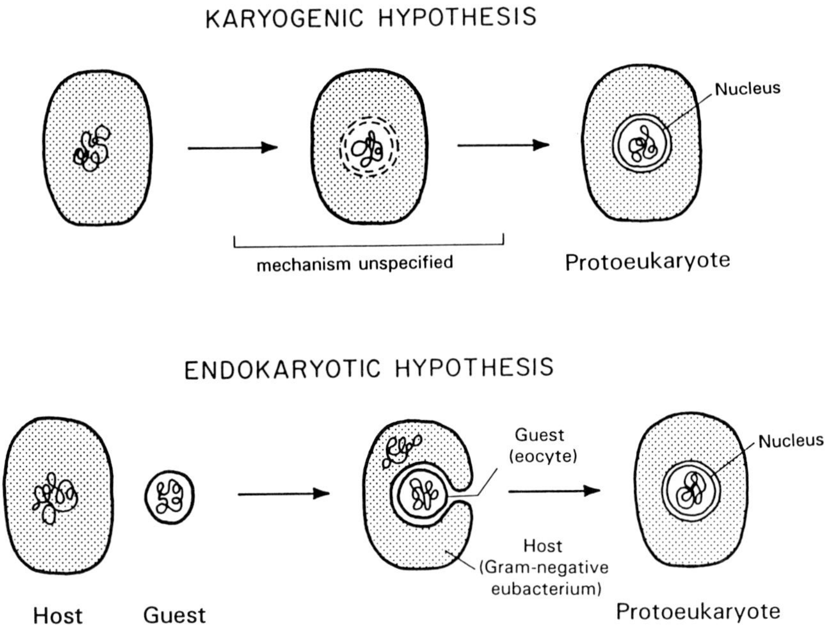

Fig. 9. Competing hypotheses for the evolution of the nucleus (Lake & Rivera, 1994).

al. (1994), argued on the basis of sequences of the 70 kDa heat shock protein (HSP70), that the *Endokaryotic Hypothesis* is supported. This raised an alternative explanation for some irregularities in the distribution of ester lipids. It has long been puzzling that Bacteria, Halobacteria, and eukaryotes share ester-linked fatty acids and functional fatty acid synthetases (Kamekura & Kates, 1988) and yet there is no phylogenetic indication that these three groups are closely related. However, if the fatty acid were introduced to eukaryotes through their host, and if that host was a Bacterium with ester lipids, then these data would best fit the eocyte tree described in this review. Clearly, deciding between both theories will be important for any complete understanding of the origin of eukaryotes.

We are just starting to fathom the diversity of life on earth. Although the relationship of the nuclear genes of eukaryotes with those of prokaryotes have been controversial, it now appears that eukaryotes are very likely the sister group of the eocyte prokaryotes. The tremendous progress that is being made toward determining the relationships among bacteria and eukaryotes gives one optimism that at least some of the major events in the tree of life may soon be understood.

ACKNOWLEDGEMENTS

We thank NSF for supporting this research with a grant to one of us (JAL).

REFERENCES

Baldauf, S. L. & Palmer, J. D. (1990). Evolutionary transfer of the chloroplast tufA gene to the nucleus. *Nature*, **344**, 262–3.

Barns, S. M., Fundyga, R. E., Jeffries, M. W. & Pace, N. R. (1994). Remarkable archael diversity detected in a Yellowstone National Park hot spring environment. *Proceedings of the National Academy of Sciences, USA*, **91**, 1609–13.

Bourne, H. R., Sanders, D. A. & McCormick, F. (1991). The GTPase superfamily: conserved structure and molecular mechanism. *Nature*, **349**, 117–27.

Brown J. W., Daniels, C. J. & Reeve, J. N. (1989). Gene structure, organization, and expression in archaebacteria. *CRC Critical Reviews in Microbiology*, **16**, 287–338.

Burggraf, S., Stetter, K. O., Rouviere, P. & Woese, C. R. (1991). *Methanopyrus kandleri: an archeal methanogen unrelated to all other known methanogens. Systematic Applied Microbiology*, **14**, 346–51.

de Queiroz, K. & Gauthier, J. (1990). Phylogeny as a central principle in taxonomy – phylogenetic definitions of taxon names. *Systematic Zoology*, **39**, 307–22.

Eldridge, N. & Cracraft, J. (1986). *Phylogenetic Patterns and the Evolutionary Process*. Columbia University Press, New York.

Fuhrman, J. A., McCallum, K. & Davis, A. A. (1992). Novel major archaebacterial group from marine plankton. *Nature*, **356**, 148–9.

Gerbi, S. A. (1985). Evolution of ribosomal DNA, in *Molecular Evolutionary Genetics*, R. J. MacIntyre, ed., Chapter 7, Plenum Publishing, New York.

Giovannoni, S. J., Fritschgi, T. B., Moyer, C. L. & Field, K. G. (1990). Genetic diversity in Sargasso sea bacterioplankton. *Nature*, **345**, 60–3.

Gogarten, J. P., Kibak, H., Dittrich, P., Taiz, L., Bowman, E. J., Bowman, B. J., Manolson, M. F., Poole, R. J., Date, T., Oshima, T., Konishi, J., Denda, K. & Yoshida, M. (1989). Evolution of the Vacuolar H^+-ATPase: implications for the origin of Eukaryotes. *Proceedings of the National Academy of Sciences, USA*, **86**, 6661–5.

Goodwin, T. W. (1980). *The Biochemistry of the Carotenoids*, Volume I, Chapman and Hall, London and New York.

Gray, M. W. (1992). The endosymbiont hypothesis revisited. *International Review of Cytology*, **141**, 233–357

Gray, M. W. (1993). Origin and evolution of organelle genomes. *Current Opinion in Genetics and Development*, **3**, 884–90.

Gupta, R. S., Aitken, K., Falah, M. & Singh, B. (1994). Cloning of *Giardia lamblia* heat shock protein HSP70 homologs: implications regarding origin of eukaryotic cells and of endoplasmic reticulum. *Proceedings of the National Academy of Sciences, USA*, **91**, 2895–6.

Hill, W. E., Dahlberg, A., Garrett, R. A., Moore, P. B., Schlessinger, D. & Warner, J. R. (eds). (1990). *The Ribosome, Structure, Function, and Evolution*. American Society of Microbiology Press, Washington, DC.

Hoffman, M. (1992). Researchers find organism they can really relate to. *Science*, **257**, 32.

Huber, R. & Stetter, K. O. (1992). The *Thermotogales*: hyperthermophilic and extremely thermophilic bacteria. In *Thermophilic Bacteria*, Kristansson, J. K., ed., pp. 185–194, CRC Press, Boca Raton, Florida.

Iwabe, N., Kuma, K. I., Hasegawa, M., Osawa, S. & Miyata, T. (1989). Evolutionary relationship of Archaebacteria, Eubacteria and Eukaryotes inferred from phylogenetic trees of duplicated genes. *Proceedings of the National Academy of Sciences, USA*, **86**, 9355–9.

Jurnak, F. (1985). Structure of the 6DP domain of EF-Tu and location of the amino acids homologous to ras oncogene proteins. *Science*, **230**, 32–6.

Kamekura, M. & Kates, M. (1988). Lipids of halophilic archaebacteria. In *Halophilic Bacteria*, Rodrigues-Valera, F., ed., pp. 24–54. CRC Press, Boca Raton, Florida.

Kjems, J. & Garrett, R. A. (1988). Novel expression of the ribosomal RNA genes in the extreme thermophile and archaebacterium, *Desulfurococcus mobilis*. *EMBO Journal*, **6**, 3521–7.

la Cour, T. F., Nyborg, J., Thirup, S. & Clark, B. F. (1985). Structural details of the binding of guanosine diphosphate to elongation factor Tu from *E. coli* as studied by X-ray crystallography. *EMBO Journal*, **4**, 2385–8.

Lake, J. A. (1982). Mapping evolution with ribosome structure: intralineage constancy and interlineage variation. *Proceedings of the National Academy of Sciences, USA*, **79**, 5948–52.

Lake, J. A. (1988). Origin of the eukaryotic nucleus determined by rate-invariant analysis of rRNA sequences. *Nature*, **331**, 184–6.

Lake, J. A. (1991). The order of sequence alignment can bias the selection of tree topology. *Molecular Biology and Evolution*, **8**, 378–85.

Lake, J. A. (1994). Reconstructing evolutionary trees from DNA and protein sequences: paralinear distances. *Proceedings of the National Academy of Sciences, USA*, **91**, 1455–9.

Lake, J. A. (1995). Calculating the probability of multitaxon evolutionary trees: bootstrappers Gambit. *Proceedings of the National Academy of Sciences, USA*, **92**, 9662–6.

Lake, J. A. & Rivera, M. C. (1994). Was the nucleus the first endosymbiont? *Proceedings of the National Academy of Sciences, USA*, **91**, 2880–1.

Lechner, K. & Bock, A. (1987). Cloning and nucleotide sequence of the gene for an archaebacterial protein synthesis elongation factor Tu. *Molecular and General Genetics*, **208**, 523–8.

Lockhart, P. J., Steel, M. A., Hendy, M. D. & Penny, D. (1994). Recovering evolutionary trees under a more realistic model of sequence evolution. *Molecular Biology and Evolution*, **11**, 605–12.

Marsh, T. L., Reich, C. I., Whitlock, R. B. & Olsen, G. J. (1994). Transcription factor IID in the Archaea: sequences in the *Thermococcus celer* genome could encode a product closely related to the TATA-binding protein of eukaryotes. *Proceedings of the National Academy of Sciences, USA*, **91**, 4180–4.

Olsen, G. J. & Woese, C. R. (1989). A brief note concerning archaebacterial phylogeny. *Canadian Journal of Microbiology*, **35**, 119–23.

Penny, D. (1988). What was the first living cell? *Nature* (News and Views), **331**, 111.

Reiter, W.-D., Hudepohl, U. & Zillig, W. (1990). Mutational analysis of an archaebacterial promoter: essential role of a TATA box for transcription efficiency and start-site selection *in vitro*. *Proceedings of the National Academy of Sciences, USA*, **87**, 9509–13.

Rivera, M. C. & Lake, J. A. (1992). Evidence that eukaryotes and eocyte prokaryotes are immediate relatives. *Science*, **257**, 74–6.

Rivera, M. C. & Lake, J. A. (1996). The phylogeny of *Methanopyrus kandleri*. *International Journal of Systematic Bacteriology*, **46**, 348–51.

Rowlands, T., Baumann, P. & Jackson, S. P. (1994). The TATA-binding protein: a general transcription factor in eukaryotes and archaebacteria. *Science*, **264**, 1326–9.

Runnegar, B. (1993). Proterozoic eukaryotes: evidence from biology and geology.

In *Early Life on Earth*, Bengtson, S., ed., pp. 287–297. Columbia University Press, New York.

Smith, M. W., Feng, D. F. & Doolittle, R. F. (1992). Evolution by acquisition: the case for horizontal gene transfers. *Trends in Biochemical Science*, **17**, 489–93.

Stetter, K. O., Fiala, G., Huber, G., Huber, R. & Segerer, A. (1990). Hyperthermophilic microorganisms. *FEMS Microbiology Review*, **75**, 117–24.

Tu, J. & Zillig, W. (1982). Organization of rRNA structural genes in the archaebacterium *Thermoplasma acidophilum*. *Nucleic Acids Research*, **10**, 7231–45.

Volters, J. & Erdmann, V. A. (1989). The structure and evolution of Archaebacterial ribosomal RNAs. *Canadian Journal of Microbiology*, **35**, 43–51.

Waterman, M. S. & Eggert, M. (1987). A new algorithm for best subsequence alignments with application to tRNA–rRNA comparisons. *Journal of Molecular Biology*, **197**, 723–8.

White, R. J. & Jackson, S. P. (1992). The TATA-binding protein: a central role in transcription by RNA polymerases I, II and II. *Trends in Genetics*, **8**, 284–8.

Wich, G., Hummel, H., Jarsch, M., Bar, U. & Bock, A. (1986). Transcription signals for stable RNA genes in *Methanococcus*. *Nucleic Acids Research*, **14**, 2459–67.

Wich, G., Sibold, L. & Bock, A. (1986). Genes for tRNA and their putative expression signals in methanogens. *Systematic and Applied Microbiology*, **7**, 18–27.

ORGANELLAR EVOLUTION

MICHAEL W. GRAY AND DAVID F. SPENCER

Program in Evolutionary Biology,
Canadian Institute for Advanced Research,
Department of Biochemistry, Dalhousie University,
Halifax, Nova Scotia B3H 4H7, Canada

INTRODUCTION

According to the endosymbiont hypothesis, mitochondria and chloroplasts originated as bacteria-like organisms that were taken up by and entered into a symbiotic relationship with a nucleus-containing host cell (Margulis, 1970). Although this hypothesis has had a long and contentious history (Sapp, 1990), over the past three decades it has assumed the status of a theory. This transition has, in large part, been a consequence of the discovery that mitochondria and chloroplasts contain DNA, and the subsequent realization that the genes present in these genomes could provide a phylogenetic key to the evolutionary history of the organelles containing them (Gray & Doolittle, 1982). Over the past 30 years, an extensive store of information has accumulated about the molecular biology of mitochondrial DNA (mtDNA) and chloroplast DNA (cpDNA), and this database has provided many insights into organellar origins and evolution (Gray, 1992). Aside from the basic question of whether or not the endosymbiont hypothesis is correct, molecular biological data have been instrumental in addressing such issues as where within the bacterial lineage mitochondria and chloroplasts are likely to have originated, whether the endosymbioses that gave rise to them occurred only once (monophyletic origin) or more than once (polyphyletic origin), and what has happened to organellar genomes since their bacterial progenitors first took up residence within the eukaryotic cell.

SIZE AND FUNCTION OF ORGANELLAR GENOMES

A developing theme in organelle evolution is that organellar genomes are genetically rather conservative, while at the same time displaying considerable variation in size, physical form, overall genome organization, gene arrangement, pattern and mechanism of gene expression, and mode and tempo of evolution (Gillham, 1994; Gray, 1989*a*, 1989*b*, 1991, 1992, 1993, 1995). Structural, functional and evolutionary variation is particularly pronounced in the case of mitochondrial genomes, which range in size from 6 kilobase pairs (kbp) in malaria parasites, *Plasmodium* sp., to as much as

2500 kbp in the flowering plant, *Cucumis melo* (muskmelon). Mitochondrial DNAs also encompass genomes that at the level of primary sequence are among the most rapidly evolving (mammalian mtDNA) and the most slowly evolving (angiosperm mtDNA). Given these extremes, and considering their relatively small sizes, organellar genomes provide excellent material for a comparative genomics approach to discovering how genomes change in evolution.

In their capacity as energy-transducing organelles, chloroplasts generate ATP through photosynthesis whereas mitochondria do so through electron transport and oxidative phosphorylation. In both cases, a small number of the proteins required for organelle biogenesis and function (mostly having to do with energy production) are encoded in and expressed from the organellar DNA, which also carries the genes for some of the components of the organelle-specific protein synthesizing system that translates the organellar DNA-encoded mRNAs. Translation components specified by organellar DNA include the rRNA species of the organellar ribosome, often (but not always) some of the ribosomal proteins, and usually (but again not always) at least some or even all of the tRNAs required for organellar protein synthesis. Despite this essential genetic function of organellar DNA, most of the genes required for organellar biogenesis and function are encoded in nuclear DNA. It is presumed that most of these organellar nuclear genes have been transferred there from the bacteria-like, proto-organellar genome in the course of evolution (Gray, 1992, 1993; Schuster & Brennicke, 1994).

ORIGIN AND EVOLUTION OF CHLOROPLAST GENOMES

Primary endosymbiosis: single or multiple?

Biochemical and ultrastructural similarities first suggested an evolutionary link between chloroplasts and photosynthetic bacteria, notably blue-green algae, or Cyanobacteria. Various types of molecular evidence have been advanced in support of this contention, including similarities in patterns of gene organization (operon arrangement), gene expression (promoter, RNA polymerase and ribosome structure; nature and function of protein synthesis factors; use of *N*-formylmethionyl-tRNA as initiator in translation; antibiotic sensitivity of both transcription and translation), and gene structure (rRNA, tRNA and protein). This accumulated body of evidence (summarized in Gray, 1989*b*, 1991, 1992, 1993; Delwiche, Kuhsel & Palmer, 1995) overwhelmingly supports an origin of chloroplasts and the chloroplast genome from within the cyanobacterial phylum of (eu)bacteria, via one or more endosymbiotic events.

Among a number of criteria, pigment composition distinguishes different chloroplast types. All chloroplasts contain chlorophyll *a* as the primary photosynthetic pigment associated with photosystem I and II reaction centres; however, they differ in their content of accessory pigments. Chlorophyll *b* is present in the chl a/b chloroplasts of land plants, green algae (chlorophytes) and euglenoid algae (euglenophytes); chlorophyll *c* is found in the chl a/c chloroplasts of chromophytes (golden algae, brown algae, diatoms) and dinoflagellates; and phycobiliproteins are present in the chl a/PB chloroplasts of rhodophytes (red algae), cryptomonads and glaucocystophytes (the latter two groups also contain chlorophyll *c*). In an early formulation of the endosymbiont hypothesis as it applies to chloroplasts, Mereshkowsky (1910) suggested that chloroplasts of different pigment composition were derived from 'cyanophytes' having the corresponding pigment arrays. More recently, this idea was resurrected and restated by Raven (1970). Such a scheme implies that chloroplasts arose more than once, in a series of separate endosymbioses; i.e. that chloroplasts are polyphyletic. Because cyanobacteria contain chlorophyll *a* plus phycobiliproteins, a cyanobacteria-like endosymbiont was an obvious candidate for the direct ancestor of the chl a/PB chloroplast. More recently, newly discovered eubacteria containing chlorophylls *a* and *b* (prochlorophytes) were proposed as possible specific ancestors of chl a/b chloroplasts (Lewin, 1981), whereas the chlorophyll *a*/*c*-containing eubacterium *Heliobacterium chlorum* was singled out as a possible direct ancestor of chl a/c chloroplasts (Margulis & Obar, 1985). However, the evidence to date favours the view that there was but a single origin of chloroplasts, with diversification of accessory photosynthetic pigments taking place subsequent to the one primary endosymbiotic event that established the proto-choroplast and its genome (Gray, 1993).

Recently published phylogenetic trees, based on comparison of both small subunit (SSU) rRNA sequences (Nelissen *et al.*, 1995) and TufA (translation elongation factor Tu) amino acid sequences (Delwiche *et al.*, 1995), bring strong additional support to the idea of an early origin of all chloroplasts from within the cyanobacterial lineage. This is also seen in the tree shown in Fig. 1, in which cyanobacteria and chloroplasts are united with a bootstrap value of 100/100. Other studies have failed to support a specific affiliation of chl a/b chloroplast types with prochlorophytes and of chl a/c chloroplasts with *H. chlorum* (Gray, 1989*b*, 1991; Delwiche *et al.*, 1995).

The question of whether chloroplasts are monophyletic within the cyanobacterial radiation has been widely debated (Cavalier-Smith, 1987, 1992; Gray, 1991, 1992; Bryant, 1992; Martin, Somerville & Loiseaux-de Goër, 1992; Morden *et al.*, 1992). In the most recent studies, Nelissen *et al.* (1995) favour a monophyletic scenario, whereas Delwiche *et al.* (1995) are equivocal on this point, citing relatively low bootstrap support for either monophyly or polyphyly in various TufA trees. This is also the case in the SSU rRNA tree shown in Fig. 1, in which strict monophyly of chloroplasts is supported at a bootstrap level of only 34/100. Note, however, that in this

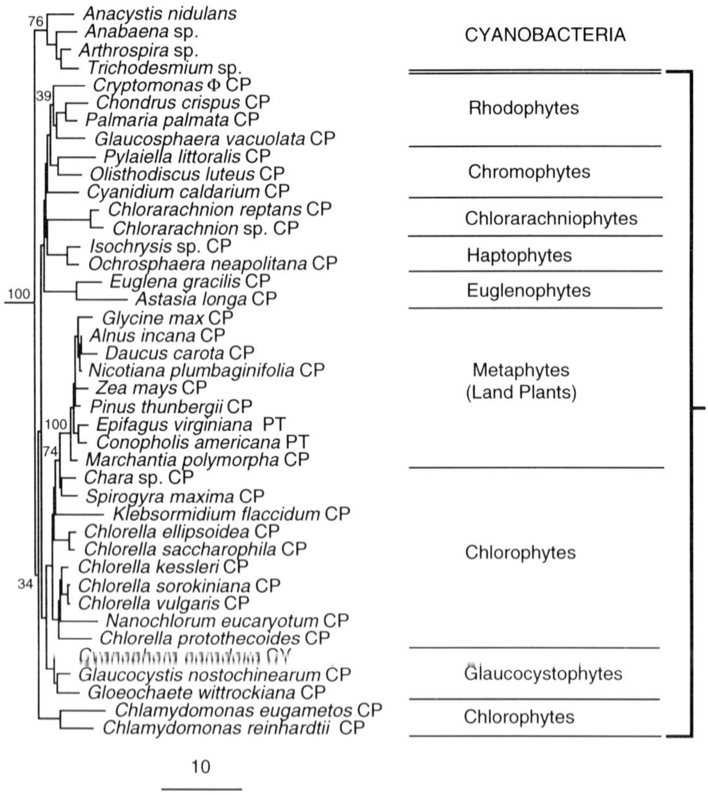

Fig. 1. Cyanobacterial/chloroplast portion of a eubacterial/organellar SSU rRNA tree. Abbreviations: CP, chloroplast; CY, cyanelle; PT, plastid (= non-photosynthetic chloroplast). The analysis employed an aligned data set of 236 published eubacterial, chloroplast/plastid and mitochondrial SSU rRNA sequences (1009 characters). The tree was generated using DNADIST from PHYLIP version 3.5 (Felsenstein, 1993) with the ML option and a transition–transversion ratio of 1.0. The dissimilarity matrix from this program was used as input to NEIGHBOR (the neighbor-joining program of PHYLIP). Bootstrap analysis (100 replicates) was performed using the programs SEQBOOT, DNADIST, NEIGHBOR and CONSENSE. The Figure shows the region of the initial distance tree leading to cyanobacteria and chloroplasts. Bootstrap values added to some of the relevant branches indicate the number of times out of 100 the taxa to the right of that node affiliate. Other analysis algorithms (transversion and simple identity-based) gave essentially the same results. The bar indicates 10 substitutions per 100 nucleotides.

particular analysis, the *Chlamydomonas* sp. chloroplast sequences behave anomalously, in that they do not branch as expected with the other green algae. Among groupings that do not appear in the consensus bootstrap tree is a clade comprising all chloroplasts except those of *Chlamydomonas* sp. and also excluding cyanobacteria (31%). Thus, if the *Chlamydomonas* sequences are excluded from consideration, bootstrap support for chloroplast monophyly increases substantially.

In this distance tree, the glaucocystophytes appear as a sister group to the remainder of the chlorophytes plus land plants (a grouping never observed by Delwiche *et al.*, 1995); however, they cluster (albeit with a very low bootstrap value, 23%) with the chromophyte/rhodophyte lineage in the consensus bootstrap tree, an affiliation consistent with both the glaucocystophyte chloroplast pigment composition (chl a/c/PB) and Rubisco type (proteobacterial; see below). Relatively high bootstrap values unite chlorophytes, with the exception of *Chlamydomonas* sp., and land plants (96%); *Chara/Spirogyra* with land plants (74%); and all land plants (100%).

The continuing volatility in different studies of deep branchings within the cyanobacterial/chloroplast clade may in part reflect the great difficulty, in these single-gene trees, of resolving lineages that may have radiated early and relatively rapidly within both the cyanobacteria *per se* and the chloroplast types derived from them. So far, phylogenetic analyses have failed to reveal a convincing affiliation between chloroplasts and a specific group within the cyanobacterial lineage, although this may simply be a consequence of the current limited sampling. Finding a single such link to all chloroplast types would help to resolve the issue of whether or not this organelle is monophyletic, although, as pointed out by Delwiche *et al.*, 1995, a polyphyletic origin of chloroplasts from closely related cyanobacteria would be very difficult, if not impossible, to discern.

In contrast to the results of the SSU rRNA and TufA analyses noted above, phylogenetic trees constructed from RbcS and RbcL amino acid sequences (representing the large and small subunits of ribulose-1,5-*bis*phosphate carboxylase/oxygenase, or Rubisco) clearly and consistently indicate a proteobacterial, not cyanobacterial, origin of these genes in chl a/P and chl a/c (but not chl a/b) chloroplasts (Gray, 1991, 1992; Delwiche *et al.*, 1995). Among possible explanations of this apparent biphyletic origin of Rubisco genes is lateral transfer of these genes from a proteobacterial source. An alternative possibility (Martin *et al.*, 1992; Liaud *et al.*, 1994) is that ancestral photosynthetic proteobacteria and cyanobacteria possessed two sets of Rubisco genes that were differentially lost in the two main chloroplast lineages (land plant plus chlorophytes versus rhodophytes plus chromophytes; see Fig. 1), subsequent to a monophyletic origin of the chloroplast. In fact, the photosynthetic α-proteobacterium *Rhodobacter sphaeroides*, whose genome does contain two different sets of Rubisco genes, provides a contemporary precedent for such a situation (Gibson, Falcone & Tabita, 1991).

These sorts of inconsistencies between various gene trees emphasize the need to integrate a variety of data in drawing conclusions about organellar origins and evolution. Shared molecular characters that support both a cyanobacterial and a monophyletic origin of chloroplasts have been discussed elsewhere (Gray, 1991, 1992, 1993; Delwiche *et al.*, 1995; Reith, 1995). As well as data drawn from genes encoded by chloroplast DNA, information derived from nuclear genes for chloroplast function can be very useful in providing phylogenetic insights. A recent example supporting a monophyletic origin of chloroplasts is the demonstration that polypeptides from the light harvesting complex of a red alga are immunologically related to those of angiosperms and chromophytes, but not to those of cyanobacteria (Wolfe *et al.*, 1994).

Complete sequences of chloroplast genomes: a new source of phylogenetic information

Determination of the complete sequence of an organellar genome permits definition of its entire genetic information content. In land plants and algae, sequenced cpDNAs have been found to contain primarily genes for photosynethesis and gene expression. Emerging data for rhodophyte and chromophyte chloroplast genomes indicate that these generally have more genes encoding proteins involved in these two functions, as well as additional genes specifying other biochemical activities. The latter include genes for biosynthesis (amino acids, fatty acids, chlorophyll, carotenoids, phycobiliproteins, thiamine), nitrogen assimilation, redox reactions, protein transport, and protein degradation. In comparing the completely sequenced, 191 kbp chloroplast genome of *Porphyra purpurea*, a red alga (approximately 250 genes), with those of land plants (120–156 kbp, approximately 120 genes), Reith (1995) noted that there were approximately 100 common genes, approximately 150 genes that are present only in *P. purpurea* cpDNA, and some 20 genes that were only found in the chloroplast genome of land plants. Loss or transfer of genes to the nucleus is presumed to account for much of the observed disparity in gene content. In the highly reduced (70 kbp) chloroplast genome of *Epifagus virginiana*, a non-photosynthetic parasitic angiosperm, virtually all of the genes required for photosynthetic function have been lost (dePamphilis & Palmer, 1990).

The availability of an increasing number of complete sequences of chloroplast genomes sets the stage for a systematic survey for phylogenetically informative characters. An illustration of this approach is shown in Table 1, which scores the presence or absence in various sequenced cpDNAs of genes encoding proteins of the chloroplast small ribosomal subunit (*rps* genes). Some 20 cpDNA-encoded genes have been characterized as homologues of 20 of the 21 *rps* genes present in the *Escherichia coli* genome (only an *rps21* homologue has so far not been identified in any cpDNA; Harris, Boynton & Gillham, 1994). Several interesting points emerge from this comparison. (i) An almost complete set of *rps* genes is encoded by *P. purpurea* cpDNA, which lacks only *rps15* in addition to *rps21* (Reith & Munholland, 1995). This presumably reflects the ancestral state of the proto-chloroplast genome. (ii) Except for *rps1*, the chloroplast genome of the chromophyte alga, *Odontella sinensis* (a centric diatom), encodes the

Table 1. *Small subunit ribosomal protein genes (rps) encoded by sequenced chloroplast genomes*

rps	Ppu [a]	Osi	Cpa	Egr	Mpo	Pth	Eva	Nta	Osa	Zma
1	■ [b]	○	○	○	○	○	○	○	○	○
2	■ [b]	■	■	■	■	■	■	■	■	■
3	■	■	■	■	■	■	■	■	■	■
4	■	■	■	■	■	■	■	■	■	■
5	■	■	■	○	○	○	○	○	○	○
6	■	■	■	○	○	○	○	○	○	○
7	■	■	■	■	■	■	■	■	■	■
8	■	■	■	■	■	■	■	■	■	■
9	■	■	■	■	○	○	○	○	○	○
10	■	■	■	○	○	○	○	○	○	○
11	■	■	■	■	■	■	■	■	■	■
12	■	■	■	■	■	■	■	■	■	■
13	■	■	■	○	○	○	○	○	○	○
14	■	■	■	■	■	■	■	■	■	■
15	○	○	○	○	■	■	○	■	■	■
16	■	■	■	○	○	○	○	■	■	■
17	■	■	■	○	○	○	○	○	○	○
18	■	■	■	■	■	■	■	■	■	■
19	■	■	■	■	■	■	■	■	■	■
20	■	■	■	○	○	○	○	○	○	○

[a]Organism names are abbreviated as follows (chloroplast genome size in bp is given in parentheses): Ppu, *Porphyra purpurea* (191 028); Osi, *Odontella sinensis* (119 704); Cpa, *Cyanophora paradoxa* (135 599); Egr, *Euglena gracilis* (142 171); Mpo, *Marchantia polymorpha* (121 024); Pth, *Pinus thunbergii* (119 707); Eva, *Epifagus virginiana* (70 028); Nta, *Nicotiana tabacum* (155 844); Osa, *Oryza sativum* (134 525); Zma, *Zea mays* (140 387). Complete genome sequences and accompanying GenBank annotation are available electronically from the World Wide Web site of the Organelle Genome Megasequencing Program (URL http://megasun.bch.umontreal.ca/ogmpproj.html; select 'Genome Bioinformatics' then 'Organelle Genome Databases').

[b]○, gene absent; ■, gene present.

same *rps* gene set as *Porphyra* cpDNA (Kowallik *et al.*, 1995), consistent with the affiliation of the chl a/c and chl a/PB algae in single gene-based phylogenetic trees (see, e.g. Fig. 1). (iii) The cpDNA of *Cyanophora paradoxa*, a glaucocystophyte alga of uncertain phylogenetic affiliation (see Delwiche *et al.*, 1995, and Fig. 1), has precisely the same set of *rps* genes as *Odontella* cpDNA (Stirewalt *et al.*, 1995), consistent with the view (Delwiche *et al.*, 1995) that *C. paradoxa* represents a very early branch within the chloroplast lineage. (Among other ancestral characters retained by the *Cyanophora* chloroplast, or 'cyanelle', is a vestigial peptidoglycan cell wall.) (iv) The *rps* genes found in the cpDNA of green algae and land plants constitute a subset of those in *Porphyra* cpDNA, indicative of differential (and substantial) loss of *rps* genes in the chl a/b line of chloroplast evolution. The same set of 12 *rps* genes is found in the cpDNAs of two monocots (*Oryza sativum*, rice and *Zea mays*, maize) and a dicot (*Nicotiana tabacum*, tobacco), compared with the 19 *rps* genes encoded by *Porphyra* cpDNA. (v) In spite of its substantial reduction in size relative to other metaphyte cpDNAs, the *Epifagus* chloroplast genome retains virtually the entire set of angiosperm *rps* genes, lacking only *rps15* and *rps16* (the latter is also absent from the chloroplast genomes of *Pinus thunbergii* (pine, a gymnosperm), and *Marchantia polymorpha* (liverwort, a bryophyte)). (vi) The cpDNA of *Euglena gracilis*, a chl a/b euglenophyte, also has a basically metaphyte *rps* composition, a character consistent with the various protein-based phylogenies (including TufA; Delwiche *et al.*, 1995) that unite *Euglena* and metaphyte chloroplasts. The chloroplast of *Euglena* and the plastid of its nonphotosynthetic relative *Astasia* regularly affiliate with the chl a/c chloroplast group in rRNA trees (as, for example, in Fig. 1), but this behavior has been ascribed to a treeing artifact caused by an extreme A+T bias in the *Euglena* chloroplast rRNA genes (Lockhart *et al.*, 1992). Although there is an increasing body of evidence grouping euglenophytes with chlorophytes and land plants at the level of the chloroplast genome, in nuclear rRNA trees euglenophytes affiliate with the trypanosomatid protozoa as an early diverging eukaryotic branch, well separated from the chlorophyte/metaphyte clade (Sogin, Edman & Elwood, 1989). This discordance between chloroplast and nuclear phylogenies is best explained by a secondary acquisition of chloroplasts through endosymbiosis between the original *Euglena* host cell and a *eukaryotic*, chl a/b alga, as originally proposed by Gibbs (1978).

Chloroplast acquisition by secondary endosymbiosis

Another characteristic that distinguishes various chloroplast types is the number of surrounding membranes. Chlorophyte and rhodophyte chloroplasts are bounded by only two membranes, those of *Euglena* and dinoflagellates are typically surrounded by three membranes, whilst chloroplasts in chromophytes and some other algae have four distinct membranes. It has been postulated that those chloroplasts having more than two membranes have been acquired in a secondary endosymbiotic event between a nonphotosynthetic host cell and a eukaryotic alga (that is, a eukaryote-eukaryote rather than eukaryote-prokaryote endosymbiosis; see Gibbs, 1981). According to this scenario, only the chloroplasts of red and green algae trace their descent directly to a primary endosymbiosis with a eubacterial – specifically cyanobacterial – endosymbiont.

The postulate of a secondary acquisition of chloroplasts in some algal groups is attractive because it can explain differences between chloroplast and nuclear phylogenies such as are seen in the case of *Euglena gracilis*, mentioned above. Until recently, an inference of secondary endosymbiosis had been difficult to prove because in most cases, all that would appear to remain of the putative endosymbiont is the chloroplast itself and one or two additional membranes, supposedly representing the plasma membrane of the endosymbiont and (if a fourth membrane is present) the phagosome membrane of the host.

In a few cases, reduction of the eukaryotic endosymbiont has not progressed to the stage of complete elimination of everything except the chloroplast and additional membrane(s). In these instances, remnants of the endosymbiont nucleus persist in the form of a distinctive structure called the *nucleomorph*, located between the two innermost and two outermost chloroplast membranes (the periplastidal compartment). Studies of nucleomorph DNA have shown that it encodes a set of eukaryotic-like rRNA genes that specify the RNA components of 80S ribosomes present in the periplastidal compartment. These nucleomorph rRNA genes are distinct from those encoded by the main nucleus, which specify the rRNA components of the cytoplasmic 80S ribosomes. These and other data have been instrumental in affirming the hypothesis of a secondary acquisition of chloroplasts in some algal groups (Gray, 1994; McFadden & Gilson, 1995). Algae for which there is now compelling evidence of a secondary endosymbiosis of this type include two photosynthetic cryptomonads, *Cryptomonas* Φ and *Pyrenomonas salina*, the non-photosynthetic cryptomonad *Chilomonas paramecium*, and a chlorarachniophyte, *Chlorarachnion*.

How many secondary endosymbioses have there been in algal evolution? This question is still being investigated, with the most recent evidence supporting a possible common origin of the chloroplast in the three flagellated cryptomonads noted above (most likely from a red algal endosymbiont), but a separate origin of the amoeboid chlorarachniophyte chloroplast (Cavalier-Smith, Allsopp & Chao, 1994). Thus, whereas chloroplasts may have had a single primary origin from a cyanobacterial endosymbiont, there have evidently been several secondary events of lateral chloroplast transfer within the eukaryotic lineage.

ORIGIN AND EVOLUTION OF MITOCHONDRIAL GENOMES

Coding function of mitochondrial DNA: genetic conservatism versus structural diversity

In general, mitochondrial genomes are smaller than chloroplast genomes and carry a correspondingly smaller array of genes. They are also much more variable in size and organization, and can display strikingly different patterns and rates of evolution (for example, mammalian versus land plant). For these reasons, it has been even harder than in the case of chloroplasts to apply molecular data to convincingly answer questions such as how many times mitochondria arose, and what the evolutionary connections are at the level of the mitochondrial genome.

For the most part, the coding function of mtDNA is directed toward the synthesis of key inner membrane components of the respiratory chain. Typically, genes encoded by mtDNA include one or more *nad* (Complex I; NADH:ubiquinone oxidoreductase), *cob* (III: ubiquinol:cytochrome *c* oxidoreductase), *cox* (IV: cytochrome *c* oxidase) and *atp* (V: ATP synthase), in addition to rRNA and tRNA genes (Table 2). Some mtDNAs, notably those

Table 2. *Respiratory chain genes in sequenced mitochondrial DNAs*

	Hsa[a]	Pan	Mpo	Cre	Pwi	Aca	Ccr
Complex I (*nad*[b])							
1	■[c]	■[d]	■	■	■	■	■
2	■	■[d]	■	■	■	■	■
3	■	■[d]	■	❍	■	■	■
4	■	■[d]	■	■	■	■	■
4L	■	■[d]	■	❍	■	■	■
5	■	■[d]	■	■	■	■	■
6	■	■[d]	■	■	■	■	■
7	❍[c]	❍	❐[c]	❍	■	■	❍
9	❍	❍	■	❍	■	■	❍
11	❍	❍	❍	❍	❍	■	❍
Complex II (*sdh*)							
2	❍	❍	❍	❍	❍	❍	■
3	❍	❍	■	❍	❍	❍	■
4	❍	❍	■	❍	❍	❍	■
Complex III (*cob*)							
	■	■	■	■	■	■	■
Complex IV (*cox*)							
1	■	■	■	■	■	■[e]	■
2	■	■	■	❍	■	■[e]	■
3	■	■	■	❍	■	■	■
Complex V (*atp*)							
1	❍	❍	■	❍	■	■	❍
6	■	■	■	❍	■	■	■
8	■	■	❍	❍	❍	❍	❍
9	❍	■	■	❍	■	■	■

[a]Organism names are abbreviated as follows (mitochondrial genome size in bp is given in parentheses): Hsa, *Homo sapiens* (16 569); Pan, *Podospora anserina* (100 314); Mpo, *Marchantia polymorpha* (186 608); Cre, *Chlamydomonas reinhardtii* (15 758); Pwi, *Prototheca wickerhamii* (55 328); Aca, *Acanthamoeba castellanii* (41 591); Ccr, *Chondrus crispus* (25 836). Complete genome sequences and accompanying GenBank annotation are available electronically from the World Wide Web site of the Organelle Genome Megasequencing Program (see Table 1 for details).

[b]Gene abbreviations are: *nad*, NADH dehydrogenase; *sdh*, succinate dehydrogenase; *cob*, apocytochrome *b*; *cox*, cytochrome oxidase; *atp*, ATP synthase.

[c]■, gene present; ❐, pseudogene; ❍, gene absent.

[d]Gene absent in certain ascomycete yeast, including *Saccharomyces cerevisiae* and *Schizosaccharomyces pombe*.

[e]A single open reading frame encodes both *cox1* and *cox2* in *A. castellanii* mtDNA (Lonergan & Gray, 1996).

Table 3. *Ribosomal protein genes in sequenced mitochondrial DNAs*

	Hsa[a]	Pan	Mpo	Cre	Pwi	Aca	Ccr
Small subunit (*rps*)							
1	○[b]	○	■[b]	○	○	○	○
2	○	○	■	○	■	■	○
3	○	○	■	○	■	■	■
4	○	○	■	○	■	■	○
7	○	○	■	○	■	■	○
8	○	○	■	○	○	■	○
10	○	○	■	○	■	○	○
11	○	○	■	○	■	■	■
12	○	○	■	○	■	■	■
13	○	○	■	○	■	■	○
14	○	○	■	○	■	■	○
19	○	○	■	○	■	■	○
Large subunit (*rpl*)							
2	○	○	■	○	○	■	○
5	○	○	■	○	■	■	○
6	○	○	■	○	■	■	○
11	○	○	○	○	○	■	○
14	○	○	○	○	○	■	○
16	○	○	■	○	■	■	■

[a]Organism names are abbreviated as in Table 2.
[b]○, gene absent; ■, gene present.

of plants and protists, also carry a set of ribosomal protein genes (Table 3), and recently genes for respiratory complex II (succinate:ubiquinone oxido-reductase) have been identified in mtDNA for the first time (Table 2; Daignan-Fornier *et al.*, 1994; Leblanc *et al.*, 1995; Viehmann *et al.*, 1996; Burger *et al.*, 1996). Although the number of respiratory and particularly ribosomal protein genes varies widely among sequenced mtDNAs, to date all well-characterized mitochondrial genomes have been found to encode at least *cob* and *cox1*, the latter specifying the largest subunit of cytochrome oxidase. Thus, in spite of the profound structural diversity of mitochondrial genomes, their fundamental function is rather conservative.

A proteobacterial origin of mitochondria

A growing body of molecular evidence supports the inference made on biochemical grounds by John & Whatley (1975) that the non-sulphur purple bacteria (Proteobacteria) are the closest contemporary bacterial relatives of mitochondria. Ribosomal RNA sequence comparisons first highlighted the α-division of the proteobacteria as the probable source of the mitochondrial ancestor (Yang *et al.*, 1985), and that identification has subsequently been narrowed to one of the two major subdivisions of the α-proteobacteria, that

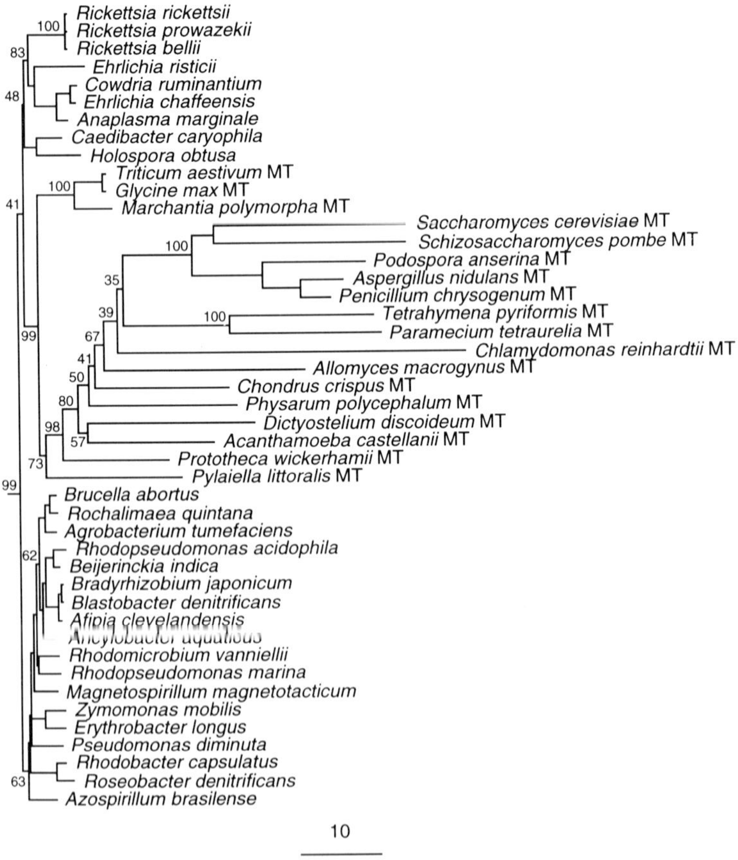

Fig. 2. α-Proteobacterial/mitochondrial portion of a eubacterial/organellar SSU rRNA tree. See Fig. 1 for details of the analysis. Abbreviation: MT, mitochondrial.

containing the obligate intracellular parasitic genera *Rickettsia, Ehrlichia* and *Anaplasma*.

A SSU rRNA tree illustrating these relationships is shown in Fig. 2 (this subtree is part of the same eubacterial/organellar tree from which the cyanobacterial/chloroplast subtree shown in Fig. 1 was extracted). The affiliation of mitochondria with the α-proteobacteria is supported by a very high bootstrap value (99/100), the same value that unites all mitochondria thus favouring a monophyletic origin of mitochondria. The mitochondrial subtree is divided into plants and non-plants, a split that also is seen consistently and persistently in these mitochondrial SSU rRNA trees (Gray, 1995), and which also draws high support in bootstrap analysis.

In the particular study shown here (Fig. 2), the association of mitochondria with *Rickettsia* and related taxa is somewhat low (41/100), due to the inclusion of *Holospora* and *Caedibacter*, two recently characterized ciliate endosymbionts (Springer *et al.*, 1993). The SSU rRNA sequences from these organisms appear to contribute some instability to this node of the tree; in fact, branching position within this particular clade suggests that *Holospora* and *Caedibacter* are deeply diverging representatives of this lineage. When these two taxa are excluded from consideration, bootstrap support rises to 73% for a specific affiliation between mitochondria and *Rickettsia* and its relatives. This association is much stronger (>95%) in other trees that include a smaller selection of mitochondria and α-proteobacteria, and that do not include *Holospora* and *Caedibacter* (Gray, 1995; D. F. Spencer, unpublished results). Precisely the same association of mitochondria with the rickettsial group of α-proteobacteria has recently been reported for phylogenetic trees based on comparison of heat shock protein (Hsp60) amino acid sequences (Viale & Arakaki, 1994; Gupta, 1995). Thus, in contrast to the situation with chloroplasts and cyanobacteria, it has been possible to associate mitochondria with a specific group of taxa within the major phylum of eubacteria from which these organelles originated.

A comparative genomics approach to mitochondrial genome evolution

Although rRNA sequence comparisons have been extremely useful in pinpointing the origin of mitochondria within the eubacteria, they have been considerably less incisive at defining relationships within the mitochondrial lineage itself. To a large extent, this is due to the striking variations in rate of sequence divergence in the mitochondrial rRNA genes of different eukaryotes. Rapid rates of divergence are manifested as very long branches in the mitochondrial clade (Fig. 2): only the relatively conservative plant mitochondrial rRNA sequences have branch lengths that approximate those of their α-proteobacterial relatives. These large rate differences can lead to a 'long branches attract' artefact in phylogenetic trees, which may underlie the robust split between the plant (slowly evolving) and non-plant (rapidly evolving) sequences seen in Fig. 2 (see Gray, 1995). Within the non-plant mitochondrial group, a few clades (e.g. fungi, ciliate protozoa) are well supported by bootstrap analysis, but most other nodes are not. As Table 4 illustrates, substantial differences in mitochondrial gene content and genome organization compound the difficulties of establishing phylogenetic connections: there is little or no evidence from these data that the land plant *M. polymorpha* and the green alga *C. reinhardtii* shared a common mitochondrial ancestor as recently as they shared a common chloroplast or nuclear ancestor.

As in the case of chloroplast genomes, complete mitochondrial genome sequences have contributed additional phylogenetically informative characters. Thus, although *C. reinhardtii* mtDNA encodes no ribosomal proteins

Table 4. *Comparison of the coding function of mitochondrial DNA*

	Marchantia polymorpha	*Chlamydomonas reinhardtii*
Size (bp)	186 608	15 758
Respiratory chain genes	17 (+1 pseudogene)	7
Ribosomal protein genes	16	0
Other ORFs	29	1
Introns	32	0
Intron ORFs	10	0
Ribosomal RNA genes		
Large subunit	+[a]	+ (9 pieces)[b]
Small subunit	+	+ (4 pieces)[b]
5S	+	−
Transfer RNA genes	29	3

[a]+, gene present; −, gene absent.
[b]Ribosomal RNA genes are fragmented and scrambled in *C. reinhardtii* mtDNA (Boer & Gray, 1988; Gray & Boer, 1988).

at all, a virtually identical set of *rps* and *rpl* genes to that encoded by *M. polymorpha* mtDNA (Oda *et al.*, 1992) is found in the mitochondrial genome of another chlorophyte alga, *Prototheca wickerhamii* (Wolff *et al.*, 1991), as well as in the mtDNA of the non-photosynthetic amoeboid protozoon, *Acanthamoeba castellanii* (Burger *et al.*, 1995; Table 3). Many of these ribosomal protein genes are arrayed in a similar manner in these three mitochondrial genomes, which is particularly striking in the case of *Marchantia* and *Acanthamoeba* (Burger *et al.*, 1995). Another strong character uniting *Marchantia* and *Prototheca* mitochondrial genomes is the presence of a distinctive mtDNA-encoded 5S rRNA gene (Wolff *et al.*, 1994). Because these and other similarities in mtDNA gene content and organization must have been extant in a common mitochondrial ancestor of *Marchantia, Prototheca* and *Acanthamoeba*, the marked differences seen in the *Chlamydomonas* mitochondrial genome are best explained by a relatively rapid and extreme evolution of the latter genome away from the ancestral pattern represented by the more conservative mtDNAs in the former three organisms.

CONCLUSIONS

Organellar genomes have provided a wealth of molecular data that have illuminated many aspects of organellar origins and evolution. Phylogenetic trees based on rRNA and protein sequences have validated the fundamental assumption of the endosymbiont hypothesis and have been particularly valuable for delineating the origin of chloroplasts and mitochondria from within the eubacteria. However, as emphasized in this review, single-gene trees have been less successful in providing an unequivocal answer to the question of whether each organelle arose only once. Relationships within each organellar lineage have also been difficult to discern on the basis of single-gene phylogenetic trees. The increasing availability of complete organellar genome sequences offers a complementary approach in our attempts to reconstruct the evolutionary history of mitochondria and chloroplasts. Information provided by this approach includes: the presence of particular genes or sets of genes; particular patterns of gene organization; intron distribution; and specific patterns and mechanisms of gene expression, including such unusual phenomena as RNA editing (Gray, 1993). Comparative data derived from complete organellar genome sequencing will be essential in our efforts to understand how and why organellar genomes, particularly mitochondrial ones, have adopted such radically different evolutionary strategies, and what mechanisms underlie these differences. Finally, complete organellar genome sequences offer the opportunity to construct phylogenetic trees from combined rather than single-protein data sets, thereby increasing substantially the number of informative characters and hence the resolution in such trees. The most profound insights into organellar evolution are likely to come from additional complete sequences of organellar genomes of unicellular eukaryotes (protists), wherein most of the biological diversity of eukaryotes exists, and most of the unexplored territory remains.

ACKNOWLEDGEMENTS

Work reported from the authors' laboratory was supported in part by a grant (MT-4124) from the Medical Research Council of Canada to M.W.G., who also acknowledges salary and interaction support from the Canadian Institute for Advanced Research (Program in Evolutionary Biology). Some of the published mitochondrial genome data reported here were generated by the Organellar Genome Megasequencing Program (OGMP), a special project of the Medical Research Council of Canada (SP-34).

REFERENCES

Boer, P. H. & Gray, M. W. (1988). Scrambled ribosomal RNA gene pieces in *Chlamydomonas reinhardtii* mitochondrial DNA. *Cell*, **55**, 399–411.

Bryant, D. A. (1992). Puzzles of chloroplast ancestry. *Current Biology*, **2**, 240–2.

Burger, G., Plante, I., Lonergan, K. M. & Gray, M. W. (1995). The mitochondrial DNA of the amoeboid protozoon, *Acanthamoeba castellanii*: complete sequence, gene content and genome organization. *Journal of Molecular Biology*, **245**, 522–37.

Burger, G., Lang, B. F., Reith, M. & Gray, M. W. (1996). Genes encoding the same three subunits of respiratory complex II are present in the mitochondrial DNA of two phylogenetically distant eukaryotes. *Proceedings of the National Academy of Sciences, USA*, **93**, 2328–32.

Cavalier-Smith, T. (1987). The simultaneous symbiotic origin of mitochondria, chloroplasts, and microbodies. *Annals of the New York Academy of Sciences*, **503**, 55–71.

Cavalier-Smith, T. (1992). The number of symbiotic origins of organelles. *BioSystems*, **28**, 91–106.

Cavalier-Smith, T., Allsopp, M. T. E. P. & Chao, E. E. (1994). Chimeric conundra: Are nucleomorphs and chromists monophyletic or polyphyletic? *Proceedings of the National Academy of Sciences, USA*, **91**, 11368–72.

Daignan-Fornier, B., Valens, M., Lemire, B. D. & Bolotin-Fukuhara, M. (1994). Structure and regulation of *SDH3*, the yeast gene encoding the cytochrome b_{560} subunit of respiratory complex II. *Journal of Biological Chemistry*, **269**, 15469–72.

Delwiche, C. F., Kuhsel, M. & Palmer, J. D. (1995). Phylogenetic analysis of *tuf*A sequences indicates a cyanobacterial origin of all plastids. *Molecular Phylogenetics and Evolution*, **4**, 110–28.

dePamphilis, C. W. & Palmer, J. D. (1990). Loss of photosynthetic and chlororespiratory genes from the plastid genome of a parasitic flowering plant. *Nature*, **348**, 337–9.

Felsenstein, J. (1993). Phylip (Phylogeny Inference Package) Version 3.5c. Distributed by the author. Department of Genetics, University of Washington, Seattle.

Gibbs, S. P. (1978). The chloroplasts of *Euglena* may have evolved from symbiotic green algae. *Canadian Journal of Botany*, **56**, 2883–9.

Gibbs, S. P. (1981). The chloroplasts of some algal groups may have evolved from endosymbiotic eukaryotic algae. *Annals of the New York Academy of Sciences*, **361**, 193–208.

Gibson, J. L., Falcone, D. L. & Tabita, F. R. (1991). Nucleotide sequence, transcriptional analysis, and expression of genes encoded within the form I CO_2 fixation operon of *Rhodobacter sphaeroides*. *Journal of Biological Chemistry*, **266**, 14646–53.

Gillham, N. W. (1994). *Organelle Genes and Genomes*, Oxford University Press, New York/Oxford.

Gray, M. W. (1989*a*). Origin and evolution of mitochondrial DNA. *Annual Review of Cell Biology*, **5**, 25–50.

Gray, M. W. (1989*b*). The evolutionary origins of organelles. *Trends in Genetics*, **5**, 294–9.

Gray, M. W. (1991). Origin and evolution of plastid genomes and genes. In *The Molecular Biology of Plastids (Cell Culture and Somatic Cell Genetics of Plants*, Vol. 7A), Bogorad L. & Vasil I. K. eds., pp. 303–30.

Gray, M. W. (1992). The endosymbiont hypothesis revisited. *International Review of Cytology*, **141**, 233–357.

Gray, M. W. (1993). Origin and evolution of organelle genomes. *Current Opinion in Genetics and Development*, **3**, 884–90.

Gray, M. W. (1994). One plus one equals one: the making of a cryptomonad alga. *American Society of Microbiology News*, **60**, 423–7.

Gray, M. W. (1995). Mitochondrial evolution. In *The Molecular Biology of Plant Mitochondria*, Levings C. S. III & Vasil I. K. eds, pp. 635–659. Kluwer Academic Publishers, Dordecht, The Netherlands.

Gray, M. W. & Boer, P. H. (1988). Organization and expression of algal (*Chlamydomonas reinhardtii*) mitochondrial DNA. *Philosophical Transactions of the Royal Society of London Series B*, **319**, 135–47.

Gray, M. W. & Doolittle, W. F. (1982). Has the endosymbiont hypothesis been proven? *Microbiological Reviews*, **46**, 1–42.

Gupta, R. S. (1995). Evolution of the chaperonin families (Hsp60, Hsp10 and Tcp-1) of proteins and the origin of eukaryotic cells. *Molecular Microbiology*, **15**, 1–11.

Harris, E. H., Boynton, J. E. & Gillham, N. W. (1994). Chloroplast ribosomes and protein synthesis. *Microbiological Reviews*, **58**, 700–54.

John, P. & Whatley, F. R. (1975). *Paracoccus denitrificans* and the evolutionary origin of the mitochondrion. *Nature*, **254**, 495–8.

Kowallik, K. V., Stoebe B., Schaffran, I., Kroth-Pancic, P. & Freier, U. (1995). The chloroplast genome of a chlorophyll *a+c*-containing alga, *Odontella sinensis*. *Plant Molecular Biology Reporter*, **13**, 336–42.

Leblanc, C., Boyen, C., Richard, O., Bonnard, G., Grienenberger, J.-M. & Kloareg, B. (1995). Complete sequence of the mitochondrial DNA of the rhodophyte *Chondrus crispus* (Gigartinales). Gene content and genome organization. *Journal of Molecular Biology*, **250**, 484–95.

Lewin, R. A. (1981). *Prochloron* and the theory of symbiogenesis. *Annals of the New York Academy of Sciences*, **361**, 325–9.

Liaud, M.-F., Valentin, C., Martin, W., Bouget, F.-Y., Kloareg, B. & Cerff, R. (1994). The evolutionary origin of red algae as deduced from the nuclear genes encoding cytosolic and chloroplast glyceraldehyde-3-phosphate dehydrogenases from *Chondrus crispus*. *Journal of Molecular Evolution*, **38**, 319–27.

Lockhart, P. J., Penny, D., Hendy, M. D., Howe, C. J., Beanland, T. J. & Larkum, A. W. D. (1992). Controversy on chloroplast origins. *FEBS Letters*, **301**, 127–31.

Lonergan, K. M. & Gray, M. W. (1996). Expression of a continuous open reading frame encoding subunits 1 and 2 of cytochrome *c* oxidase in the mitochondrial DNA of *Acanthamoeba castellanii*. *Journal of Molecular Biology*, **257**, 1019–30.

Margulis, L. (1970). *Origin of Eurkaryotic Cells*, Yale University Press, New Haven, Connecticut.

Margulis, L. & Obar, R. (1985). *Heliobacterium* and the origin of chrysoplasts. *BioSystems*, **17**, 317–25.

Martin, W., Somerville, C. C. & Loiseaux-de Goër, S. (1992). Molecular phylogenies of plastid origins and algal evolution. *Journal of Molecular Evolution*, **35**, 385–404.

McFadden, G. & Gilson, P. (1995). Something borrowed, something green: lateral transfer of chloroplasts by secondary endosymbiosis. *Trends in Ecology and Evolution*, **10**, 12–17.

Mereschkowsky, C. (1910). Theorie der zwei Plasmaarten als Grundlage der Symbiogenesis, einer neuen Lehre von der Entstehung der Organismen. *Biologisches Centralblatt*, **30**, 277–303, 321–47, 353–67.

Morden, C. W., Delwiche, C. F., Kuhsel, M. & Palmer, J. D. (1992). Gene phylogenies and the endosymbiotic origin of plastids. *BioSystems*, **28**, 75–90.

Nelissen, B., Van de Peer, Y., Wilmotte, A. & De Wachter, R. (1995). An early origin of plastids within the cyanobacterial divergence is suggested by evolutionary trees based on complete 16S rRNA sequences. *Molecular Biology and Evolution*, **12**, 1166–73.

Oda, K., Yamato, K., Ohta, E., Nakamura, Y., Takemura, M., Nozato, N., Akashi, K., Kanegae, T., Ogura, Y., Kohchi, T. & Ohyama, K. (1992). Gene organization deduced from the complete sequence of liverwort *Marchantia polymorpha* mitochondrial DNA. A primitive form of the plant mitochondrial genome. *Journal of Molecular Biology*, **223**, 1–7.

Raven, P. H. (1970). A multiple origin for plastids and mitochondria. *Science*, **169**, 641–6.

Reith, M. (1995). Molecular biology of rhodophyte and chromophyte plastids. *Annual Review of Plant Physiology and Plant Molecular Biology*, **46**, 549–75.

Reith, M. & Munholland, J. (1995). Complete nucleotide sequence of the *Porphyra purpurea* chloroplast genome. *Plant Molecular Biology Reporter*, **13**, 333–5.

Sapp, J. (1990). Symbiosis in evolution: an origin story. *Endocytobiosis and Cell Research*, **7**, 5–36.

Schuster, W. & Brennicke, A. (1994). The plant mitochondrial genome: physical structure, information content, RNA editing, and gene migration to the nucleus. *Annual Review of Plant Physiology and Plant Molecular Biology*, **45**, 61–78.

Sogin, M. L., Edman, U. & Elwood, H. (1989). A single kingdom of eukaryotes. In *The Hierarchy of Life*, Fernholm, B., Bremer K. & Jörnvall, H. eds., pp. 133–43. Elsevier Science Publishers B.V. (Biomedical Division), Amsterdam, The Netherlands.

Springer, N., Ludwig, W., Amann, R., Schmidt, H. J., Görtz, H.-D. & Schleifer, K. H. (1993). Occurrence of fragmented 16S rRNA in an obligate bacterial endosymbiont of *Paramecium caudatum*. *Proceedings of the National Academy of Sciences, USA*, **90**, 9892–5.

Stirewalt, V. L., Michalowski, C. B., Löffelhardt, W., Bohnert, H. J. & Bryant, D. A. (1995). Nucleotide sequence of the cyanelle genome from *Cyanophora paradoxa*. *Plant Molecular Biology Reporter*, **13**, 327–32.

Viale, A. M. & Arakaki, A. K. (1994). The chaperone connection to the origins of the eukaryotic organelles. *FEBS Letters*, **341**, 146–51.

Viehmann, S., Richard, O., Boyen, C. & Zetsche, K. (1996). Genes for two subunits of succinate dehydrogenase form a cluster on the mitochondrial genome of Rhodophyta. *Current Genetics*, **29**, 199–201.

Wolfe, G. R., Cunningham, F. X., Durnford, D., Green, B. R. & Gantt, E. (1994). Evidence for a common origin of chloroplasts with light-harvesting complexes of different pigmentation. *Nature*, **367**, 566–8.

Wolff, G., Plante, I., Lang, B. F., Kück, U. & Burger, G. (1994). Complete sequence of the mitochondrial DNA of the chlorophyte alga *Prototheca wickerhamii*. Gene content and genome organization. *Journal of Molecular Biology*, **237**, 75–86.

Yang, D., Oyaizu, Y., Oyaizu, H., Olsen, G. J. & Woese, C. R. (1985). Mitochondrial origins. *Proceedings of the National Academy of Sciences, USA*, **82**, 4443–7.

AN RNA VIRUS TREE OF LIFE?

EDWARD C. HOLMES[1], ERNEST A. GOULD[2] AND PAOLO M. DE A. ZANOTTO[2]

[1]*Wellcome Centre for the Epidemiology of Infectious Disease, Department of Zoology, University of Oxford, South Parks Road, Oxford OX1 3PS, UK*
[2]*NERC Institute of Virology and Environmental Microbiology, Mansfield Road, Oxford OX1 3SR, UK*

THE NATURE OF VIRAL EVOLUTION

The recent publication of three volumes dedicated to the evolutionary biology of viruses underlines the growing importance of this field (Morse, 1993, 1994*a*; Gibbs, Calisher & García Arenal, 1995). The rapid accumulation of viral sequence data means that viruses have become some of the most studied of all organisms and, in the case of RNA viruses, interest has been heightened because evolution occurs at such speed that high levels of genetic diversity can arise within the time-frame of human observation and often with severe consequences for public health. There is also a general recognition that evolutionary methodology, and in particular the reconstruction of phylogenetic trees, represents a simple and informative way in which to understand various aspects of viral biology, and most notably the ability to reconstruct the origin and spread of viral epidemics (Holmes & Garnett, 1994; Holmes *et al.*, 1995).

Some authors, however, have gone further and suggested that viruses have often been excluded from the neo-Darwinian synthesis (the rise of modern evolutionary theory which began in the 1930s) so that an underlying theme of recent studies has been to redress this balance (Morse, 1994*b*). One simple way in which viruses have been moved into the mainstream of evolutionary theory is through attempts to classify them in a strictly hierarchical manner, from species to kingdom, based on phylogenetic criteria as is accepted for other organisms (Bishop, 1994; Francki *et al.*, 1991; Ward, 1993). At lower levels, such as within viral 'genera' and 'families' or 'groups', these classifications are generally unambiguous and reflect underlying phylogenetic relationships, as reconstructed from a number of genes, as well as patterns of genome organization. Proposals of higher-order classifications which aim to place viruses into groupings above the family level have, however, been far more controversial. Two specific problems have been identified (Ward, 1993). First, it is evident that some of the major viral groups, and particularly the RNA and DNA viruses, have independent origins either from host genomes or self-replicating RNA molecules (Strauss, Strauss & Levine, 1991). But if viruses are polyphyletic how can they be placed within a single evolutionary hierarchy? Secondly, even within groups, such as the positive-strand RNA viruses, where the notion of a monophyletic origin is easier to swallow, recombination has been so frequent, even between distantly related viruses and sometimes with host cellular genes, that no single phylogenetic tree exists for the whole genome and incongruent phylogenies are recovered from different genes (Gibbs, 1995). This history of gene shuffling has given rise to the concept of 'modular evolution' (Gibbs, 1987) where different genetic segments with different functions (the modules), or individual genes, are shuffled during evolution to produce new viruses. For some authors, this means that the pattern of viral evolution is perhaps better represented as a network rather than a continually bifurcating phylogenetic tree (Rybicki, 1990).

Despite the fluid nature of their evolution, higher-order classifications of viruses continue to appear. Ward (1993) claims that the problem of virus polyphyly can be solved by placing all viruses within a kingdom and assigning the status of phylum to each group of viruses which has an independent origin. The second problem, that of networked evolution, can be overcome by reconstructing phylogenies on the 'core module of replication machinery', the polymerase sequences which all replication-competent viruses must possess and which, perhaps alone, will give rise to a single phylogenetic tree (Ward, 1993). To some this represents the most 'objective' method in which to classify RNA viruses (Dolja & Carrington, 1992) and similar analyses have been undertaken on the DNA viruses (Braithwaite & Ito, 1993). The concept of a core module of replication has also been the cornerstone of theories of viral evolution. Koonin & Dolja (1993) suggest that different functional modules shuffle around the core module during viral evolution while Dolja & Carrington (1992) propose that the progenitor of positive-strand RNA viruses may have been a self-replicating RNA module, and that divergent viral genomes evolved through the subsequent addition of modules of differing functions.

The use of polymerase sequences as a basis for viral classification is most developed with the RNA viruses. These viruses are usually divided into four major categories based on their coding and replication strategies: positive-strand RNA viruses (+RNA viruses), negative-strand RNA viruses (−RNA viruses), double-stranded RNA viruses (dsRNA viruses) and retroviruses. The first three of these groups utilize a polymerase where the template is composed of RNA, the RNA-dependent RNA polymerases (RdRp). Retroviruses, however, have both RNA and DNA stages in their life-cycle and in place of the RdRp they encode reverse transcriptase (RT), an RNA-dependent DNA polymerase. Many have claimed that the RdRp have a single evolutionary origin because they all share a small number of short sequence motifs, and in particular the Gly–Asp–Asp (GDD) tripeptide which may function in NTP-binding (Kamer & Argos, 1984). Because a number of motifs are found in both RdRp and RT there have also been suggestions that both sets of RNA polymerases have a single common ancestor (Poch *et al.*, 1989). The conservation of the RNA polymerases is thought to be due to the fact that they rely upon a number of slowly evolving host factors for replication (Strauss *et al.*, 1991). No such conservation can be found within the structural proteins of RNA viruses, such as the capsid and envelope, which are clearly subject to very different selective pressures.

A number of classification schemes have been proposed following the phylogenetic analysis of the RNA polymerases, and most particularly for the +RNA viruses which are by far the largest group of RNA viruses (Gorbalenya, 1995). From a phylogenetic analysis of conserved motifs within the RdRp, Koonin (1991) and Koonin & Dolja (1993) suggest that +RNA viruses should be placed into 11 major groups and three supergroups from which the dsRNA viruses have evolved (Fig. 1). Each of the three supergroups should be considered as a class (*Picornavirata*, *Flavivirata* and *Rubivirata*). Other evidence cited for the supergroup divisions are that members often carry apparently homologous genes arranged in the same or variable order and that similar mechanisms are used to express some of these genes (Goldbach & de Haan, 1994; Koonin, 1991; Koonin & Dolja, 1993). However, this is not always the case. For example, supergroup II (the *Flavivirata*) includes the 'flavivirus-like' viruses which possess an RNA helicase and a serine protease as well as a RdRp, and the 'carmovirus-like' viruses, spherical plant viruses which have neither an RNA helicase nor a serine protease but which contain a capsid protein like those found in supergroup I viruses (Dolja & Carrington, 1992). Furthermore, there are doubts about the validity of this classification scheme because different phylogenetic trees have been produced by other authors. In the analysis of Bruenn (1991) for example, the leviviruses (a phage lineage) are removed from the supergroup II of Koonin & Dolja (1993) and placed instead as an outgroup to the +RNA viruses. Bruenn's tree also implies that a +RNA lineage split into picornaviruses and dsRNA viruses. Goldbach & de Haan (1994) produce a tree where the +RNA viruses have a single common ancestor which arose after an initial split with the dsRNA viruses and where the −RNA viruses have arisen from a +RNA virus lineage. Finally, Strauss *et al.* (1991) propose the existence of six major 'superfamilies' of RNA viruses, some of which encompass viruses with very different morphologies. However, all six families are thought to have a single common ancestor, and Strauss *et al.* (1991) go as far as to suggest that RNA viruses as a whole may have arisen only once.

Phylogenetic analyses have also been undertaken on the sequences of other genes of RNA viruses, most notably the RNA helicases within which three major 'superfamilies' have been observed (Dolja & Carrington, 1992; Koonin & Dolja, 1993). However, there are incongruencies between the RdRp and helicase trees such that viruses possessing RdRps from one supergroup may contain helicases from different subfamilies (Dolja & Carrington, 1992).

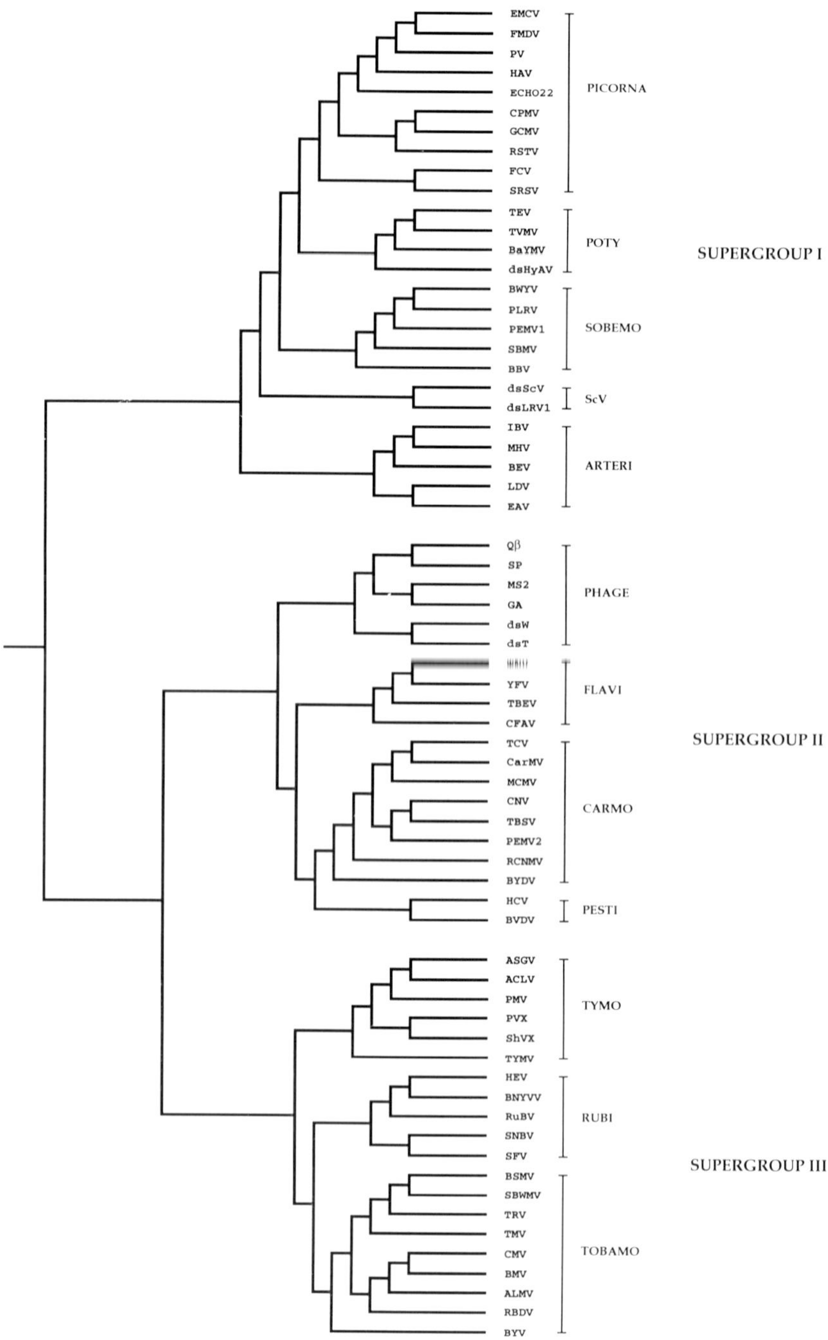

Fig. 1. The phylogenetic tree of +RNA viruses proposed by Koonin & Dolja (1993). Branch lengths are not to scale. Abbreviations for the viruses used are given in the legend of Fig. 2.

ARE RNA POLYMERASES SUITABLE PHYLOGENETIC MARKERS?

The fact that sequence alignments, as well as published trees and classification schemes of RNA viruses, differ in fundamental ways suggests that RNA polymerases might not make ideal phylogenetic markers. In particular, because these sequences are so divergent it is important to determine whether their alignments are sufficiently accurate, whether they contain enough interpretable phylogenetic signal to produce accurate phylogenetic trees, and if there is any hierarchical structure in the data. It is striking that, despite the stated aim to place viruses within the evolutionary framework already established for other organisms, the quality of much of the phylogenetic analysis performed to date only contains a veneer of the rigour that characterizes work on other taxa. For example, few attempts have been made to assess the robustness of the phylogenetic groupings obtained through bootstrap resampling even though this is a basic requirement in some evolutionary journals and is common in other areas of systematics. These sort of analyses are especially appropriate in viral taxonomy 'the classification of organisms at the edge of life' (Rybicki, 1990), because the high levels of sequence divergence mean that homology cannot be taken for granted. It is therefore informative to examine the suitability of RNA polymerases as a means of reconstructing the evolutionary history of RNA viruses.

The first question that needs to be asked with respect to RNA polymerases is whether their alignments are accurate. One simple way in which this can be assessed is to determine whether these sequences are more similar to each other than would be expected from chance alone. This is achieved by comparing the quality of their alignments to those constructed on random sequences of the same length and amino acid composition using a Monte Carlo randomization test. Sequence similarity can then be measured as the number of standard deviations (SD) above the mean value observed in comparisons involving the random sequence data. The lower the SD value, the less accurate the alignment and, for very low SD scores, the less likely that the sequences in question are actually homologous. Unfortunately, there is no hard-and-fast rule by which to interpret SD values. Scores above 15.0 for proteins of 100–200 amino acid residues are thought to signify a 'near perfect' alignment, whereas values above 5.0 indicate a 'good' alignment and values below 3.0 a 'poor' alignment (Barton, in press). However, SD values of around 7.5 have been observed in alignments from structurally unrelated proteins (Barton, in press) making it necessary to view with caution all alignments with SD scores in this range.

In order to assess the quality of alignments of RNA polymerases, Zanotto *et al.* (1995) analysed four previously published alignments using the Monte Carlo method. We reproduce the results from the analysis of two of these alignments here because they are highly informative. The first alignment consists of 80 residues incorporating five highly conserved sequence motifs from 40 RdRp and 40 RT compiled by Poch *et al.* (1989). The second alignment consists of 120 residues and includes 50 RdRp presented in Koonin & Dolja (1993) with the addition of nine other RdRp previously published by Koonin (1991). This second data set consists mainly of sequence regions around the critical residues in the RdRp core.

For both these alignments, 100 random sequences of the same length and amino acid composition as each of the RNA polymerase sequences were generated. The polymerase sequences were then aligned with each randomized sequence to generate a distribution of similarity scores from which the SD values could be calculated. The results of this analysis are presented in Figs. 2 and 3. The SD values for each pairwise comparison are presented in a 'density plot' along the grey scale. Here, the highest SD values (maximum similarity) are depicted by white squares and the lowest SD values (minimum similarity) by black squares.

It is clear that the highest SD values are found on, or near to, the diagonal and were mainly encountered in comparisons from viruses of the same family and not usually between viral families. The density plot for the Poch *et al.* (1989) data set (Fig. 2) indicates that high SD values are found within the RT sequences but that when these sequences are compared with those which encode an RdRp very low SD values (<2.5) are obtained. This indicates that there is very little sequence similarity between these two types of RNA polymerase and that, as a consequence, they may not be homologous. SD values below 2.5 were also observed in some comparisons involving the RdRp from different viral families.

The analysis of the Koonin & Dolja (1993) data set is presented in Fig. 3. Although some relatively high SD values are observed, especially when comparing the viruses assigned to putative supergroup II, most of the between-family values were low, again showing that there is little sequence similarity between these very divergent viruses. Similarly low between-family SD values were observed in the other two published sequence alignments studied by Zanotto *et al.* (1995) indicating that these results are general for RNA polymerases.

Although the levels of sequence identity between many RNA polymerases are extremely low, indicating that the underlying sequence alignments are poor and that some of these polymerases may not even be homologous, they have still been used to reconstruct phylogenetic trees. It is therefore also informative to ask whether these trees are robust by using bootstrap resampling analysis. Once again, although there is some debate as to what level of bootstrap support signifies 'significance' (Felsenstein & Kishino, 1993; Hillis & Bull, 1993; Li & Zharkikh, 1994), it is evident that we should have little confidence in groupings supported by less than 50% of replicates because they are not found in the majority of trees.

For the same two RNA polymerase data sets used in the Monte Carlo analysis, 100 bootstrap trees were reconstructed with the neighbour-joining and parsimony methods. The results are presented in Figs. 4 and 5. There is a striking lack of support for the higher-order groupings. In the Poch *et al.* (1989) data set (Fig. 4) only a few viruses could be grouped with greater than 50% bootstrap support in the neighbour-joining tree and no clusters which included both RT and RdRp sequences were obtained. A greater number of clusters were observed in the parsimony analysis but, once again, no clear

---

Fig. 2. (*overleaf*) Monte Carlo analysis of the Poch *et al.* (1989) data set. Figure taken from Zanotto *et al.* (1995). For each pairwise comparison, the number of standard deviations (SD) above the mean value found in comparisons with random sequences were calculated using the MULTALIGN program (Barton & Sternberg, 1987). The SD values were then depicted as a density plot on the grey scale from white squares (highest SD Values) to black squares (lowest SD values). Although several tones of the grey scale are available only four representative values are shown for clarity. The abbreviations for the viruses and other agents analysed are given below. Because different authors use different abbreviations, the Koonin & Dolja (1993) system was used to denote the RdRp sequences, and the Poch *et al.* (1989) system to denote RT sequences. *RdRp sequences*: Bacteriophage MS2, MS2; Bacteriophage Ga, GA; Bacteriophage Q-Beta, Qβ; Bacteriophage SP, SP; Poliovirus, PV; Encephalomyocarditis, EMCV; Foot-and-mouth disease, FMDV; Echovirus 22, ECHO 22; Hepatitis A, HAV; Coxsackievirus, CoxV; Human rhinovirus type 2, HRV2; Human rhinovirus type 14, HRV14; Feline calicivirus, FCV; Southampton, SRSV; Rice tungro spherical, RTSV; Hungarian grapevine chrome mosaic, GCMV; Cowpea mosaic, CPMV; Southern bean mosaic, SBMV; Pepper mottle, PEMV1; Black beetle, BBV; Tobacco etch, TEV; Tobacco vein mottle, TVMV; Theiler's murine encephalomyelitis, TMEV; Sindbis, Middleburg, SNBV; Pea enation mosaic, PEMV2; Semliki forest, SFV; Tobacco mosaic, TMV; Beet Necrotic yellow vein, BNYVV; Infectious bronchitis, IBV; Berne, BEV; Murine hepatitis, MHV; Barley stripe mosaic, BSMV; Soilborne wheat mosaic, SBWMV; Equine arteritis, EAV; Lactate dehydrogenase, LDV; Red clover necrotic mosaic, RCNMV; Potato virus X, PVX; Brome mosaic, BMV; Tobacco rattle, TRV; Alfalfa mosaic, ALMV; Cucumber mosaic, CMV; Raspberry bushy drawf, RBDV; Beet yellows, BYV; Carnation mottle, CarMV; Maize chlorotic mottle, MCMV; Turnip crinkle, TCV; Turnip yellow mosaic, TYMV; Barley yellow mosaic, BaYMV; Barley yellow dwarf, BYDV; Potato leaf roll, PLRV; Beet western yellow, BWYV; Tomato bushy stunt, TBSV; Cucumber necrosis, CNV; Yellow fever, YFV; Dengue serotype 4, DEN4; West Nile, WNV; Tick-borne encephalitis, TBEV; Cell fusion agent, CFAV; Hepatitis C, HCV; Bovine viral diarrhoea, BVDV; Hepatitis E, HEV; Rubella, RuBV; Semliki forest, SFV; Apple chlorotic leafspot (ACSLV), ACLV; Apple stem grooving, ASGV; Potato M, PMV; Shallot virus X, ShVX; Infectious bursal disease, IBDV; Bluetongue, BTV; Influenza A, B, InfA, InfB; Tacaribe, TacaV; Lymphocytic choriomeningitis, LCMV; Newcastle disease, NDV; Sendai, SendV; Measles, MeasV; Rabies, RabV; Vesicular stomatitis, VSV; *C. parasitica* hypovirulence virus, dsHyAV; S. *cerevisiae* virus L-A, dsScV; *Leishmania* RNA virus 1, dsLRV1; S. *cerevisiae* W RNA, dsW; S. *cerevisiae* T RNA, dsT; *RT sequences*: Human hepatitis B, HepB; Woodchuck hepatitis B, HepWo; Duck hepatitis B, HepBDu; Human endogenous C, HERVC; AKV murine leukaemia, AKVLMV; Murine Moloney Leukaemia, MoMLV; Hamster intracisternal A particle, IAPH18; Rous sarcoma, RSV; Simian Mason-Pfizer, SMPV; Murine mammary tumor, MMTV; Human endogenous retrovirus K, HERVK; Human adult T-cell leukaemia, ATLV; Human T-cell leukaemia type II, HTVLII; Bovine leukaemia, BLV; Human immunodeficiency type 1, HIV 1; Human immunodeficiency type 2, HIV 2; Caprine arthritis encephalitis, CAEV; Equine infectious anaemia, EIAV; Visna, Visna; *Drosophila* 17.6 element, 17.6; *Drosophila* 297 element, 297; *Drosophila* gypsy element, Gypsy; *Drosophila* 412 element, 412; Cauliflower mosaic, CaMV; *Dictyostelium* DIRS-1 element, Dirs; Ty912 element, TY912; *Drosophila* 1731 element, 1731; *Drosophila* copia element, Copia; Mauriceville plasmid (mtDNA), MauP; *Chlamydomonas* intron (mtDNA), RTChla; *Trypanosoma* ingi element, Ingi; *Drosophila* F factor, Ffac; Maize Cin4 element, CIN4; *Drosophila* I-factor, Ifac; Yeast class I intron (mtDNA), IntSp; Yeast class II introns (mtDNA), Int31, Int32; Mouse line-1 element, LiMd; Prosimian & human line-1 elements, LIS1, L1Hu.

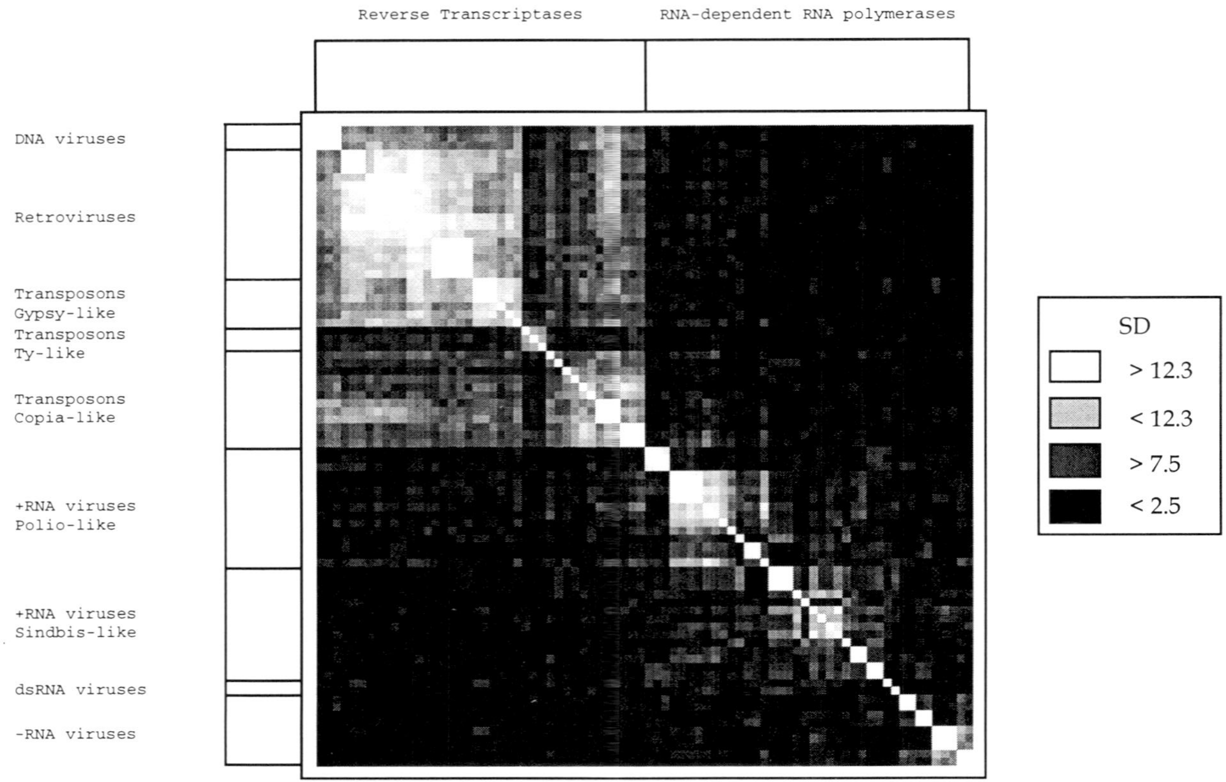
Reverse Transcriptases
RNA-dependent RNA polymerases
DNA viruses
Retroviruses
Transposons
Gypsy-like
Transposons
Ty-like
Transposons
Copia-like
+RNA viruses
Polio-like
+RNA viruses
Sindbis-like
dsRNA viruses
-RNA viruses
SD
> 12.3
< 12.3
> 7.5
< 2.5

Fig. 2.

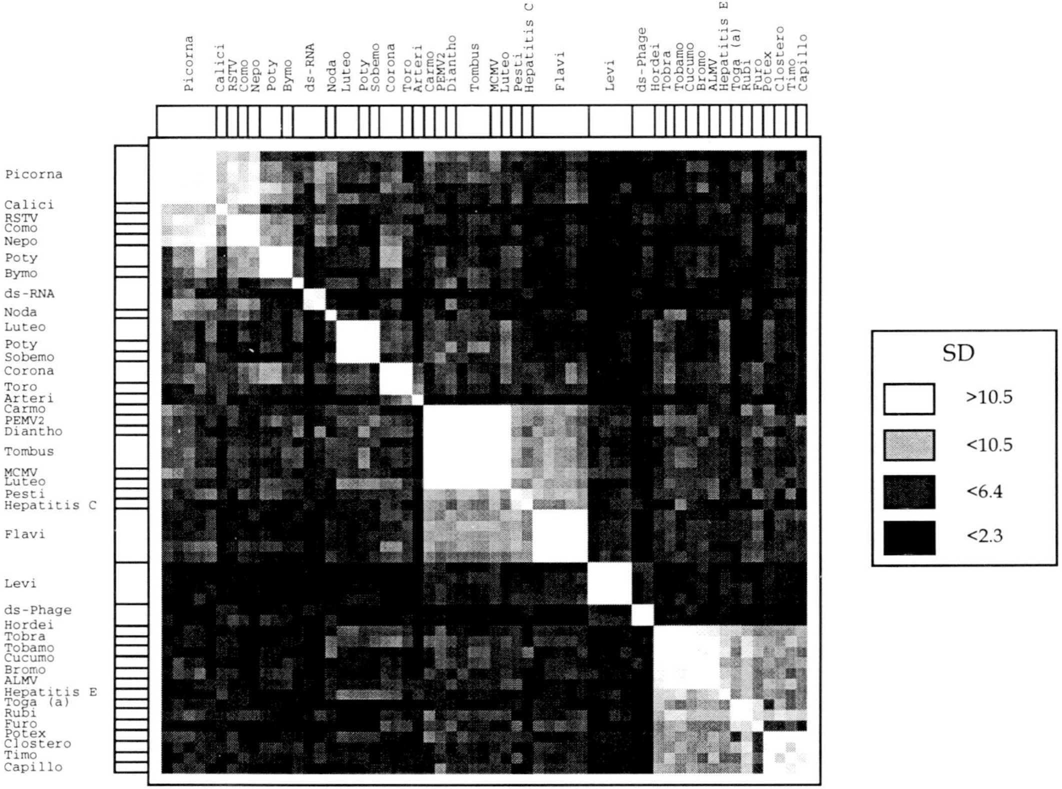

Fig. 3. Monte Carlo analysis of the Koonin & Dolja (1993) data set. Figure taken from Zanotto *et al.* (1995). (See Fig. 2 for more details.)

higher-order groupings were obtained. A similar picture is found in the bootstrap analysis of the Koonin & Dolja (1993) data set (Fig. 5). Very little resolution was observed in the neighbour-joining tree although there was some support for a putative supergroup III in the parsimony analysis. No support was obtained for any of the other higher-order groupings.

In summary, for both data sets, the bootstrap resampling analysis gave very little support to any of the previously proposed higher-order groupings of RNA viruses, with the exception of supergroup III which was only obtained with the parsimony method. Similar results were obtained in the more extensive analysis undertaken by Zanotto *et al.* (1995).

It therefore seems clear that alignments of RNA polymerases are unreliable and that we should have little confidence in the trees reconstructed from them. A final question that may be asked is whether there really is any phylogenetic structure in the data, whether a tree exists at all, or whether these sequences are merely related in a random manner. This can be done by examining the shape of the distribution of lengths of all possible bifurcating parsimony trees because data sets with phylogenetic structure produce left-skewed tree-length distributions (Hillis, 1991; Swofford, 1993). Skewness is then measured by the g1 statistic and data sets with a phylogenetic structure have g1 values significantly below zero and not significantly greater than those obtained from random data sets. The signature of phylogenetic structure can also be revealed by examining the length of the most parsimonious trees obtained. If there is phylogenetic structure in the RNA polymerase data then the most parsimonious trees reconstructed from this data will be significantly shorter than those derived from random data sets. Zanotto *et al.* (1995) performed these analyses on subsets of data representing the higher-order groupings from four published alignments of RNA polymerases and in all cases the distribution and length of parsimony trees could not be distinguished from those obtained by chance alone. This shows that there is an absence, at higher levels, of hierarchical structure in the RNA polymerase sequence data.

On following pages.

Fig. 4. Majority-rule consensus bootstrap trees for the Poch *et al.* (1989) data set using the (a) neighbour-joining and (b) parsimony methods. 100 random data sets were generated using the PHYLIP SEQBOOT program (Felsenstein, 1993). Distances between the sequences were then estimated using the PHYLIP PROTDIST program using the 'categories' distance model which accounts for the chemical similarity of the protein sequences. Phylogenetic trees were constructed using the neighbour-joining and parsimony methods (PHYLIP programs NEIGHBOR and PROTPARS, respectively). Only branches with greater than 50% bootstrap support, constructed using the PHYLIP CONSENSE program, are shown. All trees are shown as unrooted with branch lengths not drawn to scale. The abbreviations for the viruses analysed are given in the legend to Fig. 2. Figure taken from Zanotto *et al.* (1995).

Fig. 5. Majority-rule consensus bootstrap trees for the Koonin & Dolja (1993) data set. Figure taken from Zanotto *et al.* (1995). (See Fig. 4 for more details and Fig. 2 for the abbreviations used.)

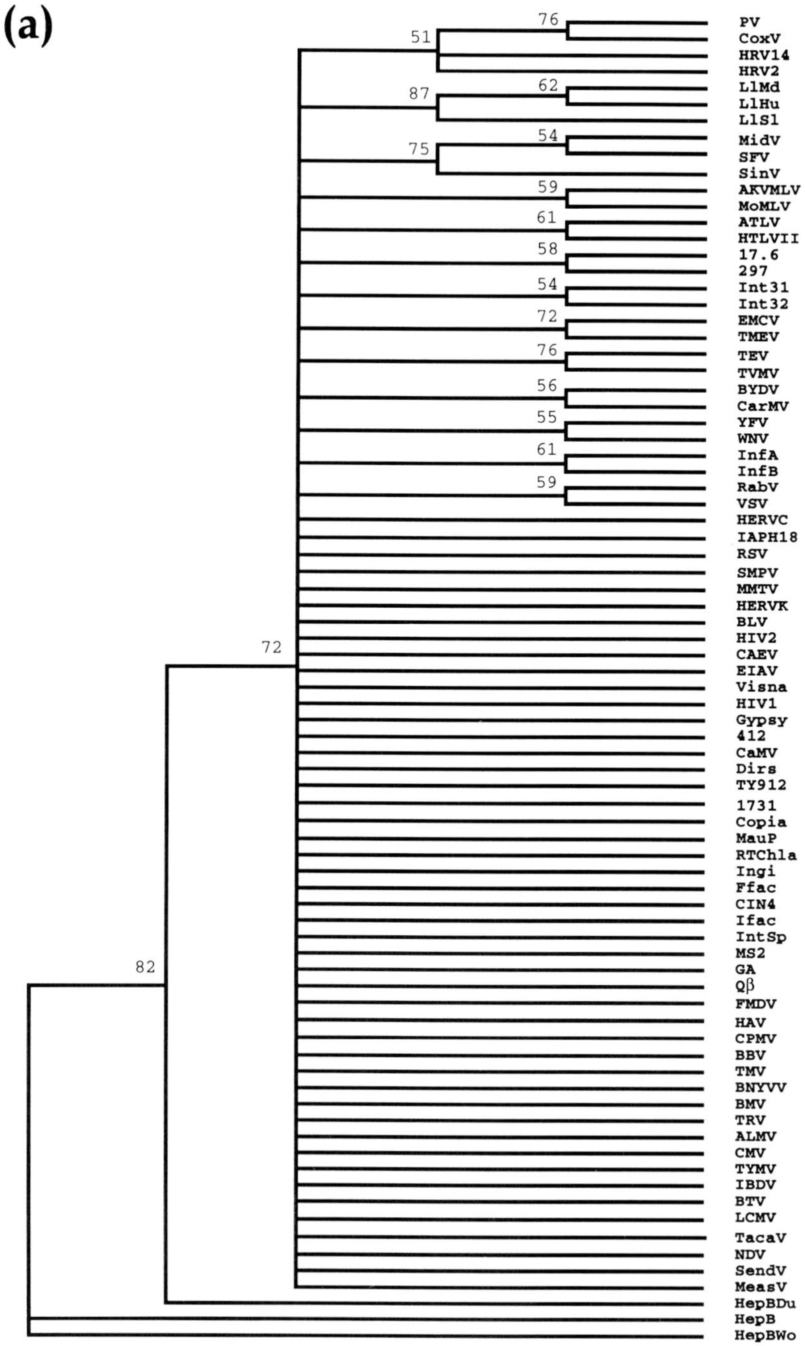

Fig. 4(a). Neighbour-joining.

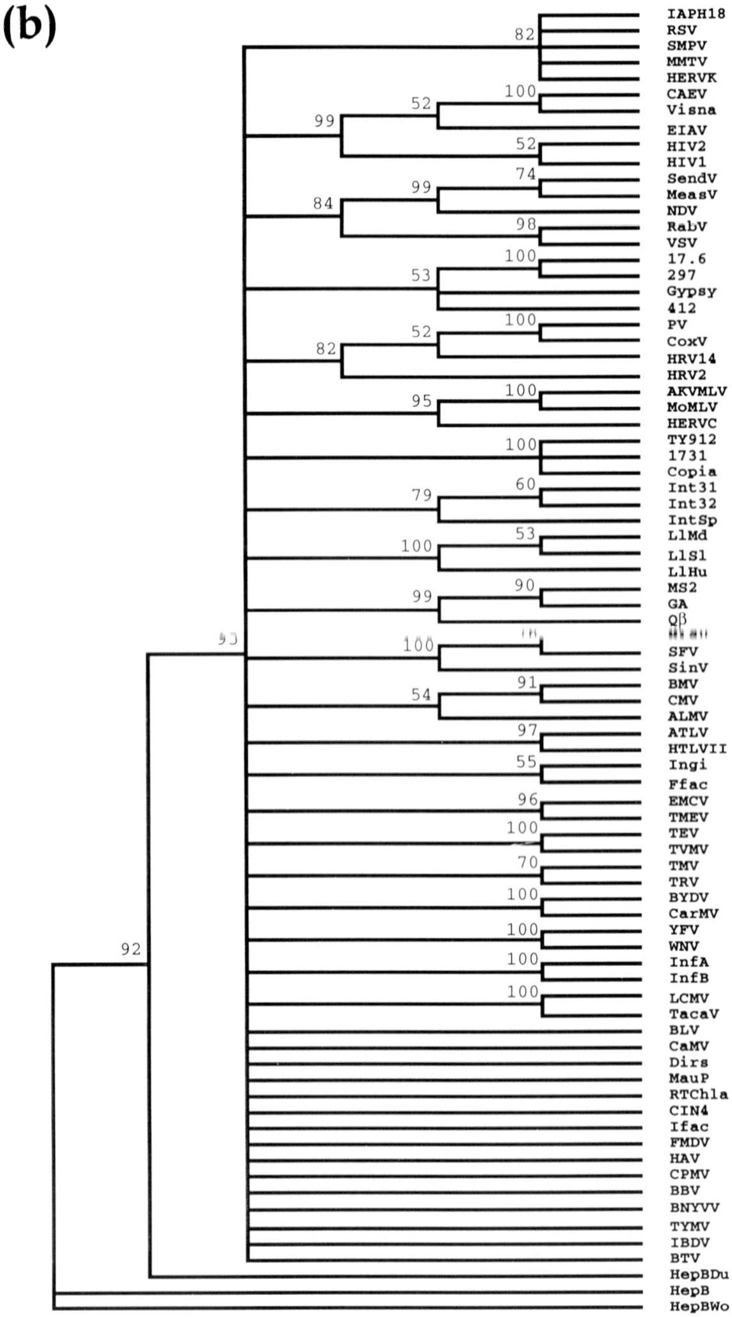

Fig. 4(b). Parsimony.

(a)

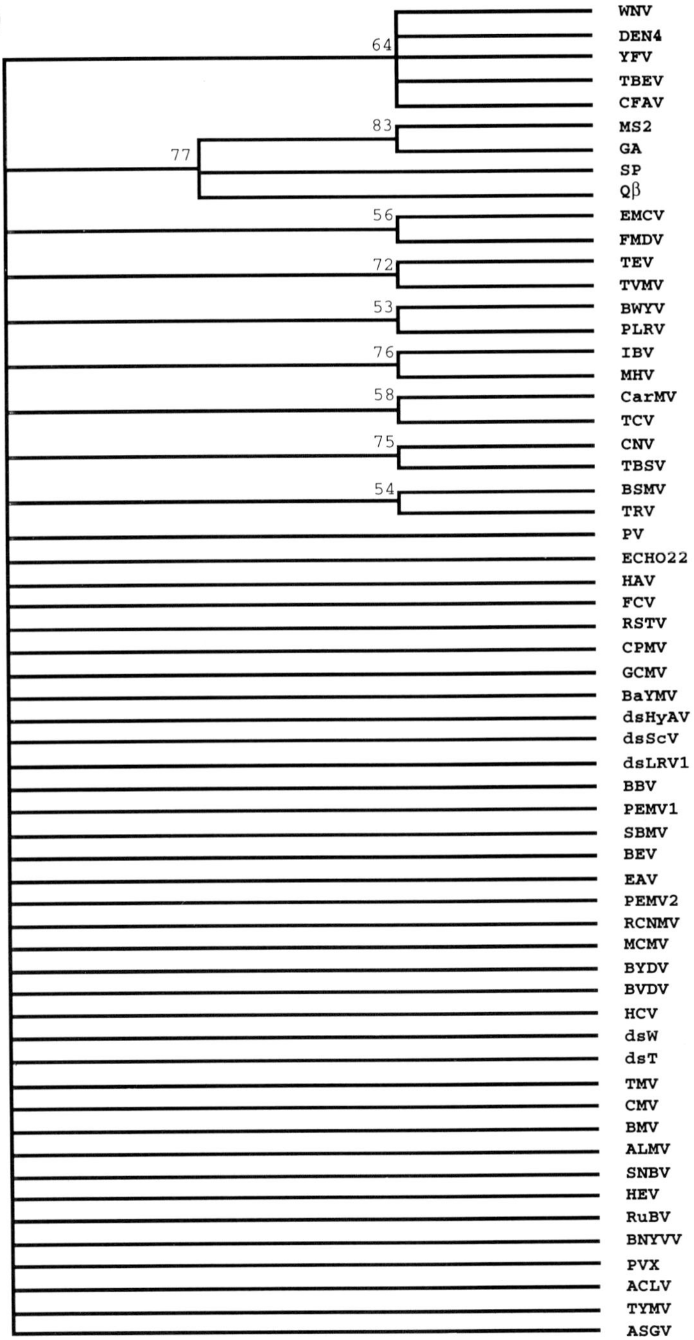

Fig. 5(a). Neighbour-joining.

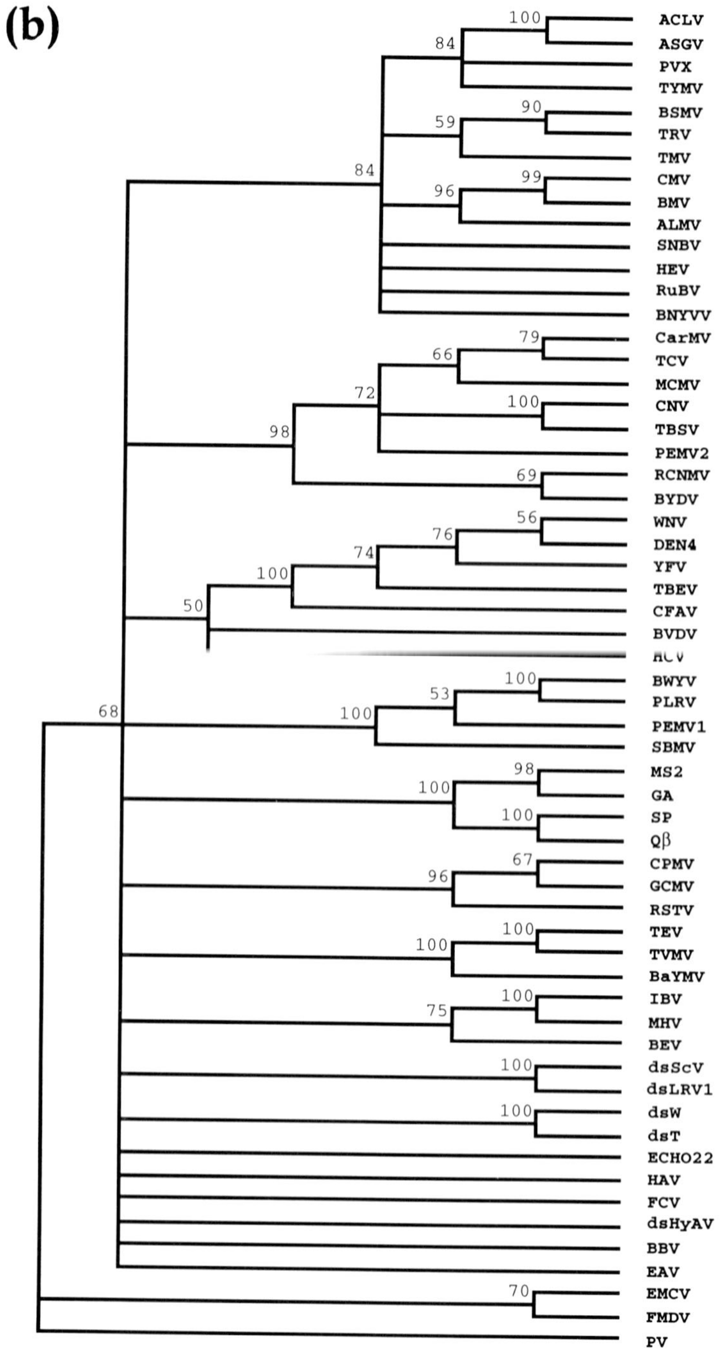

Fig. 5(b). Parsimony.

IS A HIGHER TAXONOMY OF RNA VIRUSES WORTHWHILE?

From the analyses described here, it is clear that the sequences of RNA polymerases cannot be used to construct a single reliable phylogenetic tree for all RNA viruses, nor even for all +RNA viruses. This is because of a lack of basic sequence similarity, such that alignments may be highly inaccurate, and a lack of clear phylogenetic signal as reflected in the bootstrap analysis. The extremely low SD values obtained in comparisons between the RdRp and RT sequences may mean that these polymerases are not homologous and consequently that RNA viruses, along with RNA polymerase function, have evolved independently from a number of different ancestors. The motifs that are apparently conserved among diverse RNA polymerases would then have arisen through convergent evolution, an already documented example of which is the appearance of RdRp-like motifs in components of telomerases (Lundbald & Blackburn, 1990). This observation, along with the lack of conservation in primary sequence and size among polymerases, suggests that a successfully functioning polymerase can be achieved in a number of different ways and through a number of independent evolutionary routes. However, the presence of these motifs, and particularly the keystone GDD tripeptide, has been considered by other authors as strong evidence that RNA polymerases are homologous and monophyletic. If this is indeed the case, then the phylogenetic signal which should link these sequences has been erased through time because of the extremely high rates of sequence divergence which characterize the RNA viruses.

Whatever the underlying reasons, the lack of phylogenetic structure in the RNA polymerases, the core replication module, means that they cannot reliably be used to construct classifications of RNA viruses which reflect underlying phylogenetic relationships and that the classifications proposed to date may be premature. For example, we can detect nothing at the level of the RNA polymerase sequence which suggests that +RNA viruses can be split into three major supergroups or 11 major groups. It is striking that, even within viral families, such as the *Flaviviridae*, which currently includes the flaviviruses, pestiviruses and hepatitis C virus, the very low levels of sequence similarity encountered make it extremely difficult to resolve phylogenetic relationships with any confidence. Consequently, rather than present a single resolved phylogenetic tree, we feel that it is more appropriate and realistic to depict the evolutionary relationships between RNA viruses as a set of distinct and well-supported 'subtrees', the links between which are currently unclear. For example, within the *Flaviviridae*, the flaviviruses, pestiviruses and hepatitis C virus would each be represented as unconnected subtrees. We believe that these subtrees reflect the minimal reliable component of phylogenetic history which is supported by the data.

At face value this might seem like an overly pessimistic conclusion. We believe, however, that this viewpoint has more in keeping with the true nature of viral evolution than the strait-jacket of a Linnean scheme, such as that proposed by Ward (1993). As Rybicki (1990) put it: 'rigid hierarchies cannot be imposed on organisms that appear – evolutionarily speaking – to have been rather promiscuous in sharing their characteristics around. Rather, virological systematists should adopt a flexible wait-and-see approach, and erect hierarchies only where these are unambiguously obvious'. To go further, it is also our view that the higher-order classification of RNA viruses has little practical benefit and may even be counter-productive. Classifications at this level often contain viruses with very different patterns of genome organization, that infect different hosts by different routes and with different disease outcomes. There is little to be gained from assigning a virus to a certain supergroup except perhaps a false sense of order. Furthermore, if these viruses do not have a single common ancestor and possess no obvious hierarchical structure, why try and impose one? We consider such proposals to be dangerous because they place an order and pattern onto viral evolution that probably does not exist in nature. A case in point arises in the study of Strauss *et al.* (1991) who, in claiming that 'all the RNA viruses can be classified into so few superfamilies suggests that RNA viruses probably arose very few times in the evolutionary history of life on earth', directly use their classification scheme to make an evolutionary inference. It is also possible that by placing viruses into such well-defined and arranged groups the question of viral origins may become clouded as attempts are made to explain why viruses in particular super-groups are apparently related yet have such different genome organizations and modes of operation. For example, when faced with the extremely divergent +RNA viruses which constitute supergroup II, Dolja & Carrington (1992) state that 'it is difficult to visualise the nature of a progenitor virus that gave rise to both carmo- and flavi-like viruses, as only the conserved RdRp unites these diverse groups'. Indeed, much of the literature on viral evolution is concerned with how different patterns of genome organization have arisen (for example, Gorbalenya, 1995). While an important and interesting topic, these will remain no more than scenarios until we have the basis of a reliable phylogeny.

The goal of phylogenetic systematics, the cladistic school of classification, is that classifications of organisms should reflect the one true tree of life, a desire that automatically provides an objective framework by which to proceed. However, a history of multiple origins and recombination mean that there is not one true tree of life for the RNA viruses and even within the supposedly core RNA polymerase sequences no clear hierarchical structure can be found. The truth of viral evolution is, therefore, not one which can be easily reconciled with an evolutionary hierarchy. Recognition of the complexity of viral evolution has also led to the 'quasi-species' concept – that the extremely high error rates of RNA viruses will lead to a distribution of mutants on which natural selection acts as a whole (Nowak, 1992). Taken together, this means that viruses, in some respects, have evolved in ways unlike the usual diploid-outcrossing animals on which the neo-Darwinian edifice was built. The real goal in studies of viral evolution and systematics should therefore be to encompass these differences. In this case, perhaps viruses are best left at the edge of classification.

ACKNOWLEDGEMENTS

This work was supported by grants from The Royal Society and The Wellcome Trust.

REFERENCES

Barton, G. J. (1996) Protein sequence alignment and database scanning. In *Protein Structure Prediction: A Practical Approach*, Sternberg, M. J. E., ed. IRL Press, Oxford. In press.

Barton, G. J. & Sternberg, M. J. E. (1987). A strategy for the rapid multiple alignment of protein sequences – confidence levels from tertiary structure comparisons. *Journal of Molecular Biology,* **198**, 327–37.

Bishop, D. H. L. (1994). Virus taxonomy: alive and well! *Society for General Microbiology Quarterly,* **21**, 36–8.

Braithwaite, D. K. & Ito, J. (1993). Compilation, alignment and phylogenetic relationships of DNA polymerases. *Nucleic Acids Research,* **21**, 787–802.

Bruenn, J. (1991). Relationships among the positive-strand and double-strand RNA viruses as viewed through their RNA-dependent RNA polymerases. *Nucleic Acids Research,* **19**, 217–25.

Dolja, V. V. & Carrington, J. C. (1992). Evolution of positive-strand RNA viruses. *Seminars in Virology,* **3**, 315–26.

Felsenstein, J. (1993). *PHYLIP (Phylogeny Inference Package)*. Version 3.5c. Distributed by the author. Department of Genetics, University of Washington, Seattle.

Felsenstein, J. & Kishino, H. (1993). Is there something wrong with the bootstrap on phylogenies? A reply to Hillis and Bull. *Systematic Biology,* **42**, 193–200.

Francki, R. I. B., Faquet, C. M., Knudson, D. L. & Brown, F. (1991). *Classification and Nomenclature of Viruses. Fifth Report of the International Committee on Taxonomy of Viruses*. Springer-Verlag, New York.

Gibbs, A. (1987). Molecular evolution of viruses: 'trees', 'clocks' and 'modules'. *Journal of Cell Sciences (supplement),* **7**, 319–37.

Gibbs, A., Calisher, C. H. & García Arenal, F. (eds.) (1995). *Molecular Basis of Virus Evolution*. Cambridge University Press, Cambridge.

Gibbs, M. J. (1995). The Luteovirus supergroup: rampant recombination and persistent partnerships. In *Molecular Basis of Virus Evolution*, Gibbs, A., Calisher, C. H. & García Arenal, F., eds., pp. 351–368. Cambridge University Press, Cambridge.

Goldbach, R. & de Haan, P. (1994). RNA viral supergroups and evolution of RNA viruses. In *The Evolutionary Biology of Viruses*, Morse, S. S., ed., pp. 105–119. Raven Press, New York.

Gorbalenya, A. E. (1995). Origin of RNA viral genomes. In *Molecular Basis of Virus Evolution*, Gibbs, A., Calisher, C. H. & García Arenal, F., eds., pp. 44–66. Cambridge University Press, Cambridge.

Hillis, D. M. (1991). Discriminating between phylogenetic signal and random noise in DNA sequences. In *Phylogenetic Analysis of DNA Sequences*, Miyamoto, M. M. & Cracraft, J., eds., pp. 278–294. Oxford University Press, Oxford.

Hillis, D. M. & Bull, J. J. (1993). An empirical test of bootstrapping as a method for assessing confidence in phylogenetic analysis. *Systematic Biology,* **42**, 182–92.

Holmes, E. C. & Garnett, G. P. (1994). Genes, trees and infections: molecular evidence in epidemiology. *Trends in Ecology and Evolution,* **9**, 256–60.

Holmes, E. C., Nee, S., Rambaut, A., Garnett, G. P. & Harvey, P. H. (1995). Revealing the history of infectious disease epidemics using phylogenetic trees. *Philosophical Transactions of the Royal Society of London Series B,* **349**, 33–40.

Kamer, G. & Argos, P. (1984). Primary structural comparison of RNA-dependent polymerases from plant, animal and bacterial viruses. *Nucleic Acids Research,* **12**, 7269–82.

Koonin, E. V. (1991). The phylogeny of RNA-dependent RNA polymerases of positive-strand RNA viruses. *Journal of General Virology,* **72**, 2197–206.

Koonin, E. V. & Dolja, V. V. (1993). Evolution and taxonomy of positive-strand RNA viruses: implications of comparative analysis of amino acid sequences. *Critical Reviews in Biochemistry and Molecular Biology,* **28**, 375–430.

Li, W.-H. & Zharkikh, A. (1994). What is the bootstrap technique. *Systematic Biology,* **43**, 424–430.

Lundbald, V. & Blackburn, E. H. (1990). RNA-dependent polymerase motifs in EST1: tentative identification of a protein component of an essential yeast telomerase. *Cell,* **60**, 529–30.

Morse, S. S. (ed.) (1993). *Emerging Viruses*. Oxford University Press, Oxford.

Morse, S. S. (ed.) (1994*a*). *The Evolutionary Biology of Viruses*. Raven Press, New York.

Morse, S. S. (1994*b*). Toward an evolutionary biology of viruses. In *The Evolutionary Biology of Viruses*, Morse, S. S., ed., pp. 1–28. Raven Press, New York.

Nowak, M. A. (1992). What is a quasispecies? *Trends in Ecology and Evolution,* **7**, 118–21.

Poch, O., Sauvaget, I., Delarue, M. & Tordo, N. (1989). Identification of four conserved motifs among the RNA-dependent polymerase encoding elements. *EMBO Journal,* **8**, 3867–74.

Rybicki, E. (1990). The classification of organisms at the edge of life or problems with virus systematics. *South African Journal of Sciences,* **86**, 182–6.

Strauss, E. G., Strauss, J. H. & Levine, A. J. (1991). Virus evolution. In *Fundamental Virology*, Fields, B. N., Knipe, D. M., Chanock, R. M., Hirsch, M. S., Melnick, J. L., Monath, T. P. & Roizman, B., eds., 2nd edn., pp. 167–190. Raven Press, New York.

Swofford, D. L. (1993). *Phylogenetic Analysis Using Parsimony (PAUP)*. Version 3.1.1. University of Illinois, Champaign.

Ward, C. W. (1993). Progress towards a higher taxonomy of viruses. *Research in Virology,* **144**, 419–53.

Zanotto, P. M. de A. (1995). Aspects of the molecular evolution of Baculoviruses and Flaviviruses. DPhil Thesis, University of Oxford.

Zanotto, P. M. de A., Gibbs, M. J., Gould, E. A. & Holmes, E. C. (1996). A Reevaluation of the higher taxonomy of viruses based on RNA polymerases. *Journal of Virology* (in press).

EVOLUTION OF HEPATITIS C VIRUS

PETER SIMMONDS

Department of Medical Microbiology, University of Edinburgh, Teviot Place, Edinburgh EH8 9AG, UK

Although recently discovered, hepatitis C virus (HCV) has been the subject of extensive genetic analyses. These range from investigations of virus evolution within a single infected individual, to the analysis of sequence similarities in certain highly conserved regions of the genome to classify HCV amongst the families and superfamilies of positive stranded RNA viruses (cf. Holmes *et al.*, this volume). Between these two extremes are the use of nucleotide sequences to investigate the sources and epidemiology of HCV transmission to the current proposals to classify HCV into 'genotypes' on the basis of nucleotide sequence comparisons since this cannot be currently done using antigenic differences ('serotypes'). What constitutes the 'Evolution of HCV', the title of this chapter, is a topic that can be therefore addressed at several different levels.

In this review, I will first outline the evidence that we currently have that HCV is related to pestiviruses and flaviviruses in the family *flaviviridae*. The rest of the review describes the use of sequence comparisons to investigate sources of infection and to reconstruct specific transmission events. Increasing knowledge of the rate of sequence change of HCV combined with our recent understanding of routes of transmission of HCV provide an insight into the mechanisms by which the currently described genotypes of HCV may have evolved.

INTRODUCTION

HCV has been identified as the main causative agent of post-transfusion non-A, non-B hepatitis (Choo *et al.*, 1989; Kuo *et al.*, 1989). The identification of HCV led to the development of diagnostic assays for infection, based upon either the detection of antibody to recombinant polypeptides expressed from cloned HCV sequences, or the direct detection of virus RNA sequences by amplification with reverse transcriptase – polymerase chain reaction (RT – PCR) using primers specific for the HCV genome. Routine serological screening of blood donors has now greatly reduced the occurrence of post-transfusion hepatitis C. Assays for antibody are also important diagnostic tools and have been used to investigate the prevalence of HCV in different risk groups, such as haemophiliacs or other recipients of blood products, and intravenous drug users (IVDUs), and to carry out epidemiological studies of HCV transmission.

The complete genomic sequence of HCV has been obtained for several isolates, revealing both its overall organization, and its relationship to other RNA viruses. It has also been possible to deduce possible methods of replication by analogy with related viruses. HCV contains a positive sense RNA genome approximately 9400 bases in length. In overall genome organization and presumed method of replication, it is most similar to members of the *flaviviridae*, particularly in coding for a single polyprotein that is then cleaved into a series of presumed structural and non-structural proteins (Fig. 1). The roles for these different proteins have been inferred by comparison with related viruses, and by *in vitro* expression of cloned HCV sequences in prokaryotic and eukaryotic systems (Table 1).

The genome of HCV has two non-coding regions at the 5′ and 3′ ends of the genome (5′NCR, 3′NCR; Fig. 1). These regions are highly structured, in which internal base-pairing produces a complex set of stem loop structures that are thought to interact with various host cell and virus proteins during replication and translation (Tsukiyama Kohara *et al.*, 1992; Brown *et al.*, 1992). The 5′NCR has recently been shown to act as an internal ribosomal entry site (IRES), that directs initiation of translation to an internal methionine codon several hundred bases downstream from the 5′ end of the genome (Tsukiyama Kohara *et al.*, 1992; Wang, Sarnow & Siddiqui, 1994). This strategy is reminiscent of that used by the *picornaviridae*.

Structurally, HCV is a small enveloped virus that has recently been visualized by electron microscopy (Kaito *et al.*, 1994). The exceptionally low buoyant density in sucrose (1.08–1.11 g/cm^3; Miyamoto *et al.*, 1992) has been attributed to the existence of heavily glycosylated external membrane glycoproteins in the virus envelope, and through its binding to low density beta-lipoprotein in plasma.

Very recently, two distinct RNA viruses have been discovered in a new world primate, a tamarin (*Sanguinis* spp.). This monkey species had previously been shown to harbour an infectious agent causing chronic hepatitis after inoculation with plasma from a surgeon (G.B.) who had developed a chronic hepatitis of unknown aetiology (Simons *et al.*, 1995). Parts of the genome of the two viruses (provisionally termed GBV-A and GBV-B) show measurable sequence similarity to certain regions of HCV. For example, a 200 amino acid sequence of part of NS-3 of GBV-A and GBV-B shows 47% and 55% sequence similarity with the homologous region in HCV (positions 1298–1497 in the HCV polyprotein; Choo *et al.*, 1991), and 43.5% sequence the same as each other. Similarly in NS-5, the region around the active site of the RNA dependent RNA polymerase (RdRp) (including the canonical GDD motif; positions 2662–2761 in HCV) shows 36% and 41% sequence similarities, and 43% between GBV-A and -B (Simons *et al.*, 1995).

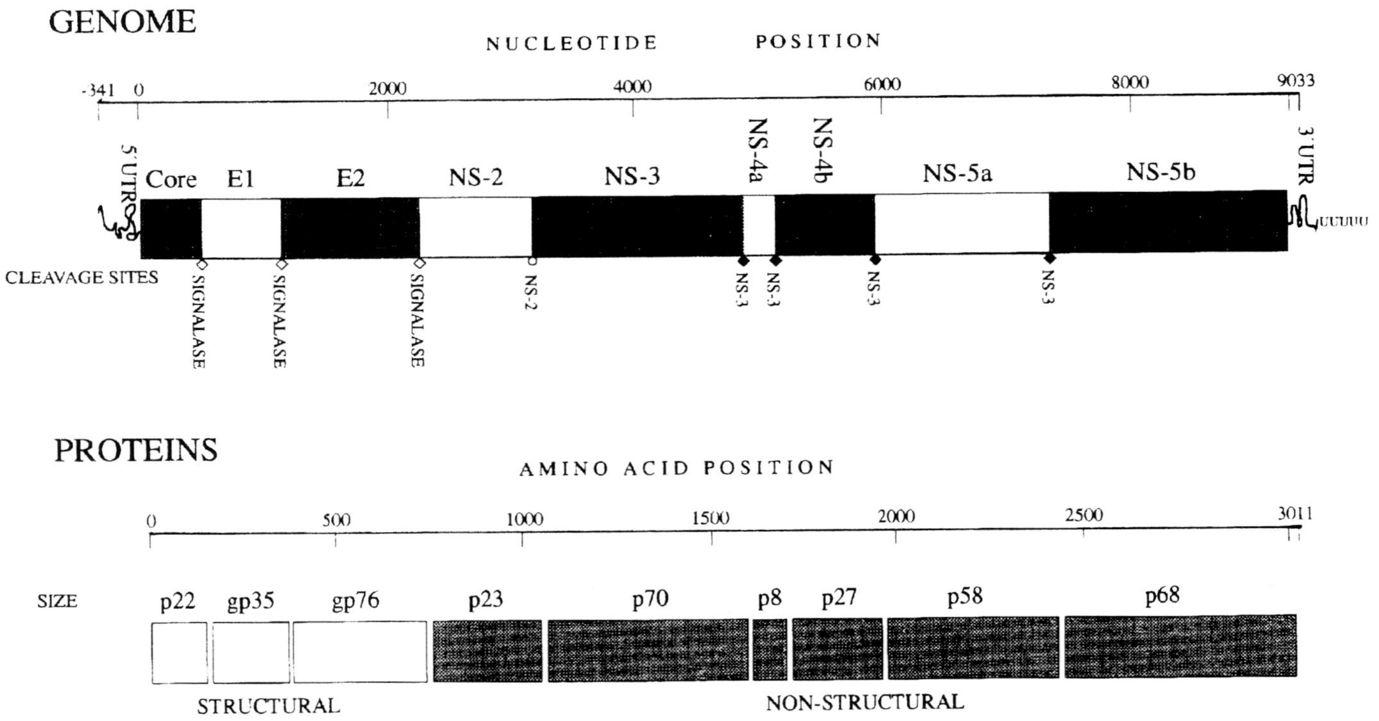

Fig. 1. Organization of HCV genome, showing cleavage sites and relative sizes of HCV-encoded proteins. Nucleotide and amino acid positions numbered as in Choo *et al*. (1991).

Table 1. *Properties of HCV proteins*

Protein	Nucleotide position[a]	Size[b]	In Virion	Antigen[c]	Function
Core	1–573	191	Yes	c22-3	Thought to form the virus nucleocapsid, and shows RNA-binding activity. Highly conserved among HCV genotypes
E1	574–1149	192	Yes		Sequence predicts a membrane anchored glycoprotein, with several potential N-linked glycosylation sites. Highly variable between genotypes
E2/NS-1	1150–2187	327 (?)	Yes (?)		Another membrane bound glycoprotein, with prominent 'hypervariable' region. The homologue in pestiviruses is a virion component, but is non-structural in flaviviruses
NS-2	2431–3078	316	No		Proteinase (Zn^{2+}-dependent)
NS-3	3079–4971	631	No	c33c	Multi-functional protein with protease activity, and with sequence motifs suggesting a helicase activity
NS-4a	4972–5133	54	No	c100-3	Co-factor for NS-3 protease
NS-4b	5134–5916	261	No		Unknown
NS-5a	5917–7260	448	No	NS-5	Unknown
NS-5b	7261–9033	591	No		Probably the RNA-dependent RNA polymerase necessary for genome replication

[a]Amino acid positions numbered as in Choo *et al.* (1991).
[b]Size expressed in number of amino acid residues.
[c]Origin of HCV antigens used in current second and third generation assays for antibody to HCV.

The degree of relatedness HCV, the GB agent, and other positive stranded RNA viruses can be shown using phylogenetic analysis of NS-5 (and homologues in other viruses) encoding the RdRp. Analysis of a 100 amino acid sequence surrounding this motif reveals relatively close clustering between HCV and GBV-A and GBV-B, while pestiviruses such as bovine viral diarrhoea virus (BVDV), and flaviviruses are distinct (Fig. 2). Remarkably, a series of plant viruses that are structurally unrelated to HCV or any other mammalian virus have RdRp amino acid sequences that are as similar to those of HCV as are the flaviviruses. Although the sequence similarity in the genomic region encoding the RdRp of positive stranded RNA viruses has usually been interpreted as indicating their evolutionary relatedness (Koonin *et al.*, 1991), it is also possible that it reflects no more than

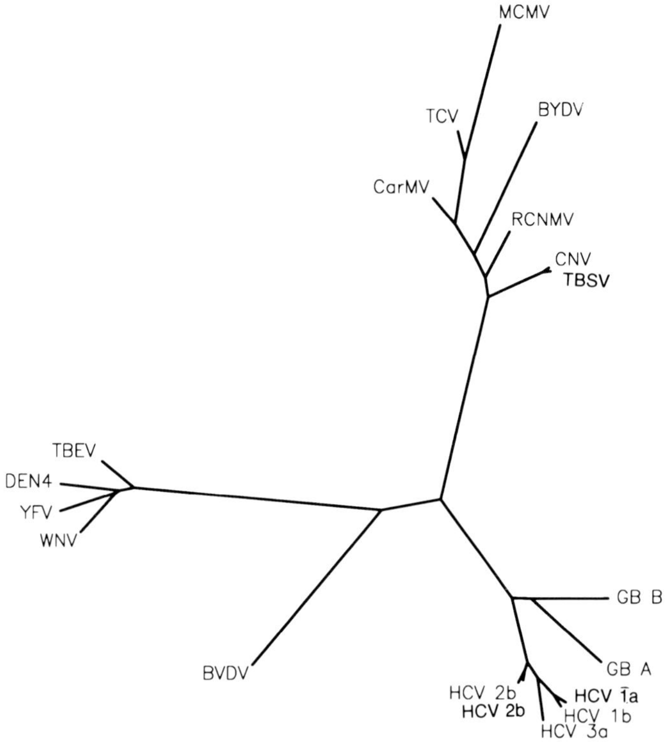

Fig. 2. Phylogenetic analysis of amino acid sequences of the region of NS-5 encoding the RdRp (positions 2662–2761 in HCV-1; Choo *et al.*, 1991). Homologous sequences from flaviviruses (TBEV: tick borne encephalitis virus; YFV: yellow fever virus; WNV: West Nile virus; DEN4: Dengue virus, serotype 4), plant viruses (CarMV: Carnation mottle virus; TCV: turnip crinkle virus; MCMV: maize chlorotic mottle virus; BYDV: barley yellow dwarf virus; RCNMV: red clover necrotic mosaic virus; CNV: cucumber necrosis virus; TBSV: tomato bushy stunt virus), pestivirus (BVDV: bovine viral diarrhoea virus); see Koonin (1991) for sources of non-HCV sequences. GBV-A, GBV-B and HCV (genotypes 1a, 1b, 2a, 2b and 3a shown) were aligned using the program CLUSTAL, and phylogenetic analysis carried out using the programs PROTDIST (PAM matrix), NEIGHBOR and DRAWTREE in the PHYLIP package (Felsenstein, 1993).

the end product of a process of convergent evolution (cf. Holmes *et al.*, this volume). For example, it is possible that many of the conserved motifs within the enzyme (e.g. the GDD sequence found in all virus RNA polymerases) represent the only or the most probable amino acid sequence with the required catalytic properties to enable nucleic acid transcription. Many unrelated viruses might therefore evolve an enzyme with such conserved elements even though they are not evolutionarily related to each other. The usefulness of evolutionary analysis in determining the relationship between highly divergent viruses.

SEQUENCE VARIABILITY OF HCV

RNA viruses replicate their genetic material using virus-encoded RNA dependent RNA polymerases. Compared with the DNA-dependent DNA polymerases found in eukaryotic cells, virus polymerases are relatively small, simple enzymes that lack accessory functions such as proof-reading. As a result, replication of RNA viruses is relatively error-prone, and is associated with rapid genetic change. How rapid this is in the case of HCV can be measured by sequence comparisons of HCV over the course of infection within an individual. For example, the complete genome sequence of HCV in an experimentally infected chimpanzee was compared with a second sample collected 8 years later, and in a separate study, a comparison was made between partial nucleotide sequences of a patient infected with HCV over a 13-year interval (Table 2). For the 9412 bases of the genome obtained in the first study, 111 nucleotide differences were observed between sequences at the beginning and the end of the study period, and 123 differences over 4923 nucleotides in the patient. Nucleotide changes occurred throughout the genome, with similar rates of sequence drift in the regions encoding non-structural proteins in both the chimpanzee and the hepatitis C patient. Only E2 and NS-2 gene showed evidence for higher rates of change, although in the case of E2, much of this resulted from particularly frequent substitutions in a small region at the amino terminus of the protein (hypervariable region; see below); other parts of E2 showed rates of change similar to other proteins.

Throughout the genome, most nucleotide changes were 'silent', i.e. did not affect the sequence of the encoded amino acid sequence. The relative infrequency of non-silent changes is found in most coding sequences of all organisms, and reflects constraints upon the extent to which proteins may vary in sequence and yet remain functional. In the case of HCV, it is possible to analyse separately the overall rate of sequence change (0.144% and 0.199% site^{-1} year^{-1}; Table 2) into rates of silent and non-silent substitution. The rate of sequence change at silent sites (approximately 0.46% and 0.65% site^{-1} year^{-1} in the chimpanzee and hepatitis C patient respectively) is 7 to 8 times higher than at non-silent sites (0.070% and 0.078% site- year-), and reflects the relative constraints upon these two types of change.

Constraints of a different kind must limit variability in the 5′NCR (0–2 changes). As described above, both the 5′NCR and 3′NCR form stem/loop structures that may be important in IRES and RNA replication functions. Nucleotide substitutions may be deleterious in regions where the RNA interacts with host ribosomal sequences or enzymes, while substitutions in the stem structure will destabilize internal base-pairing. Evidence for the importance of stem/loop structures comes from the observation of frequent co-variant changes amongst 5′NCR sequences of different HCV genotypes, in which substitutions in the stem regions of the 5′NCR are frequently

Table 2. *Sequence change of HCV*

Region	Infected chimpanzee 8.2 years (Okamoto *et al.*, 1992*a*)			Hepatitis C patient 13 years (Ogata *et al.*, 1991)			Between genotypes (1b-2a)
	Length[a]	Sequence changes[b]	Substitution rate[c]	Length[a]	Sequence changes	Substitution rate	Sequence divergence[d]
5′NCR	341	0	<0.035%	276	2	0.056%	0.07
non-structural							
Core	573	5 (2)	0.106%	420	6 (2)	0.109%	0.183 (0.089)
E1	576	10 (6)	0.212%	450	11 (2)	0.188%	0.406 (0.469)
E2/NS-1	1038	23 (15)	0.270%	960	44 (23)	0.353%	0.312 (0.289)
non-structural							
NS-2	831	14 (2)	0.205%	504	15 (3)	0.229%	0.409 (0.401)
NS-3	1827	20 (3)	0.133%	1284	25 (5)	0.150%	0.296 (0.189)
NS-4	1194	4 (1)	0.041%	n.d.			0.329 (0.261)
NS-5	2991	34 (13)	0.139%	1029	20 (5)	0.150%	0.335 (0.291)
				n.d.			
3′NCR	141	1	0.297%				0.619
TOTAL	9412	111 (42)	0.144%	4923	123 (40)	0.192%	0.318 (0.274)

[a]Lengths of sequence for each gene. More recent data has led to the identification of further cleavage sites (NS-4a and NS-4b, NS-5a and NS-5b) and minor differences in the position of cleavage sites of other genes (see Table 1). As these small differences do not materially affect the calculated values for divergence presented here, the original data are reproduced. Partial sequences for 5′NCR, core, E1, E2/NS-1, NS-2, NS-3 and NS-5 were obtained in the second study (Ogata *et al.*, 1991).
[b]Number of nucleotide sequence changes (amino acid changes in parentheses).
[c]Substitution rate (percentage) expressed as number of changes per nucleotide site per year.
[d]Uncorrected nucleotide sequence distance (amino acid sequence distances in parentheses) between complete genomic sequences of type 1b (HCV-J; Kato *et al.*, 1990) and type 2a (HC-J6; Okamoto *et al.*, 1991); genome divided as in first column (Okamoto *et al.*, 1992*a*).

accompanied by compensatory changes in the opposite nucleotide to maintain base pairing (Simmonds *et al.*, 1993*b*; Smith *et al.*, 1995).

SEQUENCE VARIABILITY IN THE ENVELOPE REGION

HCV encodes two proteins that are typical of envelope glycoproteins of other viruses. They both have hydrophobic leader sequences, a series of hydrophobic residues at the carboxyl terminus to act as a membrane anchor, and numerous potential sites for N-linked glycosylation on the presumed external domains of the proteins (Fig. 1, Table 1). The rate of sequence change in E2 and possibly in E1 has been found to be higher than the average for other parts of the genome (Table 2). However, for E2, most of the changes are concentrated in a relatively short hypervariable region of approximately 86–120 nucleotides at the amino terminus of the mature protein. This region has also been found to be highly variable between individuals infected with the same genotype, to change rapidly upon disease progression in persistently infected individuals, and to include a higher proportion of non-silent (i.e. amino acid changing) substitutions than found elsewhere in the genome (Weiner *et al.*, 1992; Hijikata *et al.*, 1991; Kato *et al.*, 1992).

As E2 is likely to lie on the outside of the virus, it may be a target for the humoral immune response to HCV. An attractive theory that explains both the high degree of envelope sequence variability and the persistent nature of HCV infection, is that changes in E2 alter the antigenicity of the virus to allow 'immune escape' from neutralizing antibodies (Weiner *et al.*, 1992). In this model, continuing virus replication is a race between diversification of HCV and the efforts of the host immune system to respond to changes by developing neutralizing antibody with an ever greater range of reactivity. The hypervariable region might be a particularly important target for antibody-mediated neutralization, and therefore be under the greatest pressure to change.

In summary, longitudinal studies such as the one described above provide information on the rate of sequence change *in vivo*, and on the constraints that limit variation in different parts of the genome. These and other studies (Abe, Inchauspe & Fujisawa, 1992; Cuypers *et al.*, 1991) therefore form the basis for the use of nucleotide sequence comparisons as an epidemiological tool to investigate HCV transmission.

MOLECULAR EPIDEMIOLOGY

Investigations of HCV transmission at the nucleotide sequence level differ considerably in the methodologies used. At one extreme is the use of highly variable regions of the genome for sequence comparisons, particularly the hypervariable region in E2. As described above, this region differs greatly between individuals, and the finding of similar sequences between individuals could be construed as evidence for epidemiological relatedness. Similarity or identity of sequences in the hypervariable region has thus been used as evidence for mother to child transmission of HCV (Weiner *et al.*, 1993), and for a common source of infection amongst recipients of a contaminated batch of anti-D immunoglobulin (Hohne, Schreier & Roggendorf, 1994). Similarly, sequence identity in E1 has been used as evidence for mother to child transmission (Inoue *et al.*, 1991).

An alternative approach is the use of phylogenetic analysis to reconstruct the evolutionary histories of the variants being compared. A simple but effective method is to first estimate the pairwise distances between sequences and then to cluster such distances into an evolutionary tree. There are numerous methods that estimate pair-wise distances between sequences, and which construct evolutionary trees from sequence distances, that differ in the assumptions they make about the evolutionary process (Nei, 1987). Sequence analyses based upon phylogenetic trees have the advantage over simple sequence comparisons in that confidence limits can be obtained for observed groupings of sequences (such as epidemiologically related clusters of infection). The most frequently used method for this is bootstrap re-sampling where a large number of trees are reconstructed on sets of randomly drawn (with replacement) sites from the original data (Felsenstein, 1988). Alternatively, probability values for particular groupings of sequences may be implicit in the evolutionary model used (e.g. maximum likelihood).

The above methods have been used for analysis of HCV sequences found amongst individuals with common sources of infection. For example, phylogenetic analysis of a 222 base fragment of the NS-5 region was used to investigate a possible single source infection cluster amongst recipients of anti-D immunoglobulin in 1977 (Power *et al.*, 1994). Anti-rhesus D immunoglobulin (anti-D) is prepared from human plasma containing high levels of anti-rhesus D antibody. Although the risk of transmission of viruses by these preparations has been considered to be low, a statistically higher than expected rate of infection was recently found in blood donors who had received Irish manufactured anti-D immunoglobulin in 1977 (Power *et al.*, 1994).

DNA sequences amplified from the NS-5 region from samples collected in 1993 from eleven of the implicated 1977 anti-D recipients were compared with those of epidemiologically unrelated individuals infected with the same genotype (type 1b) (Fig. 3). Sequences from the anti-D recipients clustered together, along with the sequence of the contaminating virus recovered from the 1977 batch (B250) and from the original HCV-infected donor from whose plasma the blood product was manufactured. Maximum likelihood analysis of the dataset showed that the branch leading to the group of sequences containing the implicated recipients and batch B250 was

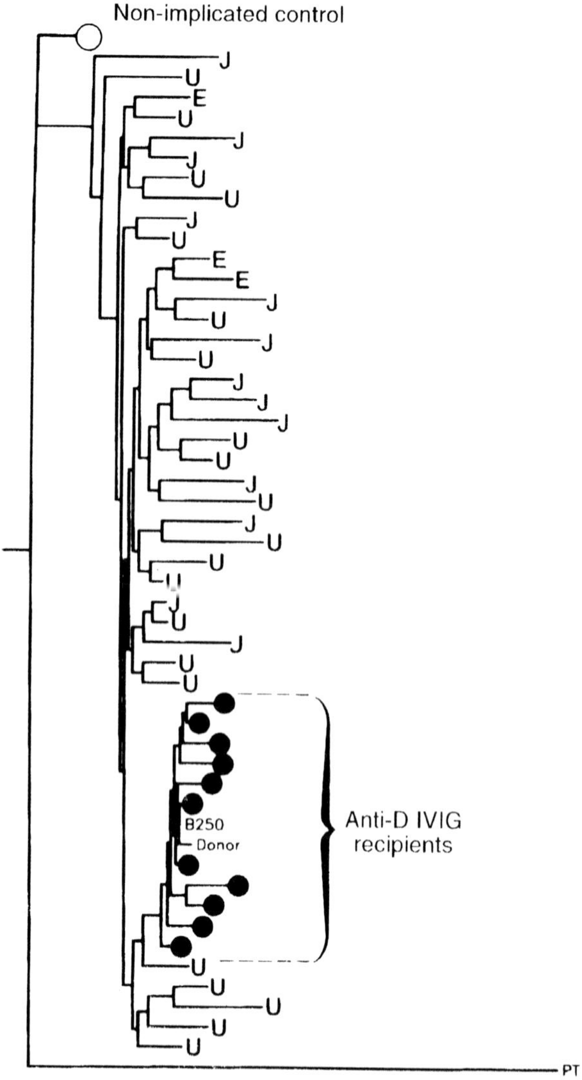

Fig. 3. Phylogenetic relationships between sequences from the NS-5 region from those exposed to implicated batch of anti-D immunoglobulin in 1977 (●) with those epidemiologically unrelated type 1b variants from Japan (J), USA (U) and Europe (E). B250: NS-5 sequence of HCV recovered from batch B250 of anti-D immunoglobulin (Ig); Donor: sequence of variant infecting suspected donor to plasma pool used to manufacture batch B250. Phylogenetic analysis was carried out on a segment (222 base pairs; positions 7975–8196) of the NS-5 gene amplified, sequenced and analysed as previously described (Simmonds *et al.*, 1993*a*). Sequence distances were calculated using the program DNAML in a dataset containing HCV-PT (type 1a) as an outgroup. Type 16 control sequences were obtained from published sources (Simmonds *et al.*, 1993*a*; Lau *et al.*, 1995).

statistically significant ($p < 0.05$), consistent with transmission of HCV from a common source in the anti-D recipients.

RATE OF HCV SEQUENCE CHANGE

In the above example, the rate of sequence change shown in the virus from the anti-D recipients in this study was slow enough to preserve evidence of relatedness over the 17 years from infection. The diversity of NS-5 sequences amongst recipients of anti-D immunoglobulin can be represented by a frequency histogram of individual sequence distances obtained by pairwise comparison of each sequence to each other in the dataset (in the case of a larger set of 23 recipients described here, 253 pairwise comparisons) (Fig. 4*a*). Sequence divergence values were 'corrected' by allowing for multiple substitutions at the same site to produce an evolutionary distance. The correction factor is necessary particularly when comparing highly divergent sequences, where it becomes increasingly probable that new changes may occur at a site already changed and therefore not be scored when computing sequence distances. Algorithms for correcting the effect of unseen multiple substitutions vary but generally provide comparable results. For the analyses presented below, Jukes-Cantor distances were used in the MEGA package (Kumar, Tamura & Nei, 1993).

Values ranged from zero (sequences identical) to 0.049, with a median of 0.020. As infection was known to have occurred 17 years previously, the (corrected) median rate of sequence change in the group as a whole was 0.059% $site^{-1}$ $year^{-1}$ (Fig. 4*a*). The rate of sequence change could also be estimated by analysis of the range of sequence distances between each recipient with the NS-5 sequence recovered from the infective batch of anti-D immunoglobulin archived in 1977. A median value of 0.011 corresponds to a rate of sequence change of 0.065% $site^{-1}$ $year^{-1}$, similar to the value obtained above. The previous longitudinal studies based on single patients (Okamoto *et al.*, 1992*a*; Ogata *et al.*, 1991) and cross-sectional studies (Cuypers *et al.*, 1991) produced values generally higher than this, but within the range of values found amongst the anti-D recipients.

Assuming the rate of sequence change to be constant, it is possible to use this 'clock' to estimate the time of divergence of epidemiologically unrelated type 1b variants. These sequences were obtained from sequence databases and are derived from a range of HCV-infected blood donors and hepatitis patients in Europe, USA, and Asia. The distribution of pairwise distances between the 40 sequences (780 pairwise comparisons) was markedly different from those found amongst the anti-D recipients (Fig. 4*b*), with a median value of 0.058. Using rates of sequence divergence calculated above, this suggests that type 1b originated from a common ancestor around 50 years ago (1940–1945).

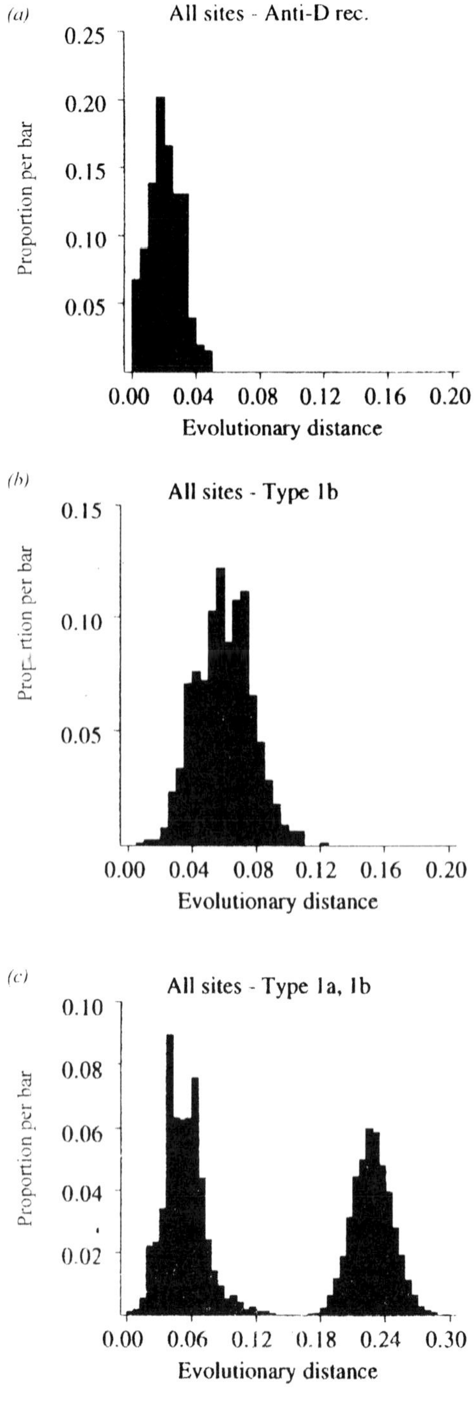
(a)
All sites - Anti-D rec.
0.25
0.20
0.15
0.10
0.05
Proportion per bar
0.00 0.04 0.08 0.12 0.16 0.20
Evolutionary distance
(b)
All sites - Type 1b
0.15
0.10
0.05
Proportion per bar
0.00 0.04 0.08 0.12 0.16 0.20
Evolutionary distance
(c)
All sites - Type 1a, 1b
0.10
0.08
0.06
0.04
0.02
Proportion per bar
0.00 0.06 0.12 0.18 0.24 0.30
Evolutionary distance

In the previous analyses, evolutionary distances were calculated using every nucleotide in each sequence. However, nucleotide changes that alter the encoded amino acid sequence are usually less frequent than those occurring at silent sites. For example, in the NS-5 region such non-silent changes are observed at a frequency approximately one-fifth of the value expected if mutations were entirely random. For example, over the 17 years following infection with anti-D immunoglobulin, the median pairwise distance for silent sites between infected recipients was 0.056, corresponding to a rate of change of 0.165% $site^{-1}$ $year^{-1}$, nearly three times higher than the analysis based upon total sites (0.059% $site^{-1}$ $year^{-1}$; see above). A 'clock' based on silent sites is in many cases preferable, as silent nucleotide changes are generally under less constraint than those that change amino acids. Using this approach, we have calculated the median synonymous pairwise distance between epidemiologically unrelated type 1b sequences to be 0.210 (compared with 0.058 for all sites), placing the time divergence of this genotype slightly further back at 64 years. Although in this particular case the difference is small, it becomes more significant for comparisons between more divergent sequences, where higher rates of multiple substitution occur at silent sites and which could be greatly underestimated by analyses based upon all sites.

Using these two clocks, it is also possible to analyse the diversity of sequences within other HCV genotypes. This analysis is currently being completed, but results so far indicate some variation. For example, sequence variability amongst type 3a variants is more restricted than other genotypes, suggesting a more recent dissemination. Amongst 27 sequences, median pairwise distances for total and silent sites are 0.042 and 0.156 respectively, predicting a time of divergence of 36–47 years ago. In contrast, types 2a and 2b are more diverse, originating around 70–90 years ago. However, all currently described genotypes show evidence of relatively recent spread into the currently identified risk groups for HCV infection, and one paralleling the widespread use of blood transfusion and other parenterally delivered medical treatments. The more restricted variability of type 3a is perhaps a reflection of its association in Europe with drug abuse, a route of transmission that has become common more recently than other parenteral routes.

Fig. 4. Distribution of sequence distances between (*a*) recipients of anti-D immunoglobulin; n=23; median value 0.020; (*b*) Epidemiologically unrelated individuals infected with genotype 1b; n=40; median value 0.058; (*c*) Individuals infected with type 1a (n=83) or 1b (n=40). Two distributions are observed, corresponding to the set of pairwise comparisons within 1a or 1b (median value 0.051) and to the set of pairwise comparisons between 1a and 1b (median 0.225); note change of scale.

ORIGINS OF HCV GENOTYPES

As with most viruses, we have no historical record of human infection with HCV, or information on its prevalence in different human populations in the past. We are therefore almost entirely reliant on inferences based upon the current distribution and diversity of genotypes in different geographical regions to illuminate their history and origins.

The first indication of the genetic heterogeneity of HCV came from comparisons of variants of HCV obtained from Japan with that of the prototype HCV variant obtained from the USA, HCV-PT (Choo *et al.*, 1991). The complete genome sequences of HCV-J (Kato *et al.*, 1990) and -BK (Takamizawa *et al.*, 1991) from Japan showed 92% similarity to each other, but only 79% with HCV-PT. At that time, the former were referred to as the 'Japanese' type (or type II), while those from the USA (HCV-PT and -H) were classified as type I. However, far more divergent variants of HCV have since been found in Japan (Okamoto *et al.*, 1991, 1992*b*) and elsewhere (Mori *et al.*, 1992; Chan *et al.*, 1992), leading to the adoption of an extended classification of HCV into types and subtypes.

Phylogenetic analysis of a 222-base fragment of NS-5 from a series of HCV-infected individuals in Europe, North and South America and the Far East showed HCV to be structured into six approximately equally divergent groups of sequences, each of which was composed of more closely related groups (Fig. 5). Similar relationships also exist between variants if analysis is carried out in E1 (Bukh, Purcell & Miller, 1993), NS-4 (Bhattacherjee *et al.*, 1995) or core regions (Bukh, Purcell & Miller, 1994), and it currently appears that any of these could be used for classification into genotypes.

A current proposal for the nomenclature of HCV genotypes is to divide HCV into types, corresponding to the main branches in the phylogenetic tree, and subtypes corresponding to the more closely related sequences within most of the major groups (Enomoto *et al.*, 1990; Simmonds *et al.*, 1993*a*, 1994). The types have been numbered 1 to 6, and the subtypes a, b and c, in both cases in order of discovery. Therefore, the sequence cloned by Chiron is assigned type 1a, HCV-J and -BK are type 1b, HC-J6 is type 2a and HC-J8 is type 2b.

The distinction between type and subtype was originally made on the basis of phylogenetic analysis, where two levels of branching were found upon analysis of NS-5, NS-3 and core sequences (Enomoto *et al.*, 1990; Chan *et al.*, 1992). For genotypes 1 to 6, the sets of sequence distances between sequences of the same subtype, between different subtypes and finally between different major genotypes, form three non-overlapping ranges. For example, sequence divergences of the 222-base fragment in NS-5 between members of the same subtype ranged from 0 to 12%, between subtypes from 12 to 22% and between types from 28 to 44% (Simmonds *et al.*, 1993*a*). Similar, non-overlapping ranges can also be computed for other regions of

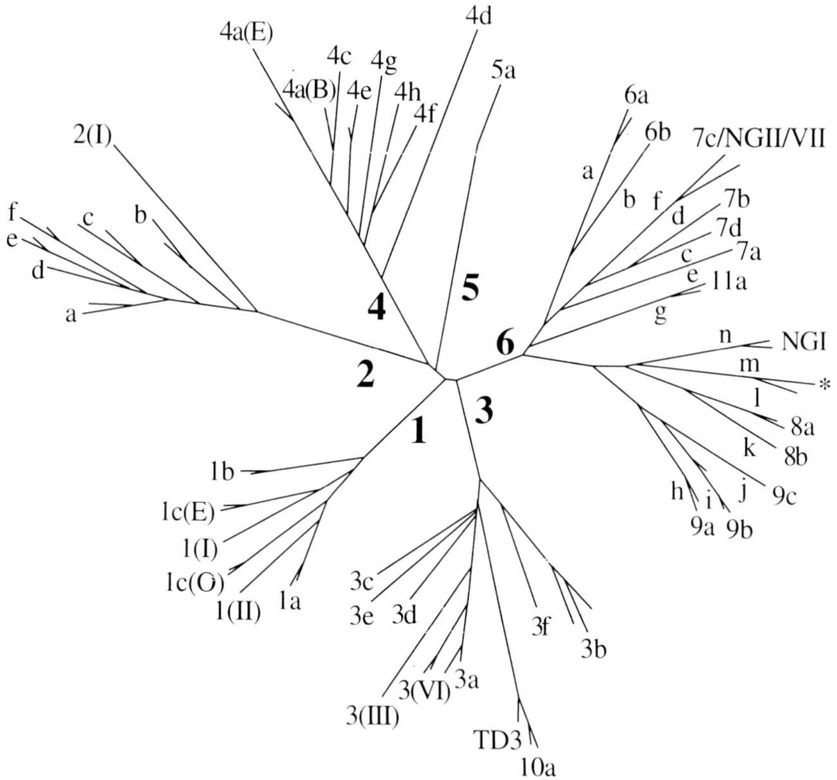

Fig. 5. Phylogenetic analysis of nucleotide sequences in part of the HCV NS-5 region. Individual sequences (numbered dots, identified in Simmonds *et al.*, 1993*a*) are separated by branch lengths proportional to their reconstructed evolutionary distances. Six main groups of sequence variants are shown (1–6); The numbering of the phylogenetic groups follows the consensus nomenclature for HCV (Simmonds *et al.*, 1994), although the final nomenclature of the highly divergent variants in the type 6 group (labelled 6a, 7a, 7b etc) has yet to be agreed. The division of HCV into six clades is supported by bootstrap resampling (100 replicates), producing values ranging from 90% (clade 3 and 6), 96% (clade 4), 98% (clade 1) to 100% (clades 2 and 5).

the genome whose values reflect their degree of variability. For example, the ranges for the three categories are slightly larger for E1 (Bukh *et al.*, 1993; Stuyver *et al.*, 1994) and smaller (and partially overlapping between type and subtype) for the core region (Bukh *et al.*, 1994; Mellor *et al.*, 1995).

There is a remarkable similarity in the pattern of sequence changes between those found among genotypes of HCV with those that accumulate over relatively short observation periods within a single infected individual. For example, regions that were unchanged over the 8.2 year study period, such as the 5′NCR are also the most conserved between genotypes of HCV (Table 2). Conversely, the greatest degree of sequence divergence is found in those regions that changed most rapidly in the chimpanzee, such as E2 and

E1. The main difference is one of scale, different genotypes of HCV may show up to 32% sequence divergence, whereas only 1.2% of nucleotides changed over 8 years in the chimpanzee.

DIVERGENCE OF HCV SUBTYPES

For the reasons discussed in the previous section, it is tempting to imagine that the genotypes of HCV represent nothing more than variants of HCV that diverged a long time ago. Analysis of median pairwise distances could therefore be used to estimate their time of divergence. For example, comparison of a large set of type 1a sequences in the NS-5 region with those of type 1b produces a bi-modal distribution of values (Fig. 4*c*). The peak nearest the origin corresponds to the set of pairwise comparisons within type 1a sequences superimposed upon those within type 1b. The median value (0.051) is similar to that obtained from comparing type 1b sequences alone (0.057; Fig. 4*b*).

The second peak is composed of the distribution of pairwise comparisons between 1a and 1b giving median values of 0.225 (all sites; Fig. 4*c*) and 0.722 (silent sites; data not shown) suggesting times of divergence of 191 and 219 years, respectively. The lineages of many genotypes of HCV are split into subtypes in a way similar to that of 1a and 1b. Type 1 itself is now known to contain numerous subtypes that diverge at a point close to that of 1a and 1b (for example, type 1c; Fig. 5). Similarly, there are seven subtypes of type 2 shown in Fig. 5, as well as others described elsewhere. Type 3 originally comprised two subtypes, but more recently another 7–8 have been characterized. Remarkably, most of the common subtypes are similarly divergent from each other; for example, the median distance between type 2a and 2b sequences over the 222 bp fragment of NS-5 is 0.243 compared with the value of 0.225 for 1a/1b. As a result, the distributions of distances between most subtypes throughout all genotypes can be superimposed to form a composite distribution (Simmonds *et al.*, 1993*a*), with values in the range from 0.149 to 0.349 and a median value of 0.235 (Mellor *et al.*, 1995).

The underlying reasons for this apparent uniformity in divergence values remains unclear. If we take 200 years or possibly longer as the estimate of the time of divergence of these subtypes, then it may be possible to discover an underlying epidemiological explanation for the accelerated process of diversification of lineages at that time. For example, significant long distance travel between continents became more frequent 200–300 years ago, and may perhaps have transported HCV to new communities in Africa and India/SE Asia where currently the greatest diversity of HCV subtypes is observed (Mellor *et al.*, 1995).

An alternative explanation is that the level of diversity that exists between subtypes represents a barrier beyond which further sequence change occurs much more slowly. For example, median synonymous pairwise distances of

0.7–1 that exist between subtypes predict that most silent sites have changed at least once since the time of divergence. Further diversification requires amino acid replacements, which accumulate more slowly because of strong selection pressure against nucleotide changes that alter the properties of the encoded HCV proteins. For this reason, the above estimates of the times of divergence of subtypes might be much greater than 200 years ago. How much earlier cannot be predicted because we do not know what constraints exist upon sequence change in NS-5 (or any other protein), and in any case these are likely to vary greatly between individual amino acid residues depending upon how critical they are to the function of the overall enzymatic activity of the protein.

Another factor that influences sequence distances that has not yet been taken into account is the difference in the numbers of transitions (G <-> A, U <-> C) compared with transversions (G or A <-> C or U). For example, amongst the anti-D recipients, three times more transitions were observed than would be expected by chance, so clearly this type of substitution would saturate earlier than other types of sequence change as sequences diverged. This would affect the calculation of evolutionary distances, and consequently increase estimates of times of divergence. However, even correcting for this may be inadequate, as different types of transition vary in frequency. For example, there are more C <-> U changes than G <-> A, and so the former would saturate sooner than the latter. In summary, as the methods for calculation of evolutionary distances are continually refined, this will generally have the effect of pushing back the time of divergence of the subtypes of HCV.

DIVERGENCE OF HCV GENOTYPES

The degree of sequence variability between the six major genotypes of HCV is similar to that found in other flaviviruses. For example, the envelope proteins of four serotypes of dengue fever virus differ from each other by 26% to 40%, similar to the divergence of the E1 or E2 proteins of different HCV genotypes (34–40% and 26–29%, respectively). In the case of dengue and other flaviviruses, the degree of sequence variability is sufficient to lead to considerable antigenic differences, leading to the use of the term 'serotype' to describe variants that differ in neutralization properties from each other. Although there is no *in vitro* assay to measure neutralization of HCV, antigenic differences have been demonstrated by enzyme-linked immunosorbent assays between recombinant proteins or peptides from different genotypes (McOmish *et al.*, 1993; Simmonds *et al.*, 1993*c*; Tsukiyama Kohara *et al.*, 1993; Mondelli *et al.*, 1994).

As with other RNA viruses, it is difficult to place a specific geographical or temporal origin for the common ancestor of the existing HCV genotypes. However, the preponderance of types 3 and 6 in South East Asia, and types

1, 2 and 4 in central and west Africa provide some clues, particularly as their presence is associated with a diversity of subtypes not observed in Europe, USA or the Far East (see above), and which may be associated with their long term presence. How old virus types are can only be speculated upon; there is considerable amino acid divergence among them but the rate of non-silent substitution is not known. Unfortunately, a time of divergence based upon silent substitutions cannot be estimated as distances are too large to be meaningful.

CONCLUSION

This review has attempted to describe the process of evolution in an RNA virus, based upon the evidence currently available from extant viruses. There is currently no evidence to suppose that HCV is in any way untypical of other RNA viruses and many of the analytical techniques used have been or could be applied to the study of other virus families. The uses of sequence analysis range from its use as a tool to study transmissions of HCV to its global epidemiology. With the large amount of sequence information currently available for HCV, there is an unparalleled opportunity to investigate the process of evolution, and the relation between virus nucleotide sequence diversity and variation in phenotype.

REFERENCES

Abe, K., Inchauspe, G. & Fujisawa, K. (1992). Genomic characterisation and mutation rate of hepatitis C virus isolated from a patient who contracted hepatitis during an epidemic of non-A, non-B hepatitis in Japan. *Journal of General Virology*, **73**, 2725–9.

Bhattacherjee, V., Prescott, L. E., Pike, I., Rodgers, B., Bell, H., El-Zayadi, A. R., Kew, M. C., Conradie, J., Lin, C. K., Marsden, H., Saeed, A. A., Parker, D., Yap, P. L. & Simmonds, P. (1995). Use of NS-4 peptides to identify type-specific antibody to hepatitis C virus genotypes 1, 2, 3, 4, 5 and 6. *Journal of General Virology*, **76**, 1737–48.

Brown, E. A., Zhang, H., Ping, H.-L. & Lemon, S. M. (1992). Secondary structure of the 5′ nontranslated region of hepatitis C virus and pestivirus genomic RNAs. *Nucleic Acids Research*, **20**, 5041–5.

Bukh, J., Purcell, R. H. & Miller, R. H. (1993). At least 12 genotypes of hepatitis C virus predicted by sequence analysis of the putative E1 gene of isolates collected worldwide. *Proceedings of the National Academy of Sciences, USA*, **90**, 8234–8.

Bukh, J., Purcell, R. H. & Miller, R. H. (1994). Sequence analysis of the core gene of 14 hepatitis C virus genotypes. *Proceedings of the National Academy of Sciences, USA*, **91**, 8239–43.

Chan, S.-W., McOmish, F., Holmes, E. C., Dow, B., Peutherer, J. F., Follett, E., Yap, P. L. & Simmonds, P. (1992). Analysis of a new hepatitis C virus type and its phylogenetic relationship to existing variants. *Journal of General Virology*, **73**, 1131–41.

Choo, Q. L., Kuo, G., Weiner, A. J., Overby, L. R., Bradley, D. W. & Houghton,

M. (1989). Isolation of a cDNA derived from a blood-borne non-A, non-B hepatitis genome. *Science*, **244**, 359–62.

Choo, Q. L., Richman, K. H., Han, J. H., Berger, K., Lee, C., Dong, C., Gallegos, C., Coit, D., Medina Selby, R., Barr, P. J., Weiner, A. J., Bradley, D. W., Kuo, G. & Houghton, M. (1991). Genetic organization and diversity of the hepatitis C virus. *Proceedings of the National Academy of Sciences, USA*, **88**, 2451–5.

Cuypers, H. T. M., Winkel, I. N., van der Poel, C. L., Reesink, H. W., Lelie, P. N., Houghton, M. & Weiner, A. (1991). Analysis of genomic variability of hepatitis C virus. *Journal of Hepatology*, **13**, S15–19.

Enomoto, N., Takada, A., Nakao, T. & Date, T. (1990). There are two major types of hepatitis C virus in Japan. *Biochemical and Biophysical Research Communications*, **170**, 1021–5.

Felsenstein, J. (1988). Phylogenies from molecular sequences: inferences and reliability. *Annual Review of Genetics*, **22**, 521–65.

Felsenstein, J. (1993). *PHYLIP Inference Package version 3.5*, Department of Genetics, University of Washington, Seatle.

Hijikata, M., Kato, N., Ootsuyama, Y., Nakagawa, M., Ohkoshi, S. & Shimotohno, K. (1991). Hypervariable regions in the putative glycoprotein of hepatitis C virus. *Biochemical and Biophysical Research Communications*, **175**, 220–8.

Hohne, M., Schreier, E. & Roggendorf, M. (1994). Sequence variability in the env-coding region of hepatitis C virus isolated from patients infected during a single source outbreak. *Archives in Virology*, **137**, 25–34.

Inoue, Y., Miyamura, T., Unayama, T., Takahashi, K. & Saito, I. (1991). Maternal transfer of HCV. *Nature*, **353**, 609.

Kaito, M., Watanabe, S., Tsukiyamakohara, K., Yamaguchi, K., Kobayashi, Y., Konishi, M., Yokoi, M., Ishida, S., Suzuki, S. & Kohara, M. (1994). Hepatitis C virus particle detected by immunoelectron microscopic study. *Journal of General Virology*, **75**, 1755–60.

Kato, N., Hijikata, M., Ootsuyama, Y., Nakagawa, M., Ohkoshi, S., Sugimura, T. & Shimotohno, K. (1990). Molecular cloning of the human hepatitis C virus genome from Japanese patients with non-A, non-B hepatitis. *Proceedings of the National Academy of Sciences, USA*, **87**, 9524–8.

Kato, N., Ootsuyama, Y., Tanaka, T., Nakagawa, M., Nakazawa, T., Muraiso, K., Ohkoshi, S., Hijikata, M. & Shimotohno, K. (1992). Marked sequence diversity in the putative envelope proteins of hepatitis C viruses. *Virus Research*, **22**, 107–23.

Koonin, E. V. (1991). The phylogeny of RNA-dependent RNA polymerases of positive-strand RNA viruses. *Journal of General Virology*, **72**, 2197–206.

Kumar, S., Tamura, K. & Nei, M. (1993). *MEGA: Molecular Evolutionary Genetics Analysis, version 1.0*, Pennsylvania State University, Pennsylvania.

Kuo, G., Choo, Q. L., Alter, H. J., Gitnick, G. L., Redeker, A. G., Purcell, R. H., Miyamura, T., Dienstag, J. L., Alter, M. J., Stevens, C. E., Tegtmeier, F., Bonino, F., Columbo, M., Lee, W.-S., Kuo, C., Berger, K., Schuster, J. R., Overby, L. R., Bradley, D. W. & Houghton, M. (1989). An assay for circulating antibodies to a major etiologic virus of human non-A, non-B hepatitis. *Science*, **244**, 362–4.

Lau, J. Y. N., Mizokami, M., Kolberg, J., Davis, G. L., Prescott, L. E., Ohno, T., Perrillo, R. P., Lindsay, K. L., Gish, R. G., Kohara, M., Simmonds, P. & Urdea, M. S. (1995). Application of hepatitis C virus genotyping systems in United States patients with chronic hepatitis C. *Journal of Infectious Diseases*, **171**, 281–9.

McOmish, F., Chan, S.-W., Dow, B. C., Gillon, J., Frame, W. D., Crawford, R. J., Yap, P. L., Follett, E. A. C. & Simmonds, P. (1993). Detection of three types of hepatitis C virus in blood donors: Investigation of type-specific differences in serological reactivity and rate of alanine aminotransferase abnormalities. *Transfusion*, **33**, 7–13.

Mellor, J., Holmes, E. C., Jarvis, L. M., Yap, P. L., Simmonds, P. & International Collaborators (1995). Investigation of the pattern of hepatitis C virus sequence diversity in different geographical regions: implications for virus classification. *Journal of General Virology*, **76**, 2493–507.

Miyamoto, H., Okamoto, H., Sato, K., Tanaka, T. & Mishiro, S. (1992). Extraordinarily low density of hepatitis C virus estimated by sucrose density gradient centrifugation and the polymerase chain reaction. *Journal of General Virology*, **73**, 715–18.

Mondelli, M. U., Cerino, A., Bono, F., Cividini, A., Maccabruni, A., Arico, M., Malfitano, A., Barbarini, G., Piazza, V., Minoli, L. & Silini, E. (1994). Hepatitis C virus (HCV) core serotypes in chronic HCV infection. *Journal of Clinical Microbiology*, **32**, 2523–7.

Mori, S., Kato, N., Yagyu, A., Tanaka, T., Ikeda, Y., Petchclai, B., Chiewsilp, P., Kurimura, T. & Shimotohno, K. (1992). A new type of hepatitis C virus in patients in Thailand. *Biochemical and Biophysical Research Communications*, **183**, 334–42.

Nei, M. (1987). *Molecular Evolutionary Genetics*, Columbia University Press.

Ogata, N., Alter, H. J., Miller, R. H. & Purcell, R. H. (1991). Nucleotide sequence and mutation rate of the H strain of hepatitis C virus. *Proceedings of the National Academy of Sciences, USA*, **88**, 3392–6.

Okamoto, H., Okada, S., Sugiyama, Y., Kurai, K., Iizuka, H., Machida, A., Miyakawa, Y. & Mayumi, M. (1991). Nucleotide sequence of the genomic RNA of hepatitis C virus isolated from a human carrier: comparison with reported isolates for conserved and divergent regions. *Journal of General Virology*, **72**, 2697–704.

Okamoto, H., Kojima, M., Okada, S.-I., Yoshizawa, H., Iizuka, H., Tanaka, T., Muchmore, E. E., Ito, Y. & Mishiro, S. (1992*a*). Genetic drift of hepatitis C virus during an 8.2 year infection in a chimpanzee: variability and stability. *Virology*, **190**, 894–9.

Okamoto, H., Kurai, K., Okada, S., Yamamoto, K., Lizuka, H., Tanaka, T., Fukuda, S., Tsuda, F. & Mishiro, S. (1992*b*). Full-length sequence of a hepatitis C virus genome having poor homology to reported isolates: comparative study of four distinct genotypes. *Virology*, **188**, 331–41.

Power, J. P., Lawlor, E., Davidson, F., Yap, P. L., Kenny-Walsh, E., Whelton, M. J. & Walsh, T. J. (1994). Hepatitis C viraemia in recipients of Irish intravenous anti-D immunoglobulin. *Lancet*, **344**, 1166–7.

Simmonds, P., Holmes, E. C., Cha, T. A., Chan, S.-W., McOmish, F., Irvine, B., Beall, E., Yap, P. L., Kolberg, J. & Urdea, M. S. (1993*a*). Classification of hepatitis C virus into six major genotypes and a series of subtypes by phylogenetic analysis of the NS-5 region. *Journal of General Virology*, **74**, 2391–9.

Simmonds, P., McOmish, F., Yap, P. L., Chan, S. W., Lin, C. K., Dusheiko, G., Saeed, A. A. & Holmes, E. C. (1993*b*). Sequence variability in the 5′ non coding region of hepatitis C virus: identification of a new virus type and restrictions on sequence diversity. *Journal of General Virology*, **74**, 661–8.

Simmonds, P., Rose, K. A., Graham, S., Chan, S. W., McOmish, F., Dow, B. C., Follett, E. A. C., Yap, P. L. & Marsden, H. (1993*c*). Mapping of serotype-specific, immunodominant epitopes in the NS- 4 region of hepatitis C virus (HCV) – use of type-specific peptides to serologically differentiate infections with HCV type 1, type 2, and type 3. *Journal of Clinical Microbiology*, **31**, 1493–503.

Simmonds, P., Alberti, A., Alter, H. J., Bonino, F., Bradley, D. W., Brechot, C.,

Brouwer, J. T., Chan, S. W., Chayama, K., Chen, D. S., Choo, Q. L., Colombo, M., Cuypers, H. T. M., Date, T., Dusheiko, G. M., Esteban, J. I., Fay, O., Hadziyannis, S. J., Han, J., Hatzakis, A., Holmes, E. C., Hotta, H., Houghton, M., Irvine, B., Kohara, M., Kolberg, J. A., Kuo, G., Lau, J. Y. N., Lelie, P. N., Maertens, G., McOmish, F., Miyamura, T., Mizokami, M., Nomoto, A., Prince, A. M., Reesink, H. W., Rice, C., Roggendorf, M., Schalm, S. W., Shikata, T., Shimotohno, K., Stuyver, L., Trepo, C., Weiner, A., Yap, P. L. & Urdea, M. S. (1994). A proposed system for the nomenclature of hepatitis C viral genotypes. *Hepatology*, **19**, 1321–4.

Simons, J. N., Pilot-Matias, T. J., Leary, T. P., Dawson, G. J., Desai, S. M., Schlauder, G. G., Muerhoff, A. S., Ersker, J. C., Buijk, S. L., Chalmers, M. L., Van Sant, C. L. & Mushahwar, I. K. (1995). Identification of two flavivirus-like genomes in the GB hepatitis agent. *Proceedings of the National Academy of Sciences, USA*, **92**, 3401–5.

Smith, D. B., Mellor, J., Jarvis, L. M., Davidson, F., Bhattacherjee, V., Kolberg, J. A., Urdea, M. S., Yap, P. L., Simmonds, P. & International Collaborators (1995). Variation of the hepatitis C virus 5′-non coding region: implications for secondary structure, virus detection and typing. *Journal of General Virology*, **76**, 1749–61.

Stuyver, L., Vanarnhem, W., Wyseur, A., Hernandez, F., Delaporte, E. & Maertens, G. (1994). Classification of hepatitis C viruses based on phylogenetic analysis of the envelope 1 and nonstructural 5b regions and identification of five additional subtypes. *Proceedings of the National Academy of Sciences, USA*, **91**, 10134–8.

Takamizawa, A., Mori, C., Fuke, I., Manabe, S., Murakami, S., Fujita, J., Onishi, E., Andoh, T., Yoshida, I. & Okayama, H. (1991). Structure and organization of the hepatitis C virus genome isolated from human carriers. *Journal of Virology*, **65**, 1105–13.

Tsukiyama Kohara, K., Iizuka, N., Kohara, M. & Nomoto, A. (1992). Internal ribosome entry site within hepatitis C virus RNA. *Journal of Virology*, **66**, 1476–83.

Tsukiyama Kohara, K., Yamaguchi, K., Maki, N., Ohta, Y., Miki, K., Mizokami, M., Ohba, K., Tanaka, S., Hattori, N., Nomoto, A. & Kohara, M. (1993). Antigenicities of group I and group II hepatitis C virus polypeptides – molecular basis of diagnosis. *Virology*, **192**, 430–7.

Wang, C. Y., Sarnow, P. & Siddiqui, A. (1994). A conserved helical element is essential for internal initiation of translation of hepatitis C virus RNA. *Journal of Virology*, **68**, 7301–7.

Weiner, A. J., Geysen, H. M., Christopherson, C., Hall, J. E., Mason, T. J., Saracco, G., Bonino, F., Crawford, K., Marion, C. D., Crawford, K. A. *et al.* (1992). Evidence for immune selection of hepatitis C virus (HCV) putative envelope glycoprotein variants: potential role in chronic HCV infections. *Proceedings of the National Academy of Sciences, USA*, **89**, 3468–72.

Weiner, A. J., Thaler, M. M., Crawford, K., Ching, K., Kansopon, J., Chien, D. Y., Hall, J. E., Hu, F. & Houghton, M. (1993). A unique, predominant hepatitis C virus variant found in an infant born to a mother with multiple variants. *Journal of Virology*, **67**, 4365–8.

PROBLEMS WITH MOLECULAR DIVERSITY IN THE EUKARYA

MITCHELL L. SOGIN, JEFFREY D. SILBERMAN, GREGORY HINKLE AND HILARY G. MORRISON

Program in Molecular Evolution, Marine Biological Laboratory, Woods Hole, MA 02543, USA

INTRODUCTION

For more than 3.5 billion years, communities of microorganisms have dominated evolution of life in our biosphere. The first unicellular creatures must have been more similar to the Bacteria or the Archaea than to cells with complex internal architecture and nuclear encoded genomes. Accordingly, phylogeneticists describe the Eukarya as a derived lineage with the Protista being ancestral to the 'higher' kingdoms of Fungi, Plantae and Animalia (Whittaker, 1969). The age of protists, and hence the origin of eukaryotes, is still undetermined. Unambiguous fossil evidence for eukaryotic algae dates back to at least 1.9 billion years ago (bya) (Lipps, 1993) but earlier origins of protists are not excluded (Knoll, 1992). More uncertain is the exact relationship of the first protists with the Archaea and the Bacteria.

Protists are an eclectic assemblage of (predominantly unicellular) eukaryotes (Sleigh, 1989; Margulis *et al.*, 1990). They inhabit diverse environments, or parasitize other protists, fungi, plants or metazoans. Their phenotypic variation far exceeds that seen in other eukaryote kingdoms (Ragan & Chapman, 1978). Memberships and relationships within the Protista are difficult to describe. There is no trait that unifies protists to the exclusion of all other eukaryotes, and there is no agreement about the relative importance of distinct characters in tests of evolutionary hypotheses. Rather than delineating a cohesive evolutionary group, the Protista describe levels of organization and represent paraphyletic lines of descent (Sogin *et al.*, 1989).

The fossil record, so important for determining the temporal branching patterns of multicellular organisms, is absent for most microbial groups and provides no clues about relationships between the primary kingdoms. Micro-paleontologists cite the great abundance of skeletal protists (Lipps, 1993) and 3.5 billion year old cyanobacterial-like stromatolites (Schopf, 1983, see also this volume). However, soft-bodied taxa that represent the earliest diverging protist lineages are not well preserved and it is all but impossible to differentiate between microfossils of archaeal versus bacterial origins. Until a time machine is invented, our only means to study early evolution of these divergent microbial taxa is to exploit molecular phylogenies based upon comparisons of genes and genomes. Comparisons of functionally independent conserved coding regions make possible phylogenetic frameworks that span life's evolutionary history (Woese, 1987; Sogin, 1994).

We are convinced of the authenticity and objective nature of gene trees and we assume they mirror the evolutionary history of genomes from the considered organisms (Zuckerkandl & Pauling, 1965). Yet, as more data become available we are presented with ostensible contradictions that cannot be readily explained by continuously dichotomous phylogenetic branching patterns. Such observations lead to arguments about right versus wrong trees and raise concerns about our underlying assumptions (Swofford & Olsen, 1990). Is it even reasonable to expect all gene trees to be congruent and to provide an accurate chronological accounting of evolutionary history from studies of molecular evolution? Aberrant rates of genetic change for different genes in different lineages confuse attempts to correlate genotypic change with the geological record and often times lead to inference of incorrect and hence incongruent phylogenies. The metaphor of a molecular clock is imperfect. At best, the length of line segments in molecular phylogenies is proportional to the product of average mutation rate in a particular gene and the time of divergence separating any two taxa. Since rates of accepted substitutions are not constant, it is sometimes very difficult to infer 'the correct' dichotomous branching pattern, much less assign accurate geological dates.

More radical explanations for incongruent gene phylogenies attack even more fundamental assumptions. Most molecular evolutionists cast hypotheses within a tree-like paradigm; few consider net-like relationships that could have been created by lateral gene transfer and/or anastamosing genomes. Descriptions of the evolutionary history of life in this biosphere must consider the potential of lateral gene transfer against a background of dichotomous branching trees.

This is a time of transition for the field of molecular evolution. Today, we rely upon fragmentary data to decide between contradictory evolutionary hypotheses. In the very near future, more comprehensive solutions to these questions will come from comparative genomics.

RIBOSOMAL RNAS AND THE REINVENTION OF EUKARYOTIC PHYLOGENETICS

More than any other gene family, studies of ribosomal RNAs have reshaped our understanding of the universal tree of life (Woese & Fox, 1977; Woese, Kandler & Wheelis, 1990). Among the many revelations, we now recognize three primary lineages; the Eukarya and two prokaryotic domains, the Archaea and the Bacteria (Fig. 1(*a*)). Perhaps even more surprising is the unexpected diversity of eukaryotic ribosomal RNAs. Differences between 16S-like rRNA sequences from protists are much greater than those seen within or between the Archaea and the Bacteria (Sogin, 1994). A progression of independent protist branches, some as ancient as the divergence between the two prokaryotic domains, preceded the nearly coincidental separation of the complex eukaryotic assemblages. Trichomonads, diplomonads and microsporidia are amitochondriate protists that are basal to all other eukaryotes included in rRNA studies (Leipe *et al.*, 1993). These organisms are without mitochondria (but see Doolittle, this volume), peroxisomes, or typical Golgi. Together with pelobionts, oxymonads, retortamonads and hypermastigids, these amitochondriate taxa comprise the Archaezoa (Cavalier-Smith, 1986). This branching pattern suggests that the Eukarya may be older than once thought possible; given the amitochondriate phenotype of early-diverging lineages, the presence of oxygen is not prerequisite to forming a nucleus.

Most extant early diverging amitochondriate protists have adopted parasitic lifestyles that may have played a role in their survival during unfavourable environmental periods. However, parasitism alone cannot explain this pattern. Free-living diplomonads related to *Giardia lamblia* are known, while other parasitic species including *Acanthamoebae, Plasmodium berghei* and *Pneumocystis carinii* diverge relatively late in eukaryotic rRNA phylogenies. Although subsequent branching patterns are better resolved, comparisons of small subunit ribosomal RNAs have not unambiguously determined the relative order of divergence for the Archaezoan lineages. Biased nucleotide compositions in the rRNA genes of some taxa and an inadequate number of informative sites compromise the resolving power of rRNAs in this region of the tree (Leipe *et al.*, 1993).

Because of their basal position in universal phylogenies, we use trichomonads, microsporidia and diplomonads as outgroups to root the eukaryal phylogeny shown in Fig. 1(*b*). This allows the inclusion of a larger number of unambiguously aligned sites than is possible in universal rRNA trees. The topology in Fig. 1(*b*) is largely congruent with likelihood (Olsen *et al.*, 1994) and parsimony analyses (Swofford, 1991). By almost any measure, protist diversity dwarfs that of the plant, animal and fungal worlds. The divergence of the primitively amitochondriate protists is followed first by amoebae-flagellates and then by a polytomy for the acellular slime moulds, euglenozoa (euglenoids and kinetoplastids) and later-diverging eukaryotes. The many lineages that separate after this radiation but prior to the separation of plants, animals and fungi have an amoeboid phenotype. Although amoeboid lineages are clearly polyphyletic, there are sub-groups of amoebae. The early-branching amoebae–flagellates form a single clade, as do the acanthamoebae and entamoebae. Other amoebae, however, are represented by independent lineages in rRNA based trees. It seems unlikely that amoebae represent a great 'way station' on the path to becoming a more complex

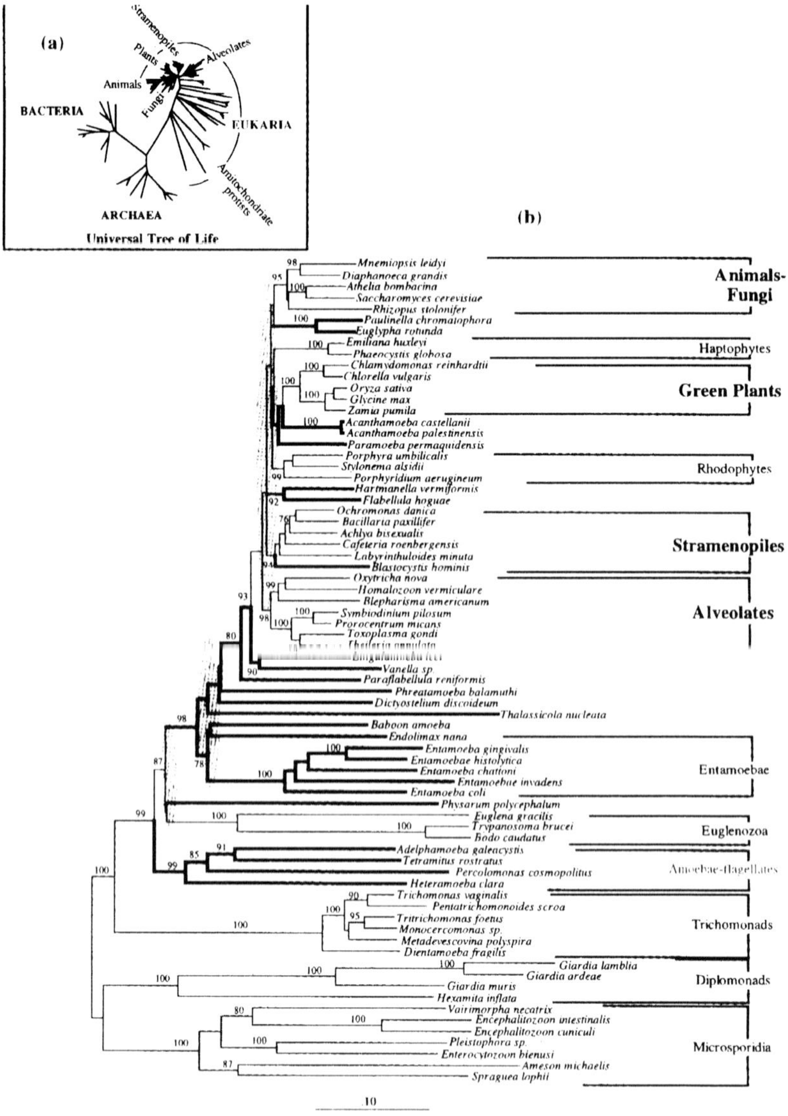

Fig. 1(*a*) and 1(*b*). Universal and eukaryote phylogenetic trees. Fig. 1(*a*) is an 'unrooted' universal phylogeny. Structural similarities for 900 sites that can be unambiguously aligned in a data set of more than 800 eukaryotes and 100 prokaryotes were computed and converted to evolutionary distances (Elwood *et al.*, 1985) using the Kimura two parameter model. The neighbour-joining method (Saitou & Nei, 1987) was used to infer the unrooted universal phylogeny shown in 1(*a*). Evolutionary distances are proportional to length of line segments separating taxa. Figure 1(*b*) is a 'rooted' phylogeny for eukaryotes. We selected sequences from an aligned database of more than 900 eukaryotic 16S-like rRNAs. Structural similarities for sites that can be unambiguously aligned for these taxa (approximately 1200 sites) were eukaryote. Incomplete taxon sampling of flagellates could explain this pattern and implies that many protist lineages including undiscovered forms are as yet not represented in the rRNA phylogenetic frameworks.

A massive radiation occurred relatively late in the evolutionary history of the Eukarya. The major evolutionary assemblages that diverged nearly simultaneously include the animals–fungi, the green plants–green algae, the Stramenopiles and the Alveolates. These four late-evolving groups and several other independent protist lineages define the 'crown' of the eukaryal subtree (Knoll, 1992). Two of these groups are newly recognized 'kingdoms' as diverse and complex as the traditional 'higher kingdoms' of Fungi, Animalia or Plantae. The Alveolates include ciliates, apicomplexans and dinoflagellates (Gajadhar *et al.*, 1991; Patterson & Sogin, 1993). The Stramenopiles include diatoms, brown algae, yellow-brown algae, golden-brown algae, water moulds, slime nets and many non-photosynthetic, heterotrophic flagellates that branch near the base of this complex assemblage (Leipe *et al.*, 1994). All members of this group have flagella with tripartite flagellar hairs not found in any other eukaryotes. The flagellar hairs may have conferred a major ecological advantage for these organisms by allowing thrust reversal when swimming and thus enhanced ability to entrap prey. How we interpret the evolution of plastids is affected by the basal position of non-photosynthetic heterotrophs within stramenopiles, the deep branching of euglenoids in the eukaryal subtree, and the phylogenetic positions of green plants–green algae, the red algae and the dinoflagellates in the eukaryal crown. There must have been multiple primary or secondary endosymbiotic events leading to the formation of several eukaryotic alga groups.

No one has resolved the exact branching order of the crown groups from

---

Fig. *Continued*

converted to evolutionary distances. We employed 'minimum evolution' (Rzhetsky & Nei, 1992) to infer a tree in which the horizontal component of separation represents the evolutionary distance between taxa. The scale bar corresponds to 10 changes/100 positions. Amoebae are indicated by heavy lines. Because of their early branching in universal phylogenies, sequences that correspond to microsporidia, diplomonads and trichomonads were used as outgroups to root the eukaryote subtree; however, their relative order of divergence is not determined. Nodes that correspond to important unresolved radiations in the eukaryote framework are indicated by shaded ovals. Bootstrap support for topological elements in the tree are based upon 1000 replicates of a neighbour-joining analysis. Bootstrap values below 75% are not shown. Bootstrap analyses have gained widespread acceptance for estimating levels of support in phylogenetic branching patterns. The significance of these 'statistical' techniques can be controversial. Bootstrap values are affected by the number of resamplings, the underlying evolutionary models, the fraction of sites that change on a given segment separating two nodes (Hillis & Bull, 1993), and the number of deep interior nodes separated by short evolutionary distances (Leipe *et al.*, 1994). Many of the low bootstrap values in Fig. 1(*b*) are indicative of the tight-ordered branching pattern displayed by the succession of independent amoebae lineages.

analyses of rRNA sequences or other gene families. We can only demonstrate a convincing common and unique evolutionary history for fungi–animals relative to other eukaryotes (Wainright *et al.*, 1993). The assignment of an absolute date to the separation of the crown groups is tempting. The 'molecular clock hypothesis' has become a metaphor in which the accumulation of genetic change is related to scales of time (Wilson, Ochman & Prager, 1987). But the clock is imprecise. Substitution rates in actin or histone coding regions differ by as much as tenfold in separate evolutionary lineages (Bhattacharya, Stickel & Sogin, 1991; Sadler & Brunk, 1992). The more proper metric for comparative molecular biology is one of relative order or relative age rather than of absolute time. This interpretation adopts the quantitative and objective attributes of the 'molecular clock' without falling prey to its failings. Phylogenetic frameworks inferred from molecular data are hypotheses of branching order based upon the extent of genetic relatedness rather than reflections of time. The massive radiation displayed by the crown groups is every bit as important and mysterious as the invention of the eukaryotic cell. This explosion of biodiversity may reflect survivorship from ancient catastrophic events, or it could reflect new modes of genome organization. Perhaps genetic innovations led to more complicated patterns of development.

Corresponding phylogenetic studies of actin genes describe eukaryal groups similar to those obtained in rRNA trees. Amoebae are an important exception. Rather than being distributed throughout the tree, phylogenetic branches corresponding to most amoebae are very short and seem to emanate from the middle of the unrooted actin tree shown in Fig. 2. There are several possible explanations for this contrasting view of actin versus rRNA evolution. The most plausible hypothesis is that functional constraints on amoeboid movement demand molecular convergence in amoeboid actins. Molecular convergence in distance-based phylogenetic reconstructions would predict the observed foreshortened line segments for amoeba actins in Fig. 2. Perhaps surveys for other actin paralogues or functional tests by cell biologists will someday explain why actins are so similar in taxa that are extremely divergent in the rRNA trees. Contradictions between the rRNA and actin trees are but one of many examples where obvious aberrations in rates of change for different molecules lead to very different evolutionary hypotheses.

DIVERSITY OF EUKARYAL PROTEIN SEQUENCES

Despite advances in eukaryal phylogenetics and the rapid growth of molecular databases, the mystery about the origin of protists and major complex groups of eukaryotes looms ever larger. Ancestral roots within the prokaryotic world are unknown and, except for the apparent lack of introns, we have only sketchy details about the genotype of early-diverging, amitochon-

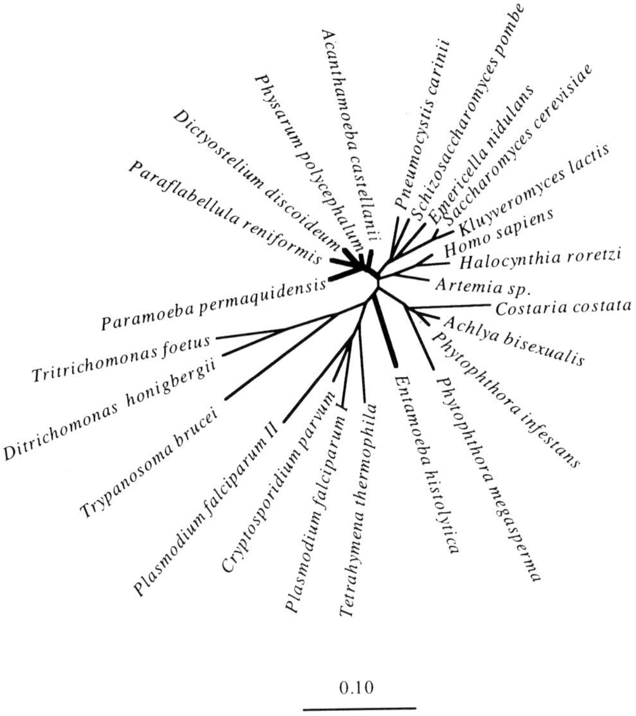

Fig. 2. Actin phylogeny. Reverse transcriptase and PCR were used to isolate actin coding regions from diverse amoebae (Bhattacharya, Stickel & Sogin, 1993). The tree is based on structural distances between first and second codon positions of actin genes (662 sites). The distance which corresponds to 10 changes per 100 nucleotide positions is indicated by the scale. Amoebae are represented by heavy line segments.

driate protists. Phylogenies based upon comparisons of a small number of other coding regions [transcription factor TFIIB (Ouzounis & Sander, 1992), DNA-dependent RNA polymerase (Klenk *et al.*, 1990), proteinase of the ubiquitin-associated proteosome (Puhler *et al.*, 1994), ATPase (Gogarten *et al.*, 1989), elongation factors for protein synthesis (Iwabe *et al.*, 1989) and tRNA synthetase (Brown & Doolittle, 1995)] suggest a common ancestry for eukaryotes and archaebacteria. But this inference is contested by analyses of glutamine synthetase (Brown *et al.*, 1994), heat shock genes (Gupta & Singh, 1992) and putative homologies between tubulin and FtsZ, a protein involved in bacterial cytokinesis (Lutkenhaus, 1993). These studies hypothesize relationships between eukaryotes and specific bacterial groups or proffer the formation of chimerical nuclear genomes (Sogin, 1991; Forterre *et al.*, 1992; Gupta & Golding, 1993).

Most molecular evolution studies that target the origin of eukaryotes and their deepest branching patterns are based upon only a handful of genes. In most cases, crown groups represent all eukaryotes, even though their physiology, metabolism and pattern of genome organization have undergone considerable change since the invention of the nucleus. Relative to animals, fungi, and plants, the databases do not adequately represent early diverging, amitochondriate protists. For example, there are many microsporidia rRNA sequences, but protein coding regions from all microsporidia include only beta-tubulin (Edlind *et al.*, 1994), an isoleucyl-tRNA synthetase (Brown & Doolittle, 1995) and a putative elongation factor-alpha gene. Trichomonads and diplomonads are somewhat better represented by approximately ten different gene classes reported for *Trichomonas* species and twice that number for *G. lamblia*. Except for cell surface antigens and giardin, the only coding regions reported from the earliest branching eukaryotes display extraordinary conservation patterns in other eukaryotes (Baldauf & Palmer, 1993; Lange, Rozario & Muller, 1994; Nagel & Doolittle, 1994; Doolittle *et al.*, 1996). This conservation facilitates isolation through heterologous probing or PCR strategies, but creates a bias against other gene families in database submissions. Unfortunately, we are largely ignorant about other genes and their conservation in early-diverging eukaryotic lineages. Knowledge of DNA replication, transcription and translation in the deepest branches of the universal tree of life is almost nonexistent. There are no descriptions of DNA polymerases, ribosomal proteins, replication factors, DNA topoisomerases, transcriptional regulatory proteins or even histones from amitochondriate protists. Such information is essential prior to our understanding the early evolution of these fundamental processes.

This void in the molecular databases constrains our view of eukaryote evolution. Furthermore, all attempts to resolve the early protist branching patterns and the relationship between the Bacteria, Archaea and Eukarya are founded upon preconceived ideas about the conserved nature of molecular markers. It is not possible to know *a priori* which genes are optimal for testing specific evolutionary hypotheses. For example, tubulins are known to be well conserved in eukaryotes and thus have been isolated and studied in nearly all phyla including diplomonads, trichomonads and microsporidia. At best, these molecules have only confirmed previously described relationships because their substitution patterns have reached saturation.

If diplomonads, trichomonads and microsporidia represent the earliest diverging lineages, any genes from primitively amitochondriate protists that are recognizable in more recently diverged groups must represent extremely ancient functions. On the other hand, given the relative simplicity of cytoskeletal organization (compared to crown group taxa) and the lack of mitochondria, peroxisomes, and typical Golgi, the amitochondriate protists such as the microsporidia might have striking molecular similarities with either the Archaea or the Bacteria. The most expedient means to explore these questions is to inventory coding regions in one or more amitochondriate lineages. The divergence of these organisms lies close to the transition between eukaryotes and prokaryotes. Similarities to genes from prokaryotes or more recently diverged eukaryotes will paint a picture of the genetic diversity in early eukaryotes and will provide important molecular tools for resolving ancient radiations.

MOLECULAR EVOLUTION AND THE ORIGINS OF THE EUKARYA

Woese and Fox discovered the Archaea based upon differences in oligonucleotide catalogues generated by T1 ribonuclease digestion of ribosomal RNAs. The eukaryotic/prokaryotic dichotomy was replaced by three primary lines of descent derived from the Progenote. According to differences in rRNA oligonucleotide catalogues and biochemical characteristics, the Archaea were as distinct from the Bacteria as either were from the Eukarya. Woese originally argued the Archaea were a monophyletic group that was the most similar to the last common ancestor of all living organisms in unrooted universal phylogenies. This satisfied the observation that ribosomal RNAs in the Archaea seem to evolve at a much slower rate than corresponding genes in the Eukarya and the Bacteria; in a genetic sense the Archaea could be considered as the least evolved of the primary lineages. Competing proposals described the Archaea as polyphyletic with the sulphur-dependent Archaea being specifically related to the Eukarya. These opposing interpretations were alternatively ascribed to differences in alignment techniques or methods for inferring evolutionary history from comparisons of molecular sequences. More recent data in the form of amino acid insertions common to elongation factors for the Eukarya and the sulphur-dependent Archaea fuel arguments about the possible polyphyletic nature of the Archaea (Rivera & Lake, 1992). However, neither of these positions explains the origins of eukaryotes or defines the phenotype of cells with nuclear encoded genomes.

If the Eukarya is a derived lineage, it is important to determine its ancestral roots in one of the bacterial domains. Putative evolutionary orthologues have been identified for several gene families in each of the primary lineages. Each of these can be used to infer a 'universal phylogeny', but the identification of the 'root' or the last common ancestor presents a special challenge. In most phylogenetic reconstructions, outgroups can polarize branching patterns. For example, bacterial sequences can polarize the relative order of branching for eukaryal lineages in a molecular tree. However, there are no known outgroups to the universal tree of life and therefore its root (and by implication the relationship of eukaryotes to one of the bacterial primary lineages) is obscure.

This problem has been addressed through studies of ancient gene duplications for two important gene families (see also Doolittle, this volume and Gogarten, this volume). Elongation factors EF-Tu and EF-G and two classes of ATPase (alpha and beta) are present in all organisms. These gene families are products of gene duplication events that likely preceded the divergence of the primary lines of descent. Phylogenies for each of these genes should reflect the evolutionary history of genomes going back to the last common ancestor of all organisms. Since the gene duplication events preceded the divergence of primary lineages from the last common ancestor, one duplicate gene partner can polarize the branching pattern of its paralogous gene family partner in molecular reconstructions. For example, the molecular phylogenies for either EF-G or EF-Tu are unrooted. However, if the two duplicate gene families are placed together in the same analysis, the duplicate gene partner EF-Tu serves as the outgroup for polarizing the branching pattern within the EF-G tree. Similarly, EF-G serves as the outgroup for rooting the topology of its duplicate gene partner tree, EF-Tu (Iwabe *et al.*, 1989). Identical strategies were applied to analyses of the duplicate gene partners alpha and beta ATPase (Gogarten *et al.*, 1989). Analyses of both gene families place the root of universal protein phylogenies within the Bacteria. Figure 3 shows the effect of rooting universal rRNA trees within the bacterial line of descent.

As reflected in Fig. 3, the duplicate gene partner story argues that eukaryotes and archaea may have shared a common evolutionary history exclusive of Bacteria. This interpretation is supported by the observed similarities between eukaryal and archaeal RNA polymerases (Puhler *et al.*, 1989), ribosomal proteins (Hui & Dennis, 1985) and the putative presence of histone-like protein in the thermoacidophilic archaebacterium Thermoplasma acidophilum (Searcy, 1975). Comparisons of many protein sequences indicate that the Archaea and Eukarya are more closely related to each other than either is to the Bacteria. This is clearly at odds with analyses of rRNAs which suggest that the Bacteria and Archaea are more closely related. Accelerated rates of change in all nuclear rRNA coding regions (the accelerated-rate hypothesis) with a constancy or even reduction of mutation rates in protein coding regions seems like a parsimonious explanation of the deep eukaryotic branching patterns in the rRNA-based phylogenies. However, this theory is more convoluted than it initially appears. It does not explain contradictory topologies observed in phylogenies for other conserved coding regions and fails to explain evolutionary processes responsible for the origin of the eukaryotic cell.

If the rate of rRNA evolution accelerated (as indicated in Fig. 3) soon after the Eukarya diverged from Archaea, it is necessary to explain the clock-like behaviour of rRNA genes during the last billion years. The geological record provides strong evidence that the crown of the eukaryote tree corresponds to events which occurred one billion years ago (Knoll, 1992). Based upon this chronological estimate, the large evolutionary distances in rRNA frameworks that correspond to early eukaryote evolution would reflect accelerated rates of mutation in rRNA genes. The accelerated-rate hypothesis places a further constraint on eukaryotic evolution. For all

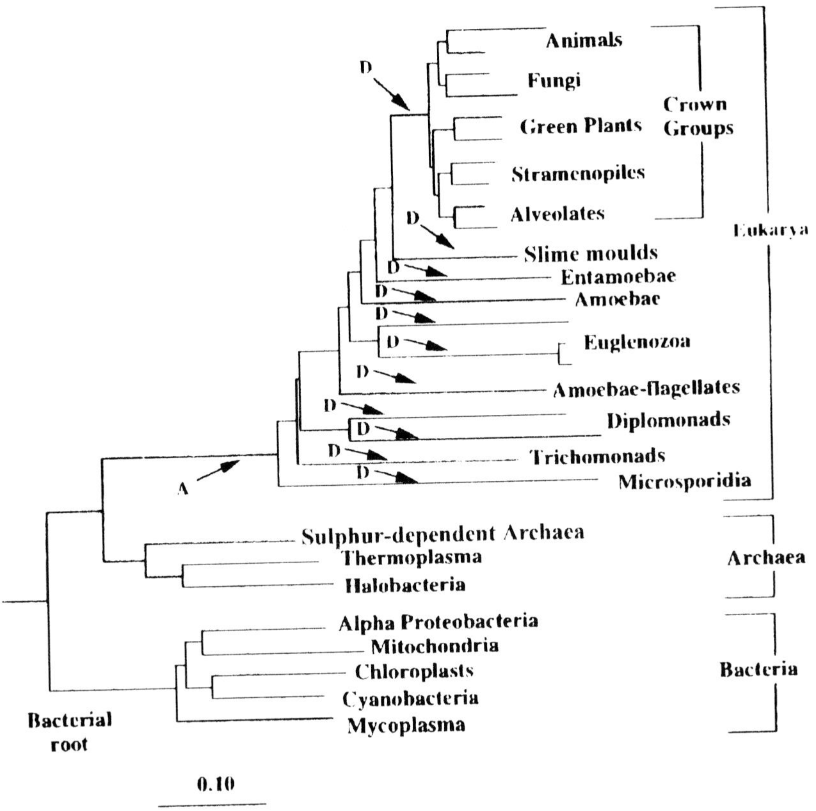

Fig. 3. Bacterial rooting of universal rRNA phylogenies. The distance matrix analysis of 16S-like rRNAs from Leipe *et al.* (1993) is shown with the root positioned within the bacterial line of descent. The horizontal component of lines separating nodes and taxa are proportional to evolutionary distances. Placement of the root is suggested by analysis of duplicate gene partners for ATPase (Gogarten *et al.*, 1989) and elongation factor genes (Iwabe *et al.*, 1989). This tree hypothesizes a common ancestor for the Archaea and the Eukarya. To explain long line segments in the Eukarya, a single acceleration of rate corresponding to 'A' has been proposed. To accommodate the one bya estimated date of divergence for eukaryotic crown groups, the accelerated-rate hypothesis requires many reductions (indicated by 'D') in rates of rRNA evolution within independent protist lineages.

branches in the eukaryote tree to exhibit approximately similar line lengths, it would be necessary to impose many independent reductions in mutation rates for rRNA genes in many independent eukaryal lineages (Fig. 3). This requirement casts doubt upon the 'most-parsimonious' solution offered by the accelerated-rate hypothesis. Furthermore, the accelerated-rate scenario does not explain hypothetical relationships between eukaryotes and sulphur-dependent Archaea based upon the 11 amino acid insertion in elongation factors shared by these two groups but missing in halophilic Archaea and all Bacteria (Rivera & Lake, 1992), nor does it address the

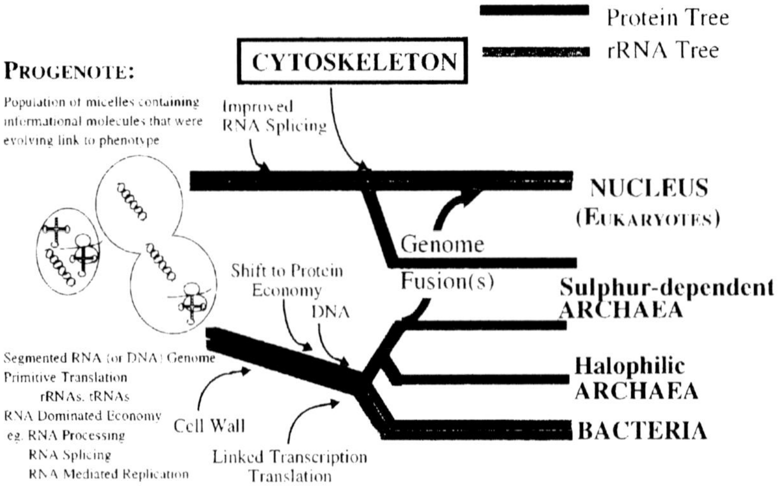

Fig. 4. Chimeric model for origins of eukaryotes. Two major lineages diverged from the theoretical Progenote. One of these lineages developed a cell wall, shifted to a protein-based metabolism and invented DNA. This led to a common ancestor for the Bacteria and the Archaea. The Archaea diverged later into Halophilic Archaea and Sulphur-dependent Archaea. The second lineage that arose from the Progenote retained an RNA-based metabolism, improved RNA splicing capacity and used primitive ribosomes to synthesize primitive cytoskeletal proteins. These aboriginal cytoskeletal proteins conferred upon this lineage the ability to engulf sulphur-dependent Archaea thus forming symbioses and eventually the nucleus. Many coding regions for eukaryotes, including elongation factors for protein synthesis, trace their origins back to the early sulphur-dependent archaeal symbionts; other genes such as the ribosomal RNAs trace their origins back to the Progenote.

proposed relationship between eukaryotes and gram-negative bacteria deduced from comparisons of heat shock patterns (Gupta *et al.*, 1994).

Something unusual must have occurred during the early evolution of eukaryotes. There seems to be a net-like rather than a tree-like history associated with the invention of the eukaryotic cell. If certain protein coding regions are considered, Archaea and Eukarya appear to be more closely related to each other than either is to the Bacteria. This relationship contradicts the rRNA phylogenies which position Archaea as specific relatives to the Bacteria. Yet other protein trees suggest ancestral roots of eukaryotes within the gram-negative bacteria or within the sulphur-dependent Archaea. Several chimeric genome models have been proposed for the origins of eukaryotes. One unusual model based upon fusions of archaeal genomes with a vestigial lineage of the RNA world is radical, yet it explains seemingly contradictory molecular analyses. The model presented in Fig. 4 emphasizes that the presence or absence of a nucleus is secondary to existence of the cytoskeleton when differentiating prokaryotes and eukaryotes. It proposes that an anucleate lineage distinct from the Archaea and the Bacteria invented sufficient cytoskeletal complexity to allow the transition from a proto-eukaryotic lineage to cells with nuclei. In this model, two primary lines of descent trace their common origins back to the Progenote rather than the three lines originally described by Woese. The Progenote was a hypothetical biological information processing system which had yet to develop a link between evolution of genotype and phenotype. In the true spirit of evolutionary biology, the scenario presented below represents maximum speculation based upon minimal data. It is set in the RNA world, but could just as well have been based upon a Progenote with DNA-based informational molecules.

The Progenote was not a single surviving individual. It was a population of micelles that could fuse or separate, allowing for the rapid exchange of macromolecular machinery. The later stages of Progenote evolution had already produced a primitive translation apparatus, but RNA-mediated events dominated metabolism. Genomes were fragmented into many nucleic acid polymers several hundred nucleotides in length. This allowed faithful replication of information molecules using replication machinery less accurate than today's polymerases. As shown in Fig. 4, two primary lineages were derived from the Progenote. The first developed a sophisticated translation apparatus capable of accurately synthesizing proteins. The RNA dominated economy of this lineage soon switched to one based upon proteins, but nucleic acids were retained as the primary information processing system. This proto-bacterial lineage invented DNA and converted fragmented informational molecules into large linear or circular chromosomes. Translation became coupled to transcription and the lineage ultimately diverged to form the archaeal and bacterial lines of descent.

The second lineage that diverged from the Progenote took a decidedly different path. It retained the fragmented genome of the Progenote but developed sophisticated RNA splicing mechanisms. Rather than switching to a protein based economy, RNA metabolism dominated and the range of RNA mediated catalysis expanded. Yet, there was still a primitive translation machinery that mediated the production of simple proteins. The most important of the proteins to evolve in this RNA based lineage were precursors of the cytoskeleton (microtubules and membrane-associated cytoskeletal proteins). Innovation of primitive cytoskeletal proteins eventually led to cells with nuclear encoded genomes.

When asked to describe the fundamental difference between eukaryotes and prokaryotes, most biologists incorrectly cite the presence versus the absence of a nucleus. The more important distinguishing feature is the presence of the eukaryotic cytoskeleton. Nuclear genome organization depends upon cytoskeletal architecture if for no other reason than to allow chromosome movement and nuclear division. The invention of the cytoskeleton conferred upon the vestigial RNA lineage the capacity to engulf other micro-organisms, a strategy not widely available to prokaryotes. This allowed entry into symbiotic relationships with bacteria. The endosymbiotic origins of mitochondria and chloroplasts are accepted facts. One or more engulfments of archaeal organisms by the proto-eukaryal lineages may have occurred even earlier. It was this combination of events which led to the formation of the earliest nucleated cells. The genome is viewed as a chimera; the archaeal component provided most of the protein coding regions while the proto-eukaryotic genome contributed genes for rRNAs and cytoskeletal proteins. Even if the model described here is incorrect, explaining cytoskeleton origins is key to understanding the ancestral roots of the Eukarya.

This hypothesis is fundamentally different from previous models which date back almost 100 years and which also propose an endosymbiotic origin of the nucleus (Margulis, 1970; Hartman, 1984; Schubert, 1988; Zillig, Palm & Klenk, 1993; Lake & Rivera, 1994). In most regards, the host lineage was completely unlike contemporary organisms and its origins likely trace back to entities that existed 3.5–4 billion years ago. The model explains sophisticated RNA splicing mechanisms in eukaryotes that are not present in the prokaryotes. It also explains Lake's observations about perceived similarity between elongation factors of sulphur-dependent archaea and eukaryotes. If the endosymbiont that give rise to the majority of genes in eukaryotes was a sulphur-dependent archaean, it would have donated its elongation factor EF-1-alpha to the eukaryal genome.

The chimerical origins of eukaryotes predict the existence of several sequence classes which will display different phylogenetic branching patterns. Coding regions such as those for cytoskeletal proteins that were invented in the vestigial RNA lineage will not be found in any prokaryotes. Coding regions invented by the Progenote will be found in all types of life, but will be most similar in the Bacteria and Archaea. If the sulphur-dependent Archaea were the endosymbionts that gave rise to the majority of eukaryal protein coding regions, there will be shared characteristics between their genes and their homologues in eukaryotes. The identification of these sequence classes will depend upon the development of complete genome sequences for phylogenetically divergent lineages.

Irrespective of whether the chimerical origins of nuclear genomes is correct, we are left with at least one more dilemma. Based upon comparisons of 57 gene families from 15 groups of divergent taxa, Doolittle *et al.* (1996) dated the crown of the eukaryote subtree at one G yr, the divergence of protists at 1.23 G yr, the separation of the Archaea and the Eukarya at 1.96 G yr, and the last common ancestor to the Bacteria and the Archaea plus Eukarya lineages at 2.1 G yr. Assuming these dates do not represent unwarranted linear extrapolations of the molecular clock, several anomalies demand extrapolation and perhaps correction. The date assigned to the crown of the eukaryotic tree is in general agreement with dates assigned by rRNA phylogenies, but the date for divergence of protists is likely to be too conservative. As pointed out by Doolittle, most of the protist sequences used in this analysis were from kinetoplastids, which are only moderately deep in the eukaryotic rRNA phylogenies. A wider sampling of protists, especially amitochondriate taxa, would provide more credibility to estimates about the age of the first protist lineages.

Molecular phylogeneticists who study rRNAs will find the relative (but not the absolute) time of divergence between the Archaea and the Eukarya to be satisfying. The divergence of the Archaea and Eukarya occurred soon after their proposed separation from the Bacteria. This is reassuring for those who adhere to Woese's original hypothesis that the last common ancestor is positioned close to the trichotomy for the three primary lines of descent. However, the time of divergence between the bacterial domains presents a major problem for interpreting the meaning of the 3.5 billion year old cyanobacterial-like stromatolites. If the assignment of 2.1 G yr to the separation of the bacterial lineages is correct, much of our evolutionary history was dominated by undescribed life forms, the most primitive of which left cyanobacterial-like stromatolite fossils. Yet the cyanobacterial phenotype is localized within bacterial radiations that occurred after the divergence of the first bacterial lineages. There may be nearly a twofold error in the extrapolation for time of divergence between the bacterial lineages. Alternatively, estimates of time of divergence between the Eukarya and Bacteria could correspond to the fusion events between sulphur-dependent Archaea and the lineage that invented the cytoskeleton. Greater accuracy in time of divergence for protein families will come when the analyses include better representation of deep branching, amitochondriate eukaryotes. However, the more important analyses will be those which provide more accurate estimates of phylogenetic branching patterns for different gene families. Such studies can only be achieved through comparative studies of genomes from diverse representatives of the Bacteria, the Archaea and the Protista.

ACKNOWLEDGEMENTS

This work was supported by the G. Unger Vetlesen Foundation and a grant from the National Institute of Health (GM32964) to M.L.S.

REFERENCES

Baldauf, S. L. & Palmer, J. D. (1993). Animals and Fungi are each other's closest relatives: congruent evidence from multiple proteins. *Proceedings of the National Academy of Sciences, USA*, **90**, 11558–62.

Bhattacharya, D., Stickel, S. K. & Sogin, M. L. (1991). Molecular phylogenetic analysis of actin genic regions from *Achlya bisexualis* (Oomycota) and *Costaria costata* (Chromophyta). *Journal of Molecular Evolution*, **33**, 4275–86.

Bhattacharya, D., Stickel, S. K. & Sogin, M. L. (1993). Isolation and molecular phylogenetic analysis of actin coding regions from *Emiliania huxleyi*, a prymnesiophyte alga, using reverse transcriptase and PCR methods. *Molecular Biological Evolution*, **10**, 689–703.

Brown, J. R. & Doolittle, W. F. (1995). Root of the universal tree of life based on ancient aminoacyl-tRNA synthetase gene duplications. *Proceedings of the National Academy of Sciences, USA*, **92**, 2441–5.

Brown, J. R., Masuchi, Y., Robb, F. T. & Doolittle, W. F. (1994). Evolutionary relationships of bacterial and archaeal glutamine synthetase genes. *Journal of Molecular Evolution*, **38**, 566–76.

Cavalier-Smith, T. (1986). The kingdom Protista: origin and systematics. *Progress in Phycology Research*, **4**. Round, F. E. and D. J. Chapman, eds.

Doolittle, R. F., Feng, D., Tsang, S., Cho, G. & Little, E. (1996). Determining divergence times of the major kingdoms of living organisms with a protein clock. *Science*, **271**, 470–7.

Edlind, T., Visvesvara, G., Li, J. & Katiyar, S. (1994). *Cryptosporidium* and microsporidial B-tubulin sequences: prediction of benzimidazole sensitivity and phylogeny. *Journal of Eukaryotic Microbiology*, **41**, 385.

Elwood, H. J., Olsen, G. J. & Sogin, M. L. (1985). The small subunit ribosomal RNA gene sequences from the hypotrichous ciliates *Oxytricha nova* and *Stylonychia pustulata. Molecular Biology Evolution*, **2**, 399–410.

Felsenstein, J. (1985). Confidence intervals on phylogenies: an approach using the bootstrap. *Evolution*, **39**, 783–91.

Forterre, P., Benachenhou-Lahfa, N., Confalonieri, F., Duguet, M., Elie, C. & Labedan, B. (1992). The nature of the last universal ancestor and the root of the tree of life, still open questions. *Biosystems*, **28**, 15–32.

Gajadhar, A. A., Marquardt, W. C., Hull, R., Gunderson, J., Ariztia Carmona, E. V. & Sogin, M. L. (1991). Ribosomal RNA sequences of *Sarcocystis muris, Theileria annulata* and *Crypthecodinium cohnii* reveal evolutionary relationships among apicomplexans, dinoflagellates, and ciliates. *Molecular Biochemical Parasitology*, **45**, 147–54.

Gogarten, J. P., Kibak, H., Dittrich, P., Taiz, L., Bowman, E. J., Bowman, B. J., Manolson, N. F., Poole, R. J., Date, T., Oshima, T., Knoishi, J., Denda, K. & Yhoshida, M. (1989). Evolution of the vacuolar H^+-ATPase: implications for the origin of eukaryotes. *Proceedings of the National Academy of Sciences, USA*, **86**, 6661–5.

Gupta, R. S. & Golding, G. B. (1993). Evolution of HSP70 gene and its implications regarding relationships between Archaebacteria, Eubacteria and Eukaryotes. *Journal of Molecular Evolution*, **37**, 573–82.

Gupta, R. S. & Singh, B. (1992). Cloning of the HSP70 gene from *Halobacterium marismortui*: relatedness of archaebacterial HSP70 to its eubacterial homologs and a model of the evolution of the HSP70 gene. *Journal of Bacteriology*, **174**, 4594–605.

Gupta, R. S., Aitken, K., Falah, M. & Singh, B. (1994). Cloning of *Giardia lamblia* heat shock protein HSP70 homologs: implications regarding origin of eukaryotic cells and of endoplasmic reticulum. *Proceedings of the National Academy of Sciences, USA*, **91**, 2895–9.

Hartman, H. (1984). The origin of the eukaryotic cell. *Speculations in Science and Technology*, **7**, 77–81.

Hillis, D. M. & Bull, J. J. (1993). An empirical test of bootstrapping as a method for assessing confidence in phylogenetic analysis. *Systems in Biology*, **42**, 182–92.

Hui, J. & Dennis, P. P. (1985). Characterization of the ribosomal RNA gene cluster in *Halobacterium cutirubrum. Journal of Biological Chemistry*, **260**, 529–33.

Iwabe, N., Kuma, K.-I., Hasegawa, M., Osawa, S. & Miyata, T. (1989). Evolutionary relationships of Archaebacteria, Eubacteria and Eukaryotes inferred from phylogenetic trees of duplicated genes. *Proceedings of the National Academy of Sciences, USA*, **86**, 9355–9.

Klenk, H.-P., Palm, P., Lottspeich, F. & Zillig, W. (1990). Component H of the DNA-dependent RNA polymerase of Archaea is homologous to a subunit shared by the three eukaryl nuclear RNA polymerases. *Proceedings of the National Academy of Sciences, USA*, **89**, 407–10.

Knoll, A. H. (1992). The early evolution of eukaryotes: a geological perspective. *Science*, **256**, 622–7.

Lake, J. A. & Rivera, M. C. (1994). Was the nucleus the first endosymbiont? *Proceedings of the National Academy of Sciences, USA*, **91**, 2880–1.

Lange, S., Rozario, C. & Muller, M. (1994). Primary structure of the hydrogenosomal adenylate kinase of *Trichomonas vaginalis* and its phylogenetic relationships. *Molecular Biochemistry Parasitology*, **66**, 297–308.

Leipe, D. D., Gunderson, J. H., Nerad, T. A. & Sogin, M. L. (1993). Small subunit ribosomal RNA of *Hexamita inflata* and the quest for the first branch in the eukaryotic tree. *Molecular Biochemistry Parasitology*, **59**, 41–8.

Leipe, D. D., Wainright, P. O., Gunderson, J. H., Porter, D., Patterson, D. J., Valois, F., Himmerich, S. & Sogin, M. L. (1994). The stramenopiles from a molecular perspective: 16S-like rRNA sequences from *Labyrinthuloides minuta* and *Cafeteria roenbergensis*. *Phycologia*, **33**, 369–77.

Lipps, J. H. (1993). *Fossil Procaryotes and Protists*. Blackwell Scientific Publications, Boston.

Lutkenhaus, J. (1993). FtsZ ring in bacterial cytokinesis. *Molecular Microbiology*, **9**, 403–9.

Margulis, L. (1970). *Origin of Eukaryotic Cells*. Yale University Press, New Haven.

Margulis, L., Corliss, J. O., Melkonian, M. & Chapman, D. J. (1990). *Handbook of Protoctista*. Jones and Bartlett Publishers, Boston.

Nagel, G. M. & Doolittle, R. F. (1994). Phylogenetic analysis of the aminoacyl-tRNA synthetase. *Journal of Molecular Evolution*, **40**, 487–98.

Olsen, G., Matsuda, H., Hagstrom, R. & Overbeek, R. (1994). fastDNAml: a tool for construction of phylogenetic trees of DNA sequences using Maximum Likelihood. *Cabios*, **10**, 41–8.

Ouzounis, C. & Sander, C. (1992). TFIIB, an evolutionary link between the transcription machineries of archaebacteria and eukaryotes. *Cell*, **71**, 189–90.

Patterson, D. J. & Sogin, M. L. (1993). Eukaryote origins and protistan diversity. In *The Origin and Evolution of Procaryotic and Eukaryotic Cells*, Hartman, H. & Matsuno, K.,eds., pp. 13–46. World Scientific Pub. Co., River Edge, NJ.

Puhler, G., Leffers, H., Gropp, F., Palm, P., Klenk, H. P., Lottspeich, F., Garrett, R. A. & Zillig, W. (1989). Archaebacterial DNA-dependent RNA polymerases testify to the evolution of the eukaryotic nuclear genome. *Proceedings of the National Academy of Sciences, USA*, **86**, 4569–73.

Puhler, G., Pitzer, F., Zwickl, P. & Baumeister, W. (1994). Proteasomes: multisubunit proteinases common to *Thermoplasma* and eukaryotes. *Systems in Applied Microbiology*, **16**, 734–41.

Ragan, M. A. & Chapman, D. J. (1978). *A Biochemical Phylogeny of the Protists*. Academic Press, New York, San Francisco and London.

Rivera, M. C. & Lake, J. A. (1992). Evidence that Eukaryotes and Eocyte Procaryotes are immediate relatives. *Science*, **257**, 74–6.

Rzhetsky, A. & Nei, M. (1992). A simple method for estimating and testing minimum-evolution trees. *Molecular Biology of Evolution*, **9**, 945–67.

Sadler, L. A. & Brunk, C. F. (1992). Phylogenetic relationships and unusual diversity in histone H4 proteins within the *Tetrahymena pyriformis* complex. *Molecular Biology of Evolution*, **9**, 70–84.

Saitou, N. & Nei, M. (1987). The neighbor-joining method: a new method for reconstructing phylogenetic trees. *Molecular Biology of Evolution*, **4**, 406–25.

Schopf, W. J. (1983). *Earth's Earliest Biosphere*. Princeton University Press, Princeton NJ.

Schubert, I. (1988). Eukaryotic nuclei of endosymbiotic origin? *Naturwissenschaften*, **75**, 89–91.

Searcy, D. G. (1975). Histone-like protein in the procaryote *Thermoplasma acidophilum*. *Biochimica et Biophysica Acta*, **395**, 535–47.

Sleigh, M. (1989). *Protozoa and Other Protists*. Edward Arnold, London.

Sogin, M. L. (1991). Early evolution and the origin of eukaryotes. *Current Opinion Genetic Development*, **1**, 457–63.

Sogin, M. L. (1994). The origin of eukaryotes and evolution into major kingdoms. In *Early Life on Earth – A Nobel Symposium*, Bengtson, S., ed., pp. 181–192. Columbia University Press, New York NY.

Sogin, M. L., Gunderson, J. H., Elwood, H. J., Alonso, R. A. & Peattie, D. A. (1989). Phylogenetic significance of the Kingdom concept: an unusual eukaryotic 16S-like ribosomal RNA from *Giardia lamblia*. *Science*, **243**, 75–7.

Swofford, D. L. (1991). *Phylogenetic Analysis Using Parsimony* (PAUP). Version 3.0. Illinois Natural History Survey, Champaign IL.

Swofford, D. L. & Olsen, G. J. (1990). In *Phylogeny Reconstruction, Molecular Systematics*, Hillis, D. M. & Moritz, C., eds., pp. 411–501. Sinauer Associates, Sunderland MA.

Wainright, P. O., Hinkle, G., Sogin, M. L. & Stickel, S. K. (1993). The monophyletic origins of the Metazoa; an unexpected evolutionary link with Fungi. *Science*, **260**, 340–3.

Whittaker, R. H. (1969). New concepts of kingdoms of organisms. *Science*, **163**, 150–60.

Wilson, A. C., Ochman, H. & Prager, E. M. (1987). Molecular time scale for evolution. *Trends in Genetics*, **3**, 341–7.

Woese, C. R. (1987). Bacterial evolution. *Microbiological Reviews*, **51**, 221–71.

Woese, C. R. & Fox, G. E. (1977). Phylogenetic structure of the procaryotic domain: the primary kingdoms. *Proceedings of the National Academy of Sciences, USA*, **74**, 5088–90.

Woese, C. R., Kandler, O. & Wheelis, M. L. (1990). Towards a natural system of organisms: proposal for the domains Archaea, Bacteria, and Eucarya. *Proceedings of the Natural Academy of Science, USA*, **87**, 4576–9.

Zillig, W., Palm, P. & Klenk, H.-P. (1993). A model of the early evolution of organisms: the arisal of the three domains of life from the common ancestor. In *The Origin and Evolution of Procaryotic and Eukaryotic Cells*, Hartman, H. & Matsuno, K., eds., pp. 163–204. World Scientific Pub. Co., River Edge NJ.

Zuckerkandl, E. & Pauling, L. (1965). Molecules as documents of evolutionary history. *Journal of Theoretical Biology*, **8**, 357–66.

EUKARYOTIC LIFE: ANAEROBIC PHYSIOLOGY

TOM FENCHEL

Marine Biological Laboratory (University of Copenhagen)
Strandpromemaden 5, DK-3000 Helsingør, Denmark

INTRODUCTION

Inspired by the idea that life originated under anoxic conditions, as originally proposed by Oparin and Haldane, extant anaerobes have long drawn attention as possible Precambrian relicts. As far as eukaryotes are concerned, however, we know now that they acquired mitochondria very early in their evolutionary history and that they evolved and diversified mainly as aerobic organisms. This is consistent with evidence that oxic habitats occurred already 3.5×10^9 years ago, which is the age of the earliest remains of cyanobacterial mats. The majority of extant anaerobic eukaryotes have thus secondarily adapted to life without oxygen and this has happened independently within several protist lineages. Recent molecular data (in particular rRNA sequences) as well as morphological traits have, however, provided evidence that a few minor groups of amitochondrial protists (the diplomonads and probably the mastigamoebae and the retortamonads) represent a very early branch in eukaryote evolution. They are believed to be primarily amitochondrial and if so, they are also primarily anaerobic (Schlegel, 1991, 1994; Sogin, this volume). Cavalier-Smith (1987*a*) coined the collective term Archezoa for these groups. (The microsporidians, which apparently also represent an ancient branch of eukaryotes are usually included in this group. They are amitochondrial, but they are also intracellular parasites which have secondarily lost a number of cell functions and structures; it makes little sense to discuss whether they are anaerobes or aerobes and the absence of mitochondria may not be original.)

In the following, organisms are considered anaerobes if they can complete their entire life cycle in the absence of O_2. Obligate anaerobes are incapable of oxidative phosphorylation (they do not have cytochrome *c*-oxidase), while facultative anaerobes can use O_2 in respiration for electron transport phosphorylation, but they can also cope indefinitely without O_2. By strict anaerobes are meant forms which are extremely sensitive to O_2-exposure. Among the anaerobic (and some microaerobic) eukaryotes O_2-sensitivity is highly variable and it correlates with the habitat of the organisms. Probably all anaerobic eukaryotes have some enzymatic defences against oxygen radicals and many can use O_2 for the restoration of redox balance in their fermentative metabolism. Practically all anaerobic eukaryotes depend on different kinds of fermentation (using substrate phosphorylation or in some cases also electron transport phosphorylation) for energy metabolism. Only one case of anaerobic respiration is so far known: the ciliate *Loxodes* is a denitrifier under anaerobic conditions (Finlay, Span & Harman, 1983). Eukaryotic anaerobic phototrophs (capable of anoxygenic photosynthesis) have not been found. Anoxygenic photosynthesis is known from several cyanobacteria which can use sulphide as an electron donor in photosynthesis (Cohen *et al.*, 1986). Some types of eukaryotic phototrophs (especially euglenoids and diatoms) are frequently found in sulphidic habitats (e.g. together with purple sulphur bacteria and in *Beggiatoa*-mats) and so it is possible that anoxygenic photosynthesis may still be discovered among protists.

Anaerobic protists occur in a variety of anaerobic and microaerobic habitats. Aquatic sediments are typically anaerobic and usually sulphidic a few mm or cm beneath the surface and if the sediments are sufficiently porous they will harbour anaerobic protists. The deep water of stratified lakes is often seasonally or permanently anoxic as is the water column beneath the pycnocline of some stratified marine basins and fjords (e.g. the Black Sea); these habitats also host a variety of anaerobic protists. So do anaerobic or microaerobic parts of microbial mats consisting of sulphur bacteria or accumulations of decaying organic debris. The intestinal tract of animals constitutes another important habitat for anaerobic protists. The rumen of artiodactyls and the termite hindgut represent the best studied and perhaps the most diverse examples, but anaerobic protozoa probably occur in the gut or caecum of all mammals and also in, e.g. sea turtles, anurans, sea urchins and in many other animals. The human intestinal tract also harbours some anaerobic protists among which a few (*Giardia lamblia* and *Entamoeba histolytica*) are pathogens of some importance.

Although anaerobic forms have evolved within many eukaryotic lineages the diversity of eukaryotic anaerobes is very low compared to aerobic forms. Also, and in contrast to aerobic habitats, most anaerobic habitats harbour a relatively low concentration of protist individuals which together represent a relatively small biomass when compared to that formed by prokaryotes.

Adaptations among eukaryotes to life without oxygen will here be discussed primarily with reference to energy metabolism and symbiotic interactions with prokaryotes; more brief references will be given to oxygen toxicity and to some evolutionary and ecological aspects. Table 1 summarizes the taxonomic distribution of anaerobic eukaryotes and their basic properties.

ENERGY METABOLISM

Oxidative phosphorylation is the most efficient type of energy conservation and the most important characteristic of anaerobes is that they have to

Table 1. *Eukaryote groups with anaerobic representatives*

Taxonomic group	Type of metabolism	Ecology[a]
Archezoa		
Rhizomastigids	Probably primarily amitochondriate.	F
Diplomonads	Clostridial type fermentation (without	F/C
Retortamonads	H_2-evolution) in cytosol (only studied in parasitic diplomonads)	F/C
Entamoeba	Probably secondarily amitochondriate. Clostridial type fermentation (without H_2-evolution) in cytosol	C
Trichomonads (+hypermastigids and oxymonads)	All with hydrogenosomes	C
Heterolobosa		
Psalteriomonas	With hydrogenosomes. Other heteroloboseans observed in anaerobic habitats, but energy metabolism is unknown	F
Various euglenoids, dinoflag., lobose amoebae, chrysomonads, choanoflag.	Only observations of growth in anoxic habitats/cultures. Mitochondria always present, but hydrogenosomes not observed. Many may be facultative anaerobes	F
Chytrids	All with hydrogenosomes.	C[2]
Ciliates	Include obligate anaerobes with hydrogenosomes or with mitochondria without hydrogenase as well as facultative anaerobes. Hydrogenosomes have evolved independently within at least six orders	F/C

[a]F: free-living, C: (intestinal) commensals, [2]free-living chytrids are common and may also include anaerobic species.

manage with considerably less efficient ways of gaining energy from their substrates. The anaerobic eukaryotes together represent almost all types of fermentative processes known from various prokaryotes. Among the anaerobic eukaryotes, some of the fermentative pathways have evolved independently in different groups. In many cases they employ existing pathways (glycolysis, parts of the citric acid enzymes in mitochondriate anaerobes), but some forms have also acquired pathways (such as clostridial-type fermentations) and enzymes which otherwise do not occur in eukaryotes.

The amitochondriate protists

The Archezoa (including the diplomonad and retortamonad flagellates and the uniflagellate mastigamoebae) do not have membrane covered organelles. Unfortunately, regarding metabolic pathways only the parasitic diplomonad *Giardia* has so far been studied in detail (Adam, 1991; Ellis *et al.*, 1993; Williams & Lloyd, 1993). All the enzymes of metabolism are found in the cytosol. The organism depends on glucose and amino acids for substrate. Carbohydrates are broken down through the glycolytic pathway and pyruvate is oxidized to acetate (a process coupled to substrate level phosphorylation) with ferredoxin as an electron acceptor and catalysed by pyruvate:ferredoxin oxidoreductase. Various organic compounds serve as electron acceptors for the reoxidation of ferredoxin and ethanol and acetate are the main end products (Fig. 1). The iron–sulphur centre of the ferredoxin is of a clostridial [4Fe–4S] type. In the absence of other evidence it may be assumed that the free-living, bacterivorous diplomonads (e.g. *Hexamita*, *Trepomonas*) have a similar type of energy metabolism. Nothing is known about the metabolism of mastigamoebae (rhizomastigids), a group which includes only free-living forms like *Mastigamoeba*, *Mastigella* and the probably related giant amoeba *Pelomyxa*. The latter lives in the ooze of ponds and lakes; it behaves partly as an anaerobe and partly as a microaerophile according to the stage in the life cycle (Whatley & Chapman-Andresen, 1990). Like some species of *Mastigella* it harbours endosymbiotic methanogens. In this case the source of H_2 is not known; although the energy metabolism of these organisms has not been studied, hydrogen evolution is, in all other known cases, associated with membrane covered organelles. It has therefore been speculated that this substrate is provided by other, non-methanogenic endosymbionts (Van Bruggen *et al.*, 1983, 1985). No details are known about the retortamonad flagellates which occur both as intestinal commensals and as free-living species (for the diversity of amitochondriate free-living protists, see Mylnikov, 1991).

The intestinal commensal/parasitic *Entamoeba* species are also devoid of mitochondria or of mitochondria-derived organelles. They seem, however, to represent a later branch in the eukaryote phylogenetic tree (Schlegel, 1991) and by implication the loss of mitochondria is secondary (with the exception of *Entamoeba* and the above-mentioned Archezoa, all anaerobic eukaryotes seem to have retained mitochondria or at least organelles which seem to derive from mitochondria; see below). *Entamoeba* has an energy metabolism which resembles that of *Giardia* (Müller, 1988; Reeves, 1984).

Protists with hydrogenosomes

In the early 1970s, Müller and co-workers discovered a new kind of organelle, the hydrogenosome, in trichomonad flagellates (see Müller, 1980, 1993). Its main function is the oxidation of pyruvate (deriving from glycolysis in the cytosol) to acetate and H_2 and the process is coupled to energy conservation through substrate phosphorylation. The organelle does not contain cytochromes, but a pyruvate:ferredoxin oxidoreductase and a

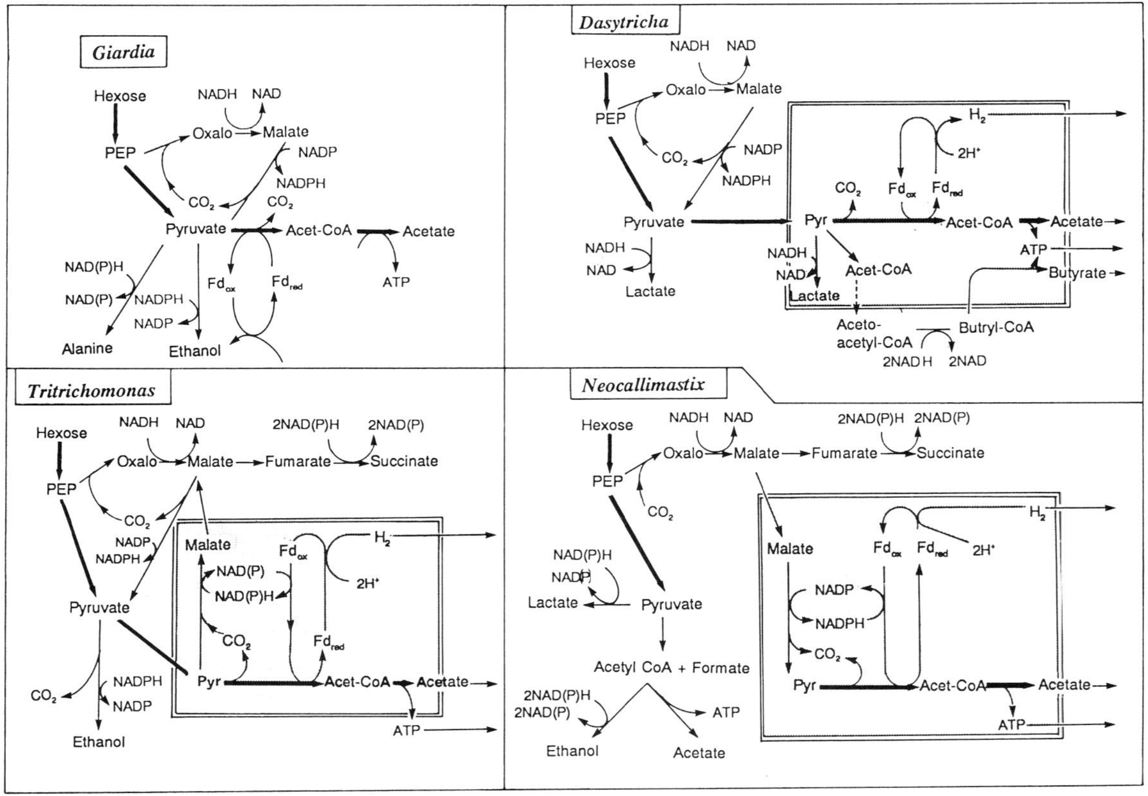

Fig. 1. Metabolic pathways in an archezoan (*Giardia*) with only cytosolic fermentation and in three forms with hydrogenosomes (shaded): a rumen trichostomatid ciliate (*Dasytricha*), a trichomonad flagellate (*Tritrichomonas*) and a rumen chytrid (*Neocallimastix*). PEP: phosphoenolpyruvate; Fd: ferredoxin (from Fenchel & Finlay, 1995).

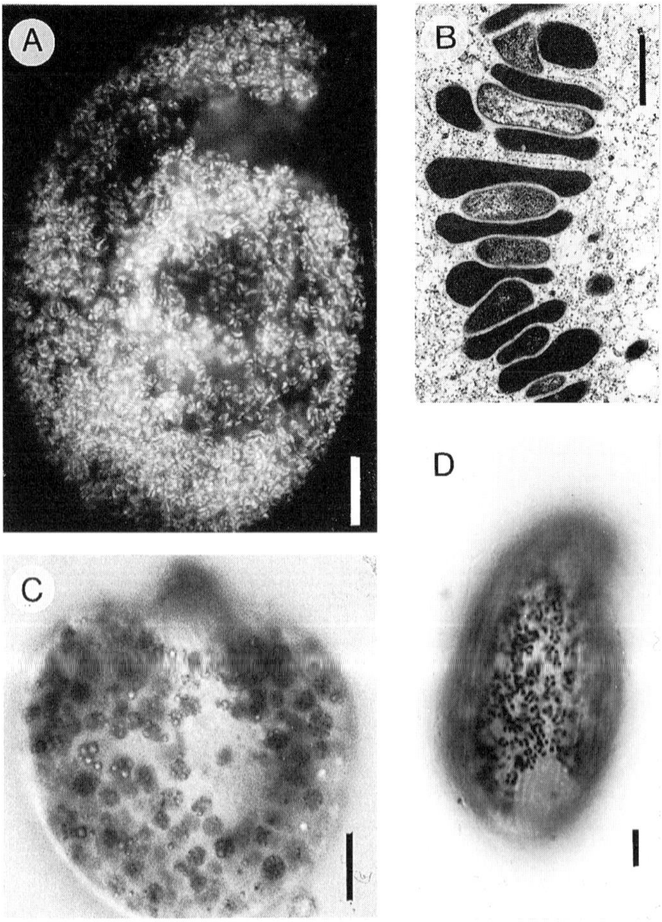

Fig. 2. A: The marine anaerobic ciliate *Plagiopyla frontata* photographed in the fluorescence microscope showing the endosymbiotic methanogenic bacteria. B: TEM photograph of assemblages of hydrogenosomes (black) and methanogenic bacteria in *P. frontata*. C: The ciliate *Strombidium purpureum* with endosymbiotic purple non-sulphur bacteria. D: *Plagiopyla frontata* with stained hydrogenosomes (glutaraldehyde fixation followed by incubation with a tetrazolium salt and H_2-gas). Scale bars: A, C–D: 10 μm; B: 1 μm.

hydrogenase. It is covered by a double membrane and it typically has an electron dense matrix; it has not yet unambiguously been demonstrated that it contains DNA. Subsequently the organelle has been found among several other anaerobic protists including ciliates, rumen chytrids and the peculiar flagellate *Psalteriomonas* (Broers *et al.*, 1990; Fenchel & Finlay, 1991*a*; Williams, 1986; Yarlett *et al.*, 1984, 1986). Hydrogenosomes are easily detected by a cytochemical staining method for hydrogenase (Fig. 2D;Finlay & Fenchel, 1989; Zwart *et al.*, 1988) and by the H_2-production (or CH_4-

production in forms with symbiotic methanogens) of pure cultures of the organisms (e.g. Fenchel & Finlay, 1992; Lloyd *et al.*, 1989).

The trichomonads include one free-living and many commensal species. They and the related hypermastigids and (probably) the oxymonads (which are both known from the hindgut of termites and cockroaches) always have hydrogenosomes and aerobic forms (with normal, cytochrome *c*-oxidase containing mitochondria) are unknown. In contrast, other major groups in which species with hydrogenosomes are found also include ordinary aerobic species. In ciliates hydrogenosomes have evolved independently within several groups (Embley & Finlay, 1994; Fenchel & Finlay, 1995).

The origin of hydrogenosomes has been debated. Müller (1980, 1988, 1993) suggested that the organelle originated as an endosymbiotic *Clostridium* in analogy with the origin of mitochondria as an endosymbiotic respiring bacterium. Conversely, Cavalier-Smith (1987*b*), Fenchel & Finlay (1995) and Finlay & Fenchel (1989) have argued that hydrogenosomes are modified mitochondria. Among the ciliates a few genera (e.g. *Cyclidium* and *Cristigera*) include both aerobic species with mitochondria and anaerobic species with hydrogenosomes (with hydrogenase). In these cases the hydrogenosomes and the mitochondria are nearly morphologically identical and both conform to the mitochondrial characteristics of the particular group of ciliates (Esteban *et al.*, 1993; Fenchel & Finlay, 1991*a*). In most other cases hydrogenosomes occur in all members of particular families or orders of ciliates. The morphology varies; in a few cases the hydrogenosomes have cristae (Fig. 3A).

In attempts to resolve the problem of the origin of hydrogenosomes, detailed structural studies of enzymes of trichomomad hydrogenosomes have been made (Gorrell *et al.*, 1984; Hrdy & Müller, 1995; Johnson *et al.*, 1990; Lahti, D'Oliveira & Johnson, 1992). Generally, these molecules proved more reminiscent of their mitochondrial than of their bacterial counterparts, but the results also show a long independent evolution of trichomonads. Unfortunately, there are no similar studies in the case of ciliates and chytrids. If hydrogenosomes represent modified mitochondria, which is strongly suggested at least in the case of ciliates and chytrids, then it remains an open question how these organelles acquired the enzymes pyruvate:ferredoxin oxidoreductase and hydrogenase.

The biochemistry of hydrogenosomes have especially been studied in detail in trichomonads, rumen ciliates and rumen chytrids and some variations have been demonstrated as recently reviewed by Müller (1993); see also Fig. 1. In some cases malate, rather than pyruvate is the substrate entering the hydrogenosome and in one case the presence of an intrahydrogenosomal lactate dehydrogenase has been demonstrated. Protozoa with hydrogenosomes produce a variety of less oxidized organic compounds (ethanol, lactate, butyrate) in addition to the principle end products: acetate and H_2 (e.g. Goosen *et al.*, 1990). The degree to which pyruvate and NADH

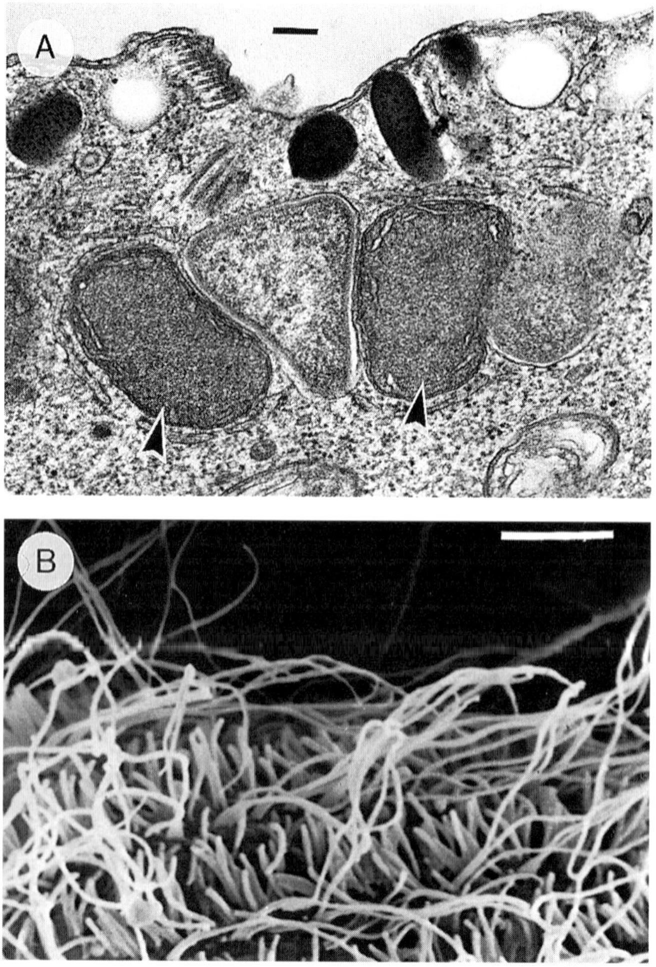

Fig. 3. A: TEM photograph of methanogenic bacteria sandwiched between hydrogenosomes (arrows) in the marine anaerobic ciliate *Metopus contortus*. The double membrane and cristae of the hydrogenosomes are visible (photograph by B. J. Finlay). B: SEM- photograph of ectosymbiotic bacteria on the surface of the anaerobic marine ciliate *Parablepharisma pellitum* (the short rods in between the longer cilia). Scale bars: A: 0.2 μm; B: 4 μm.

can be completely oxidized to acetate through H_2-excretion probably varies according to the intracellular P_{H_2} (see below).

When exposed to microaerobic environments most anaerobic protists seem to be capable of O_2-respiration (which is not coupled to energy conservation, and usually assumed to represent an O_2-detoxification mechanism; see also below) and many anaerobic eukaryotes, especially the commensal and parasitic species, typically live in microaerobic rather than in strictly anaerobic habitats. In species with hydrogenosomes, this organelle seems to be responsible for the O_2-uptake, reducing equivalents being dumped on O_2 rather than being released in the form of H_2 (Lloyd *et al.*, 1982, 1989; Paget & Lloyd, 1990; Yarlett *et al.*, 1983).

Anaerobic eukaryotes with non-hydrogenosomal mitochondria

Some obligate anaerobic protists have ordinary looking mitochondria without hydrogenase activity. This has been demonstrated in some ciliates including members of the genus *Parablepharisma* (which belongs to the most O_2-sensitive anaerobic ciliates; see Fenchel & Finlay, 1990*a*; 1991*a*) and in some flagellates (Table 1; Fenchel *et al.*, 1996). In these cases the mitochondria are likely to play a role in energy metabolism. This would be consistent with the fact that facultative anaerobic ciliates actually increase the volume and sometimes the number of individual mitochondria when grown under anoxic conditions (Bernard & Fenchel, 1996).

No investigation of the exact role of (non-hydrogenosomal) mitochondria exists in the case of obligate anaerobic ciliates. It is likely, however, that the enzymes of the citric acid cycle play a role in energy metabolism. Phosphoenolpyruvate is transformed to oxaloacetate through CO_2-assimilation in the cytosol yielding one mol ATP by substrate phosphorylation. Oxaloacetate then reoxidizes cytosolic NADH and the resulting malate enters the mitochondria and is transformed into pyruvate and fumarate. The pyruvate is then oxidized to acetate (with energy conservation through substrate phosphorylation) and this is coupled to fumarate reduction to produce succinate/propionate (with energy conservation through electron transport phosphorylation) thus maintaining mitochondrial redox balance (Fig. 4). This pathway is known from the ciliate *Tetrahymena*, which is relatively tolerant to exposure to low O_2-tensions (Shrago & Elson, 1980). The pathway is also known from many metazoans which live anaerobically for extended periods such as helminths in the intestinal tract of other animals (see Bryant, 1991). It is thus likely that the pathway also occurs in mitochondriate obligate and facultative anaerobe protists.

Facultative anaerobic protists

In the older literature there are many references to growth of aerobic protists (especially ciliates) under anaerobic conditions. These findings have since been discredited and shown to be the result of insufficient removal of traces of O_2; in fact, most aerobic protozoa are immobilised and killed within minutes when exposed to strict anoxic conditions (for references see Noland & Gojdics, 1967). However, some microaerophilic ciliates are actually capable of sustained survival under strict anoxic conditions and some can grow indefinitely under anaerobic conditions (Bernard & Fenchel, 1996). The phenomenon may also be widespread among other protist

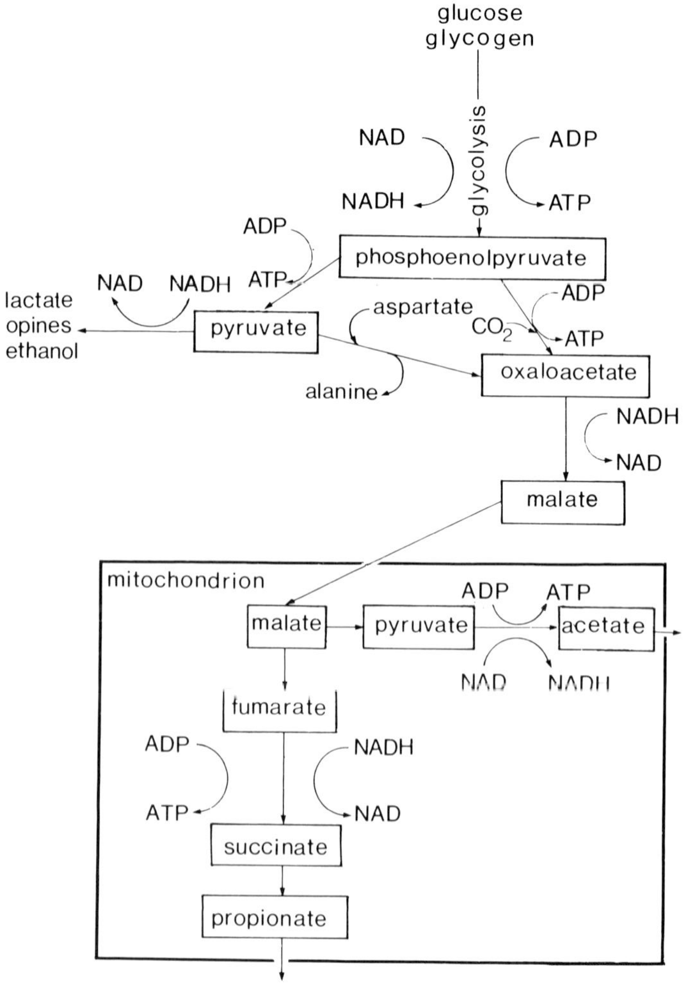

Fig. 4. Cytosolic and mitochondrial anaerobic metabolic pathways occurring in many animals which live for extended periods under anoxia and probably in most or all anaerobic protists with non-hydrogenosomal mitochondria. The mitochondrial processes use parts of enzymes of the citric acid cycle (from Fenchel &Finlay, 1995).

groups such as certain heterotrophic flagellates which are frequently found in and can be isolated from sulphidic water (Table 1; unpublished observations). It is likely that the energy metabolism is based on propionate fermentation, but this is unknown. Certainly, such facultative anaerobes present a model of how obligate anaerobes evolved from aerobes.

A special case is the freshwater ciliate *Loxodes* (and probably its marine relative *Remanella*). These ciliates are microaerophilic and in the dark they prefer a P_{O_2} of approximately 5% atm.sat. In the light, however, *Loxodes* prefers anoxia, which it survives for extended periods, but it is not known whether it will then grow and divide. It has been shown, however, that it is capable of using nitrate as an electron acceptor in a respiratory (mitochondrial) process producing nitrite as the end product (Finlay, 1985; Finlay *et al.*, 1983; Finlay, Fenchel & Gardner, 1986).

Anaerobiosis in animals

Many aquatic animals (metazoa) live in habitats which are periodically or regularly anaerobic or they are periodically forced to depend on an anaerobic metabolism in order to avoid desiccation (e.g. barnacles at low tide). Such forms are capable of survival under anaerobic conditions for shorter or for more extended periods. Those being exposed only for brief periods usually depend solely on the glycolytic pathway in the cytosol and they excrete lactate, ethanol or opines as metabolic end products. Those which endure prolonged exposure to anoxia employ CO_2-fixation and fumarate reduction in the mitochondria as explained above (Fig. 4). It has been, and still is, debated whether genuine anaerobes exist among metazoa. Candidates are especially intestinal parasites (nematodes, cestodes) and some representatives of the free-living meiofauna of sediments; especially nematodes which are often found deep down in sulphidic sediments. It is established that the energy metabolism of such organisms is fermentative throughout most of their life. All metazoa seem to have mitochondria with cytochromes. However, there is evidence to show that in some helminths the terminal respiratory enzyme is cytochrome *o* and that respiration may not be coupled to energy conservation, but only serves for O_2-detoxification (Bryant, 1991). Even so, it has not yet been possible to demonstrate that any metazoan can complete its entire life cycle without access to O_2 during at least a short period.

The barrier against the evolution of anaerobic metazoa (in the sense of complete independence of O_2) seems to be the need of oxygen for some synthetic processes rather than it being a question of energy metabolism. Such processes include the synthesis of steroids, quinone tanning and collagen synthesis. These processes probably explain why animals may irreversibly have lost the possibility of life totally without oxygen. Metazoan adaptations to anoxia is thoroughly reviewed in Bryant (1991).

SYMBIOSIS WITH PROKARYOTES

It has long been observed that many protozoa which were known or subsequently proven to be anaerobes (e.g. the flagellates of the termite hindgut, some rumen and free-living ciliates) harbour symbiotic bacteria. That such associations (including endo- as well as ectosymbionts) are particularly common among anaerobes was suggested by Fenchel, Perry & Thane (1977), and van Bruggen *et al.* (1983, 1985) subsequently showed that most of the endosymbionts are methanogens.

Since then other types of bacterial symbionts of anaerobic protozoa have been identified. The common functional significance of these associations is syntrophy; that is, that the symbionts utilise the host metabolites. Hydrogen transfer from hydrogenosomes is probably the most important aspect (at least in terms of the adaptive significance for the host), but other metabolites such as acetate or propionate are likely to be used by some types of symbionts.

Methanogenic endosymbionts

Methanogens occur only as endosymbionts and (excepting the case of *Pelomyxa*) only in species with hydrogenosomes. Methanogens are readily identified owing to the fluorescence of the coenzyme F_{420} in violet light (Fig. 2A) and by the CH_4- production of the symbiotic consortium (Fenchel & Finlay, 1992). Endosymbiotic methanogens have first of all been studied in free-living ciliates: among some 20 species of hydrogenosome-containing, anaerobic freshwater ciliates all harboured methanogens and among a similar number of marine species about half harboured methanogens (Fenchel & Finlay, 1995). Some rumen ciliates carry methanogens on their surface, but this seems to be an unstable association (Vogels, Hoppe & Stumm, 1980); more recently endosymbiotic methanogens have also been demonstrated in some rumen ciliates (Finlay *et al.*, 1994). Methanogens also occur in trichomonad flagellates from the termite hindgut (Lee *et al.*, 1987) and in the free-living flagellate *Psalteriomonas* (Broers *et al.*, 1990). Rumen chytrids do not seem to harbour methanogens; however their growth is enhanced when in co-culture with methanogens (Bauchop, 1989).

Sequencing of rRNA has shown that in free-living anaerobic ciliates symbiosis with methanogens have evolved independently within several lineages and several unrelated groups of methanogens are represented (Embley & Finlay, 1994). The degree of specialization to the endocellular life varies among methanogens from different host species. Some endosymbiotic methanogens are typical bacterial rods, but in other cases they have lost or reduced their cell wall and take an irregular shape and they are always in close physical contact with the hydrogenosomes (Fig. 3A). In the marine ciliate *Plagiopyla frontata*, the hydrogenosomes and the methanogens are disc-shaped and alternate in assemblages similar to rolls of coins (Fig. 2A). In this species, the division of the hydrogenosomes and of the methanogens is synchronized and takes place during host division; in this way the integrity of these assemblages is maintained after division. The volume fraction of methanogens in the ciliates is always 1–2% (Esteban *et al.*, 1993; Fenchel & Finlay, 1991*c*, 1995; Finlay & Fenchel, 1991).

The methanogens consume virtually all H_2 produced by the host. When methanogens are experimentally inhibited with BES (bromoethane sulphonic acid) host growth rate and yield both decrease by 25–30%; H_2 is then produced instead of CH_4, but at a somewhat lower rate than that predicted from CH_4-production of normal cells with functional methanogens (Fenchel & Finlay, 1991*b*, 1992). The implication is that H_2-transfer from the hydrogenosomes to the methanogens is important in maintaining a low intracellular P_{H_2}; otherwise H_2-production is partly inhibited and more reduced end products (relative to acetate) are produced, thus decreasing the ATP-yield of the fermentative processes. Simple calculations show that in the absence of active methanogens intracellular P_{H_2} (of a spherical ciliate with a diameter of 50 μm) will increase by a factor of about 1000 if H_2-production is assumed to be unaffected (Fenchel & Finlay, 1992).

The methanogen-anaerobic ciliate system is interesting from an evolutionary point in that it shows that this type of symbiosis (in which the endosymbiont is totally dependent on a host metabolite) is always stable (that is, the endosymbionts will always grow at the same rate as the host within a wide range of host growth rates cf Douglas, this volume). In fact, the endosymbionts behave rather like a microbial population growing in a chemostat (where the dilution rate corresponds to the growth and division of the host cells and in which the rate of substrate addition, that is metabolite production, is proportional to the growth of the host cell). Using minor modification of chemostat theory, it is thus possible to describe the symbiotic system and, e.g. to predict the symbiont biomass per host cell (Fenchel & Finlay, 1995; Finlay & Fenchel, 1992).

Other types of prokaryote symbionts

Ectosymbiotic bacteria occur on many marine anaerobic ciliates (Fig. 3B; Fenchel *et al.*, 1977) and in contrast to methanogens they also occur on some species without hydrogenosomes. They are attached to the host in depressions of the cell membrane (often in close contact to hydrogenosomes or mitochondria beneath the cell membrane), in mucous layers covering the ciliate or attached with sheaths. Circumstantial evidence suggests that they are sulphate reducers playing a similar role as the methanogens. One indication is that they occur almost exclusively on marine ciliates (the two only limnic examples are ciliates isolated from a gypsum lake). The symbiont:host biomass ratio often exceeds that of methanogens in other species and this suggests a relatively higher efficiency in energy metabolism which makes sulphate reducers the only likely candidates. In two cases (*Metopus contortus*, *Caenomorpha levanderi*) it has actually been shown that the ectosymbionts are sulphate reducers as determined by specific rRNA-probes (Fenchel & Ramsing, 1992).

There is so far only one example of the third type of endosymbionts: that of a purple non-sulphur bacterium in the marine ciliate *Strombidium purpureum* which lives in mats of purple sulphur bacteria in shallow marine sediments; the symbiont resembles *Rhodopseudomonas* as far as pigments and morphology are concerned (Fig. 2C; Bernard & Fenchel, 1994; Fenchel & Bernard, 1993*a,b*). This ciliate requires anaerobic conditions in the light; it has a photosensory behaviour congregating at wave lengths corresponding to the absorption spectrum of the symbiont (bacteriochlorophyll *a*). In the dark the ciliate seeks microaerobic environments, which is consistent with the respiratory activity of purple non-sulphur bacteria in the absence of light. This symbiosis is particularly interesting in that it offers a model for the origin of mitochondria.

OXYGEN TOXICITY

Oxygen toxicity is a complex subject and has several possible causes. Aerobic organisms have a variety of mechanisms for the protection against the effects of oxygen radicals; in anaerobes and in microaerophiles these mechanisms are partially incomplete, but they are rarely absent altogether (Morris, 1979).

The sensitivity to O_2 varies considerably among anaerobic eukaryotes and this reflects ecology rather than phylogeny. Most anaerobes are at risk to be exposed to O_2 during some part of their life and many anaerobes permanently or periodically inhabit microaerobic rather than strictly anaerobic habitats. In particular commensal and parasitic forms (e.g. *Giardia, Trichomonas*) are typically quite O_2-tolerant and can be grown at a low P_{O_2} (Lloyd *et al.*, 1982; Müller, 1988). These forms are often exposed to aerobic conditions when moving from one host individual to another and many inhabit sites in their hosts which are not strictly anaerobic (e.g. *Trichomonas vaginalis*); even the intestine and the rumen often do not constitute completely anaerobic habitats (Hillman, Lloyd & Williams, 1985).

Free-living anaerobes seem generally to be less tolerant to O_2-exposure although this is also variable; Fenchel & Finlay (1990*a*) found that among some anaerobic ciliates, most tolerated a P_{O_2} up to about 2% atm sat., but one species was more sensitive.

Anaerobic protists have several types of defences against O_2. Among them chemosensory behaviour to O_2 is important: the organisms orient themselves in oxygen gradients and thus congregate under anaerobic conditions (Fenchel & Finlay, 1990*a*). Furthermore, they consume O_2 and as long as uptake is diffusion limited the cells can maintain an anoxic intracellular environment in this way. The K_m for O_2-uptake is 1–2% atm. sat. (Fenchel & Finlay, 1990*a*; Lloyd *et al.*, 1982). This is consistent with the observation that methanogens remained active in the ciliates *Metopus* and *Plagiopyla* up to an environmental P_{O_2} of about 2% atm. sat. (Fenchel & Finlay, 1990*a*, 1992).

All tested anaerobic eukaryotes have superoxide dismutase (SOD), but not catalase or peroxidase (Yarlett *et al.*, 1987). It may be speculated that SOD is more important because O_2^- is a charged molecule and is thus trapped inside the cell whereas H_2O_2 can diffuse out of the cell.

THE ECOLOGICAL IMPLICATIONS OF ANAEROBIC METABOLISM IN PHAGOTROPHS

In anaerobic prokaryote communities the important interspecic interactions are competition (for common resources) and syntrophy (the metabolite of one species constitutes the substrate for another species). Syntrophy is important in anaerobic biota because of the metabolic diversity among anaerobic prokaryotes and because the low energetic efficiency means that most of the substrate is dissimilated rather than assimilated. With the advent of eukaryotes, phagocytosis (predation) was introduced in the world. As long as anaerobic conditions prevailed, however, this way of living had limitations because it depended on the relatively small part of the resources which had been assimilated by prey organisms. Even after the biosphere (or parts of it) had become oxic, but before eukaryotes had acquired mitochondria, the ecological role of phagotrophs would remain modest and they would largely be restricted to occupy a single level in the food chain (as consumers of bacteria).

An aerobic cell generates 32 moles of ATP for every mole of glucose while an anerobic (fermentative) one can generate only 2–4 ATP/glucose. Everything else being equal it can be shown that the growth yield and maximum growth efficiency of an anaerobic protist should be only about 25% of that of an equally sized aerobic form and this is largely the case (Fenchel & Finlay, 1990*b*). As a consequence phagotrophic food chains must be short in anaerobic ecological systems because a relatively large fraction of organic matter is dissimilated (lost) at each trophic level. In fact phagotrophic food chains of extant anaerobic habitats are short, usually represented only by a single level (as bacterivores); a very few carnivorous forms (eating other protists) do occur in some systems, but a two-step food chain seems to be the maximum realized (Fenchel & Finlay, 1995). In contrast, over evolutionary time aerobic eukaryotes diversified and grew to large forms largely as an adaptation to be able to eat larger prey or to avoid being eaten themselves. This evolution resulted in ecological systems with complex food webs and predation became a dominating type of ecological interaction. If the eukaryotes had not acquired mitochondria and oxidative phosphorylation they would have remained as a marginal group of bacterivorous microorganisms.

It is likely that the first (phagotrophic) anaerobic eukaryotes (perhaps looking somewhat like extant Archezoa) arose very early in the history of life. But prior to their acquisition of mitochondria and oxidative phosphorylation they were relegated to a marginal role in prokaryote-dominated ecological systems.

REFERENCES

Adam, R. D. (1991). The biology of *Giardia* spp. *Microbiological Reviews*, **55**, 706–32.

Bauchop, T. (1989). Biology of gut anaerobic fungi. *BioSystems*, **23**, 53–64.

Bernard, C. & Fenchel, T. (1994). Chemosensory behaviour of *Strombidium purpureum*, an anaerobic oligotrich with endosymbiotic purple non-sulphur bacteria. *Journal of Eukaryotic Microbiology*, **41**, 391–6.

Bernard, C. & Fenchel, T. (1996). Some microaerophilic ciliates are facultative anaerobes. *European Journal of Protistology*, **32** (in press).

Broers, C. A. M., Stumm, C. K., Vogels, G. D. & Brugerolle, G. (1990). *Psalteriomonas lanterna* gen. nov., sp. nov., a free-living amoeboflagellate isolated from freshwater anaerobic sediments. *European Journal of Protistology*, **25**, 369–80.

Bryant, C. (1991). *Metazoan Life without Oxygen*. London, Chapman and Hall.

Cavalier-Smith, T. (1987*a*). The origin of eukaryote and archebacterial cells. *Annals of the New York Academy of Sciences*, **503**, 17–54.

Cavalier-Smith, T. (1987*b*). The simultaneous symbiotic origin of mitochondria, chloroplasts and microbodies. *Annals of the New York Academy of Sciences*, **503**, 55–71.

Cohen, Y., Jørgensen, B. B., Revsbech, N. P. & Poplawski, R. (1986). Adaptation to hydrogen sulfide of oxygenic and anoxygenic photosynthesis among cyanobacteria. *Applied and Environmental Microbiology*, **51**, 398–407.

Ellis, J. E., Williams, R., Cole, D., Cammack, R. & Lloyd, D. (1993). Electron transport components of the parasitic protozoon *Giardia lamblia*. *FEMS Letters*, **325**, 196–200.

Embley, T. M. & Finlay, B. J. (1994). The use of small subunit rRNA sequences to unravel the relationships between anaerobic ciliates and their methanogenic endosymbionts. *Microbiology*, **140**, 225–35.

Esteban, G., Guhl, B. E., Clarke, K. J., Embley, T. M. & Finlay, B. J. (1993). *Cyclidium porcatum* n.sp.: a free-living anaerobic scuticociliate containing a stable complex of hydrogenosomes, eubacteria and archaeobacteria. *European Journal of Protistology*, **29**, 262–70.

Fenchel, T. & Bernard, C. (1993*a*). A purple protist. *Nature*, **362**, 300.

Fenchel, T. & Bernard, C. (1993*b*). Endosymbiotic purple non-sulphur bacteria in an anaerobic ciliated protozoon. *FEMS Microbiology Letters*, **110**, 21–5.

Fenchel, T. & Finlay, B. J. (1990*a*). Oxygen toxicity, respiration and behavioural responses to oxygen in free-living anaerobic ciliates. *Journal of General Microbiology*, **136**, 1953–9.

Fenchel, T. & Finlay, B. J. (1990*b*). Anaerobic free-living protozoa: growth efficiencies and the structure of anaerobic communities. *FEMS Microbiology Ecology*, **74**, 269–76.

Fenchel, T. & Finlay, B. J. (1991*a*). The biology of free-living anaerobic ciliates. *European Journal of Protistology*, **26**, 201–15.

Fenchel, T. & Finlay, B. J. (1991*b*). Endosymbiotic methanogenic bacteria in anaerobic ciliates: significance for the growth efficiency of the host. *Journal of Protozoology*, **38**, 18–22.

Fenchel, T. & Finlay, B. J. (1991*c*). Synchronous division of an endosymbiotic methanogenic bacterium in the anaerobic ciliate *Plagiopyla frontata* Kahl. *Journal of Protozoology*, **38**, 22–8.

Fenchel, T. & Finlay, B. J. (1992). Production of methane and hydrogen by anaerobic ciliates containing symbiotic methanogens. *Archives of Microbiology*, **157**, 475–80.

Fenchel, T. & Finlay, B. J. (1995). *Ecology and Evolution in Anoxic Worlds*. Oxford, Oxford University Press.

Fenchel, T. & Ramsing, N. B. (1992). Identification of sulphate reducing bacteria from anaerobic ciliates using 16S rRNA binding oligonucleotide probes. *Archives of Microbiology*, **158**, 394–7.

Fenchel, T., Perry, T. & Thane, A. (1977). Anaerobiosis and symbiosis with bacteria in free-living ciliates. *Journal of Protozoology*, **24**, 154–63.

Fenchel, T., Bernard, C., Esteban, G., Finlay, B. J., Hansen, P. J. & Iversen, N. (1995). Microbial diversity and activity in a Danish fjord with anoxic deep water. *Ophelia*, (in press).

Finlay, B. J. (1985). Nitrate respiration by Protozoa (*Loxodes* spp.) in the hypolimnetic nitrite maximum of a productive freshwater pond. *Freshwater Biology*, **15**, 333–46.

Finlay, B. J. & Fenchel, T. (1989). Hydrogenosomes in some anaerobic protozoa resemble mitochondria. *FEMS Microbiology Letters*, **65**, 311–14.

Finlay, B. J. & Fenchel, T. (1991). Polymorphic bacterial symbionts in the anaerobic ciliated protozoon *Metopus*. *FEMS Microbiology Letters*, **70**, 187–90.

Finlay, B. J. & Fenchel, T. (1992). An anaerobic ciliate as a natural chemostat for the growth of endosymbiotic methanogens. *European Journal of Protistology*, **28**, 127–37.

Finlay, B. J., Span, A. S. W. & Harman, J. M. P. (1983). Nitrate respiration in primitive eukaryotes. *Nature*, **303**, 333–6.

Finlay, B. J., Fenchel, T. & Gardner, S. (1986). Oxygen perception and O_2 toxicity in the freshwater ciliated protozoon *Loxodes*. *Journal of Protozoology*, **33**, 157–65.

Finlay, B. J., Esteban, G., Clarke, K. J., Williams, A. G., Embley, T. M. & Hirt, R. P. (1994). Some rumen ciliates have endosymbiotic methanogens. *FEMS Microbiology Letters*, **117**, 157–62.

Goosen, N. K., Van Der Drift, C., Stumm, C. K. & Vogels, G. D. (1990). End products of metabolism in the anaerobic ciliate *Trimyema compressum*. *FEMS Microbiology Letters*, **69**, 171–6.

Gorrell, T. E., Yarlett, N. & Müller, M. (1984). Isolation and characterization of *Trichomonas vaginalis* ferredoxin. *Carlsberg Research Communications*, **446**, 259–68.

Hillman, K., Lloyd, D. & Williams, A. G. (1985). Use of a portable spectrometer for the measurement of dissolved gas concentration in bovine liquid *in situ*. *Current Microbiology*, **12**, 335–40.

Hrdy, I. & Müller, M. (1995). Primary structure of the hydrogenosomal malic enzyme of *Trichomonas vaginalis* and its relationship to homologous enzymes. *Journal of Eukaryotic Microbiology* (in press).

Johnson, P. J., D'Oliveira, C. E., Gorell, T. E. & Müller, M. (1990). Molecular analysis of the hydrogenosomal ferredoxin of the anaerobic protist *Trichomonas vaginalis*. *Proceedings of the National Academy of Sciences, USA*, **87**, 6097–101.

Lahti, C. J., D'Oliveira, C. E. & Johnson, P. J. (1992). β-succinyl-coenzyme A synthetase from *Trichomonas vaginalis* in a soluble hydrogenosomal protein with an amino-terminal sequence that resembles mitochondrial presequences. *Journal of Bacteriology*, **174**, 6822–30.

Lee, M. J., Schreurs, P. J., Messer, A. C. & Zinder, S. H. (1987). Association of methanogenic bacteria with flagellated protozoa from a termite hindgut. *Current Microbiology*, **15**, 337–41.

Lloyd, D., Williams, J., Yarlett, N. & Williams, A. G. (1982). Oxygen affinities of the hydrogenosome-containing protozoa *Tritrichomonas foetus* and *Dasytricha ruminantium*, and two aerobic protozoa, determined by bacterial bioluminescence. *Journal of General Microbiology*, **128**, 1019–22.

Lloyd, D., Hillman, K., Yarlett, N. & Williams, A. G. (1989). Hydrogen production by rumen holotrich protozoa: effects of oxygen and implications for metabolic control by *in situ* conditions. *Journal of Protozoology*, **36**, 205–13.

Morris, J. G. (1979). Nature of oxygen toxicity in anaerobic microorganisms. In *Strategies of Microbial Life in Extreme Environments*, Life Science Research Report 13, Shilo, M., ed., pp. 149–162, Berlin, Dahlem Konferenzen.

Müller, M. (1980). The hydrogenosmome. In *The Eukaryotic Microbial Cell*, Gooday, G. W., Lloyd, D. & Trinci, A. P. J., eds., pp. 127–142. Cambridge, Cambridge University Press.

Müller, M. (1988). Energy metabolism of Protozoa without mitochondria. *Annual Review of Microbiology*, **42**, 465–88.

Müller, M. (1993). The hydrogenosome. *Journal of General Microbiology*, **139**, 2879–89.

Mylnikov, A. P. (1991). Diversity of flagellates without mitochondria. In *The Biology of Free-living Flagellates*, Patterson, D. J. & Larsen, J., eds. pp. 149–158. Oxford, Oxford University Press.

Noland, L. E. & Gojdics, M. (1967). Ecology of free-living protozoa. In *Research in Protozoology*, Chen, T. T., ed. vol. 2, pp. 215–266. Oxford, Pergamon Press.

Paget, T. A. & Lloyd, D. (1990). *Trichomonas vaginalis* requires traces of oxygen and high concentrations of carbon dioxide for optimal growth. *Molecular Biochemistry and Parasitology*, **41**, 65–72.

Reeves, R. E. (1984). Metabolism of *Entamoeba histolytica* Schaudinn 1903. *Advances in Parasitology*, **23**, 105–42.

Schlegel, M. (1991). Protist evolution and phylogeny as discerned from small subunit ribosomal RNA sequence comparisons. *European Journal of Protistology*, **27**, 207–19.

Schlegel, M. (1994). Molecular phylogeny of eukaryotes. *Trends in Ecology and Evolution*, **9**, 330–5.

Shrago, E. & Elson, C. (1980). Intermediary metabolism of *Tetrahymena*. In *Biochemistry and Physiology of Protozoa* (2nd ed.), Levandowsky, M. & Hutner, S. H., eds. vol. 3, pp. 287–312. New York, Academic Press.

Van Bruggen, J. J. A., Stumm, C. K. & Vogels, G. D. (1983). Symbiosis of methanogenic bacteria and sapropelic protozoa. *Archives of Microbiology*, **136**, 89–96.

Van Bruggen, J. J. A., Stumm, C. K., Zwart, K. B. & Vogels, G. D. (1985). Endosymbiotic methanogenic bacteria of the sapropelic amoebae *Mastigella*. *FEMS Microbiology Ecology*, **31**, 187–92.

Vogels, G. D., Hoppe, W. & Stumm, C. K. (1980). Association of methanogenic bacteria with rumen ciliates. *Applied and Environmental Microbiology*, **40**, 608–12.

Whatley, J. M. & Chapman-Andresen, C. (1990). Phylum Karyoblastea. In *Handbook of Protoctista*, Margulis, L., Corliss, J. O., Melkonian, M. & Chapman, D. J., eds. pp. 167–185. Boston, Jones and Bartlett Publishers.

Williams, A. G. (1986). Rumen holotrich ciliate protozoa. *Microbiological Reviews*, **50**, 25–49.

Williams, A. G. & Lloyd, D. (1993). Biological activities of symbiotic and parasitic protists in low oxygen environments. *Advances in Microbial Ecology*, **13**, 211–62.

Yarlett, N., Coleman, G. S., Williams, A. G. & Lloyd, D. (1984). Hydrogenosomes in known species of rumen entodiniomorphid protozoa. *FEMS Microbiology Letters*, **21**, 15–19.

Yarlett, N., Scott, R. I., Williams, A. G. & Lloyd, D. (1983). A note on the effect of oxygen on hydrogen production by the rumen protozoon *Dasytricha ruminantium* Shuberg. *Journal of Applied Bacteriology*, **55**, 359–61.

Yarlett, N., Orpin, C. G., Munn, E. A., Yarlett, N. C. & Greenwood, C. A. (1986). Hydrogenosomes in the rumen fungus *Neocallimastix patriciarum. Biochemical Journal*, **236**, 729–39.

Yarlett, N., Rowlands, C., Yarlett, N. C., Evans, J. C. & Lloyd, D. (1987). Respiration of the hydrogenosome-containing fungus *Neocallimastix patriciarum. Archives of Microbiology*, **148**, 25–8.

Zwart, K. B., Goosen, N. K., Schijndel, M. W., Broers, C. A. M., Stumm, C. K. & Vogels, G. D. (1988). Cytochemical localization of hydrogenase activity in the anaerobic protozoa *Trichomonas vaginalis, Plagiopyla nasuta* and *Trimyema compressum. Journal of General Microbiology*, **134**, 2165–70.

EVOLUTIONARY TRENDS IN THE FUNGI

ROYALL T. MOORE

School of Applied Biological and Chemical Sciences
University of Ulster at Coleraine
Northern Ireland BT52 1SA, UK

INTRODUCTION

Omnis vita in partes tres divisa est, to paraphrase Caesar: viruses, prokaryotes, eukaryotes; the last divided into plants, animals, and fungi. The defining characters of these six crown taxa are outlined in Table 1. The eukaryote kingdoms comprise Ecology's three-fold way of producers, consumers and breakdown organisms (Zuck, 1953). This was not always so. The massive carbon deposits of the Carboniferous indicate that the environment during this geological Period consisted, to all intents and purposes, of only producers and consumers. Corner (1964, p. 251) noted that:

> The fungi of today are those of coniferous and flowering forest that adapt themselves to other kinds of vegetation. Ferns and their like do not decay rapidly. The thickness of the Coal Measures, composed of the remains of fern-like plants, indicates that even at this stage of the world's history there were not many organisms able to remove their detritus. Sphagnum bogs are another example. Conifers, too, with resinous tissue are not so easily decayed as flowering plants. It seems indeed that it could not have been until their advent that the fungus world expanded. Flowering plants employed the animal and propagated the fungus. The flowering forest is both

Table 1. *Defining characters of the crown taxa*

Superkingdom Viridae
Membrane systems absent; either RNA or DNA present; enzymes absent or one to a few; lipids and polysaccharides usually absent; reproduce using products from host genetic material; independent synthesis of subsequently assembled parts. No free living forms
Superkingdom Prokaryotae
Nuclear material contiguous with the cytoplasm; lacking mitochondria, Golgi apparatus, chloroplasts, endoplasmic reticulum, and centriolar structures
Superkingdom Eukaryotae
Nuclear material bounded by a double membrane envelope; mitochondria always present; Golgi apparatus, chloroplasts, endoplasmic reticulum, and centriolar structures variously present. Divided into three kingdoms on the basis of nutrition:
Kingdom Plantae Autotrophic or closely related to autotrophs
Kingdom Animalia Phagotrophic heterotrophs or closely related to phagotrophs
Kingdom Fungi All lysotrophic heterotrophs

eminently edible and putrescible, wherein lies the other half of its success. Its remains do not accumulate. They are rotted down by fungus, animal, and bacterium and reincorporated in the soil.

Before proceeding further, a cautionary word about the terminology of evolutionary relationships is needed. The correct usage of homology comes from Darwin's theory of common descent (Mayr, 1982) and is the phenomenon of having a common historical origin but not necessarily the same final structure or function (for example, forelegs and wings of vertebrates); the evidence is usually fossil but molecular clocks are now being employed to calculate ancestral branching points (Bruns, White & Taylor, 1991; Berbee & Taylor, 1995; Paquin *et al.*, 1995). Paramology (Moore, 1971) applies to inferred relationships in evolutionary schemes based on contemporary forms, such as the various phylogenetic representations in prokaryotes, algae, and fungi and to those entailing molecular biology. Analogy is generally applied to similar forms that are unrelated (for example, wings of vertebrates and insects; flagella of prokaryotes and eukaryotes).

BROAD EVOLUTIONARY TRENDS IN THE FUNGI

Unlike plants and animals, which have differentiated and mostly determinant tissue systems, the somatic structure of filamentous fungi is composed of elongating hyphae (Bartnicki-Garcia, 1987; Heath, 1995) that advance out into the substrate forming a mycelium (Andrews, 1995). A mycelium is a fungal individual (Todd & Rayner, 1980; Dahlberg & Stenlid, 1995) and differs only in scale whether the hyphae are radially enlarging in a Petri dish or in nature to form a fairy ring; in fact, the world's largest organism is a mycelium of *Armillaria bulbosa* (Barla) Kile & Watling that occupies a minimum of 15 hectares, weighs in excess of 10 000 kg, and has remained genetically stable for more than 1500 years (Smith, Bruhn & Anderson, 1992). Fruit bodies and other structures are composed of organized hyphae compacted into pseudotissues.

The Fungi are characterized by: (i) a diversity of microbodies (Veenhuis & Harder, 1988; Carson & Cooney, 1990); (ii) cell walls that have a great similarity of architecture (Farkaŝ, 1990); (iii) hyphae that have a major chitin component (Girbart, 1969; Grove & Bracker, 1970; Bartnicki-Garcia & Lippman, 1989; Gooday & Schofield, 1995), extend apically (Wessels, 1986, 1990; Bartnicki-Garcia, Hergert & Gierz, 1989; Kotov & Reshetnikov, 1990; Money, 1995; Sietsma, Wösten & Wessels, 1995), and divide by centripetal invagination of the plasma membrane (Girbart, 1979; Hoch & Howard, 1980; Moore 1985*a*; Orlovich & Ashford, 1994); (iv) lomasomes: sponge-like intumescences seen on the inside of the cell wall (Moore & McAlear, 1961; Marchant & Moore, 1973; Barrett, 1986); (v) a complete absence of the Golgi organelle in the terrestrial assemblages (zygomycetes, ascomycetes, and basidiomycetes) and some of the aquatic taxa (see Moore,

1989); and (vi) nuclei in which most, if not all, gene products involved in mitosis probably have higher eukaryotic paramologues (Morris, 1990)) but which, in other ways, are exceptional. Fungi also have unique mycoviruses that are typically isometric particles 25–35 nm in diameter containing dsRNA genomes and which have no known natural vectors (Ghabrial, 1994).

The simpler nuclei of fungi have a number of characters that may be either primitive (for example, a nucleoplasm which appears bipartite in permanganate preparations and the mode of post-division separation of daughter nuclei (Bandoni, Bisalputra & Bisalputra, 1967; Moore, 1965, 1989; Moore *et al.*, 1991)) or reduced and advanced (for example, the centriole-like spindle pole body). In general, fungal nuclei are haploid, small (1 μm or so in diameter; Fig. 1), have envelope pore complexes similar to those of other eukaryotes (Allen & Douglas, 1990; Nehrbass *et al.*, 1990), and amounts of chromosomal DNA intermediate between that of bacteria and plants and animals (Sparrow, Price & Underbrink, 1972; Durán & Gray, 1989). The microtubular organizing centres (MTOC's; Oakley, 1995), chromosomes, and modes of nuclear division are unusual. There is a single spindle pole body (SPB; Fig. 2) intimately associated with the nuclear envelope (Heath, 1981; O'Donnell & McLaughlin, 1984; Rout & Kilmartin, 1990; O'Donnell, 1992) that divides during nuclear division (Lu, 1978). Mitotic chromosomes are, with rare exception, indiscernible by light microscopy, do not become organized into a metaphase plate (King & Hyams, 1982; Burns, 1988) and, at most, condense only slightly (Erhard, Barker & Green, 1988), perhaps because of the absence of H1 histones (Carter, 1978). The nuclear envelope persists during division (Wells, 1977), both in mitosis (Boekhout & Linnemans, 1982; King & Hyams, 1982; Heath *et al.*, 1982; Bourett & McLaughlin, 1986; Tanaka & Kanbe, 1986) and in meiosis (Howard & Moore, 1970; O'Donnell & McLaughlin, 1984). During division the SPB divides and its daughters generate spindle and, in mitosis, astral microtubules (mts) as they migrate around the nuclear envelope to opposing polar positions (Heath, 1981; O'Donnell, 1992). During the cell cycle, tubulin and actin undergo structural rearrangement (Kilmartin & Adams, 1984). The final parting of the isthmus between separating mitotic or meiotic daughter nuclei, karyochorisis (Moore, 1964, 1965), is effected by sequential invagination of the two envelope membranes (Moore, 1965; Motta, 1969; Egashira, Tokunaga & Tokunaga, 1972; Setliff, Hoch & Patton, 1974) similar to mitochondriokinesis and perhaps chloroplast division.

The defining characters of the major fungal phyla (Table 2) can be plotted in a progressive series of paramologies (Moore, 1989; Table 3). The oomycetes (Beakes, 1987; Dick, 1995) are by way of being algal-fungi and form a logical transition taxon between the two groups (Moore, 1971; Gunderson *et al.*, 1987; Beakes, 1989). Oomycetes have cellulose cell walls, biflagellate spores (Kole, 1965) with a leading tinsel-type flagellum and a

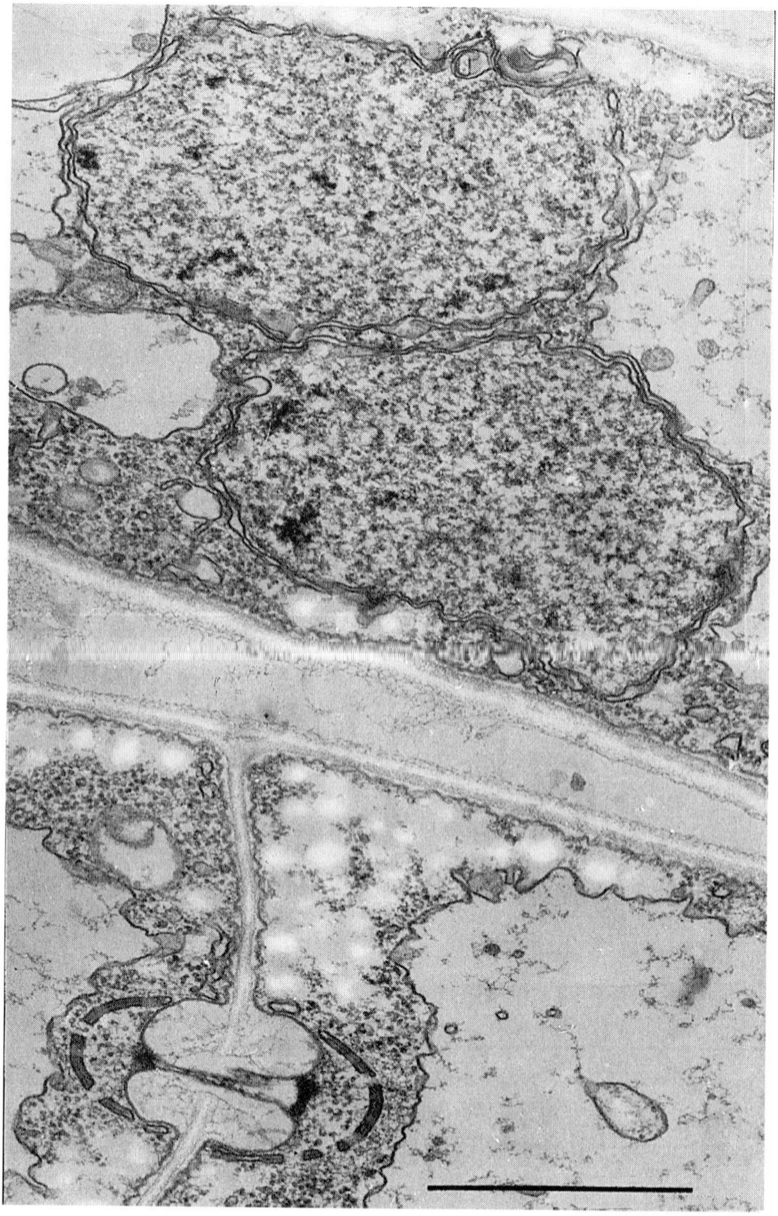

Fig. 1. Part of two dikaryotic hyphae of *Polyporus biennis* (Bull.:Fr.) Fries (from Moore, 1989). The top hypha shows the intimate association of the contrasting nuclei while the bottom hypha shows a typical dolipore with perforate parenthesomes (O_1/P_1 type septum). Scale line = 1.0 μm.

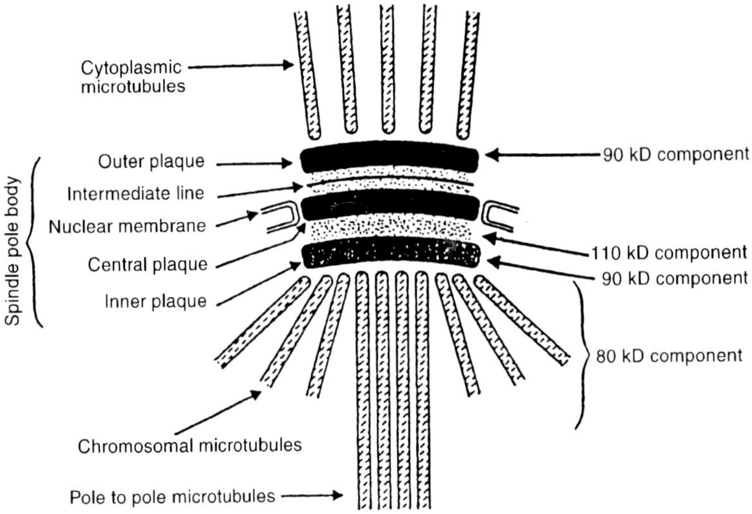

Fig. 2. Diagram of the spindle pole body (SPB). (From Rout & Kilmartin, 1990.)

trailing whip-lash flagellum, the nearly universal diaminopimelic acid lysine pathway, a typical pair of centrioles, coenocytic hyphae with occasional abscissional septa, and nuclei that are diploid (Howard & Moore, 1970). The few species of hyphochytriomycetes have a leading tinsel-type flagellum like that of the oomycetes (Olson & Fuller, 1968; Barr & Désaulniers, 1989) and are more fungal-like in having cell walls with a mixture of cellulose and chitin. The chytridiomycetes have a trailing whip-lash flagellum like that of the oomycetes (Olson & Fuller, 1968), cell walls in which chitin has replaced cellulose as the major structural component (Gooday & Schofield, 1995), the absence of the Golgi organelle in the Blastocladiales (Moore, 1968; 1989), and a shift to the aminoadipic acid lysine pathway characteristic of the rest of the fungi. The zygomycetes are the start of the terrestrial habit and are the first fungi in which the pair of centrioles, characteristic of aquatic species, is replaced by a single spindle pole body (Fig. 2). Ascomycetes and basidiomycetes are characterized by regularly septate hyphae, haplo-dikaryotic life cycles (Fig. 3), and fruiting bodies. Ascomycete septa have large pores that are guarded by Woronin bodies (Buller, 1933; Markham & Collinge, 1987) and are large enough to allow the passage of nuclei; consequently, the cells of these hyphae are functionally coenocytic (and not, in this respect, appreciatively different from the hyphae of the preceding taxa). The limited dikaryotic sporothallus is dependent ('parasitic') on the gametothallus which forms the containing fruiting body; this life cycle is analogous to that of bryophytes. Ascocarps show a general progression (Berbee & Taylor, 1995) from completely enclosed cleistothecia with scattered asci (plectomycetes), to apically open perithecia (Berbee &

Table 2. *Major taxa of the kingdom* Fungi and the subkingdom Basidiomycetia (Moore, 1994; 1996d)*

Subkingdom Mastigomycetia		
Phylum Oomycota (guest taxon)		
Phylum Hyphochytriomycota (guest taxon)		
Phylum Chytridiomycota		
Subkingdom Zygomycetia		
Phylum Zygomycota		
Phylum Trichomycota		
Subkingdom Ascomycetia		
Phylum Ascomycota		
Phylum Hemiascomycota		
Subkingdom Basidiomycetia		
Phylum Basidiomycota		
Class Hymenomycetes		
Order Agaricales	Order Coltriciales	Order Polyporales
Order Amanitales	Order Hericiales	Order Russulales
Order Boletales	Order Meruliales	Order Sistotrematales
Order Cantharellales	Order Phaeolales	Order Stereales
Order Ceratobasidiales	Order Pluteales	Order Thelephorales
Class Clavariomycetes		
Order Botryobasidiales	Order Dacrymycetales	Order Hyphodontiales
Order Clavariales	Order Hirschioporales	Order Lentinales
Order Clavuliciales	Order Hymenochaetales	Order Tulasnellales
Class Gasteromycetes		
Class Exidiomycetes		
Order Exidiales	Order Tremellales	Order Filobasidiales
Class Auriculariomycetes		
Order Auriculariales	Order Eocronartiales	Order Uredinales
Phylum Ustomycota		
Order Agaricostilbales	Order Graphiolales	Order Sporidiales
Order Atractiellales	Order Platygloeoales	Order Tilletiales
Order Exobasidiales	Order Septobasidiales	Order Ustilaginales
Order Heterogastridiales		
Subkingdom Deuteromycetia		
Phylum Deuteromycota		
Phylum Blastomycota		

*This is a practical, nomenclatural, outline designed to present a working view of the major types of fungi. The plan is broadly progressive but, because many of these taxa are undoubtedly polyphyletic, the scheme does not necessarily imply natural relationships. The evidence of molecular systematics (Gunderson *et al.*, 1987; Bruns *et al.*, 1991; Paquin *et al.*, 1995) indicates that all of the taxa except Oomycota and Hyphochytriomycota form a natural group.

Taylor, 1992) with asci arising from a basal hymenium (pyrenomycetes), to completely open apothecia of discomycetes that freely release quantities of airborne spores; the completely exposed hymenia characteristic of cup fungi and lichens are usually coloured due to pigments in the closely packed tips of paraphyses, gametothallic hyphae that separate and protect the asci. The advent of large quantities of freely liberated ascospores is accompanied by a marked decline in dispersal by asexual mitospores (sporangiospores and conidia) characteristic of all previous taxa. The advancements in this

Table 3. *Evolutionary trends in the fungi*

PHYLUM	OOMYCOTA	HYPHOCHYTRIOMYCOTA	CHYTRIDIOMYCOTA	ZYGOMYCOTA	ASCOMYCOTA	BASIDIOMYCOTA
Ploidy	2n --------------------> ln	---	2n/\ln ------------------------	----------------------------	----------------	>(n + n)
Flagella	2 ----------------------> 1	---> 1	----------------------------> 0			
Walls	Cellulose --------------	-------------------------------> chitin				
Lysine	Diaminopimelic acid -	-->	Aminoadipic acid			
Golgi	Dictyosomes (+) ---	--	----------------> (−)			
Habit	Aquatic -----------------	--	----------------------------->	Terrestrial		
Centrioles	2 ------------------------	--	----------------------------->	Single spindle pole body		
Septa	Absicissional ---------	--	--------------------------------	------------------------>	Perforate ------------>	Dolipore/ parenthesome
Meiospores	Endogenous -----------	--	--------------------------------	---------------------------	------------------------->	Exogenous
Life cycle	|< diplobiontic >|	|< ? >|	|< diplohaplobiontic >|	|< haplobiontic >|	|< haplodikaryotic	>|

In the Hyphochytriomycota the cell walls contain a mixture of cellulose and chitin. In the Chytridiomycota the Gogi dictyosome occurs in the orders Chytridiales, Harpochytridiales, and, possibly, the Monoblepharidales, but is absent from the Blastocladiales. In the Ascomycota the dikaryotic sporothallus is dependent on the gametothallus. Modified from Moore (1989).

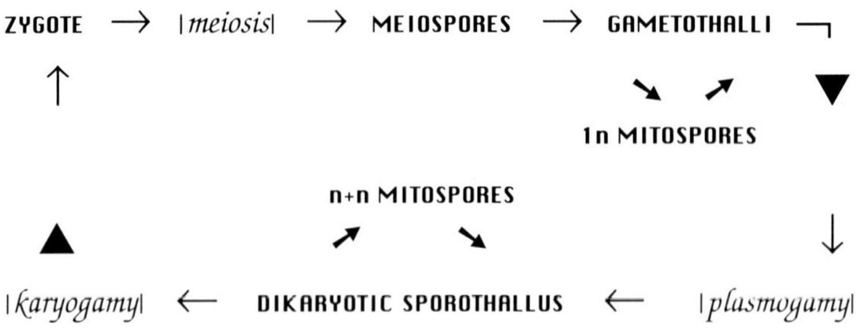

Fig. 3. Haplo-dikaryotic life cycle characteristic of ascomycetes and basidiomycetes.

terminal group of ascomycetes thus include the coincidence of large fruit-bodies, often big enough to be called 'mushrooms', with the accompanying shift to propagation mostly by meiospores with all the genetic advantages they confer. In the basidiomycetes, the dikaryotic sporothallus becomes autonomous and the cell walls and septa become lamellate (appearing light/dark/light in the electron microscope).

The paramology that links the ascomycetes to the basidiomycetes is the ontogenetic similarities of the ascus forming crozier system and the development of clamp connections on dikaryotic hyphae; the uninucleate ultimate cell with the uninucleate clamp cell, the binucleate pre-ascus penultimate cell with the binucleate hyphal tip, and the uninucleate antepenultimate cell with the uninucleate daughter hyphal cell; in both systems the uninucleate cells fuse to reestablish a dikaryotic cell with contrasting nuclei.

Only brief mention can be made of three recurring polyphyletic themes: (i) the symbiotic associations of fungi with plants (Lewis, 1987) that is mycorrhizae (Harley & Smith, 1983) and lichens (Ahmadjian & Hale, 1973), (ii) the aggregate of yeasts and yeastlike fungi (Kurtzman & Fell, 1996), and (iii) parasitism. The mycorrhizal fungus/root association ranges from broadly permissive (many fungal species have a broad host range and many plants have a diversity of mycobionts) to more or less host specific (Cullings, Szaro & Bruns, 1996). These associations afford the plant access to soil nutrients, especially phosphorus, particularly in impoverished soils, and allow the fungus to obtain carbon compounds directly. Vesicular arbuscular (VA) endomycorhizae are formed by zygomycetes (Glomales) that have evolved obligate associations with a majority of bryophytes and vascular plants (Morton, Bentivenga & Bever, 1995), including a number of important cultivated crops (Ruhle & Marx, 1979). Ectomycorrhizae are characteristic of forest trees of the north temperate zone and are formed by a number of species of Hymenomycetes and Gasteromycetes (Ruhle & Marx, 1979). Other mycorrhizal associations are formed, respectively, with the orders Ericales and Orchidales (Harley & Smith, 1983; Perotto *et al.*, 1995).

Lichens are a symbiotic association between fungi (commonly discomycetes, but also pyrenomycetes and basidiomycetes) and either unicelluar algae (usually Chlorophyta) or cyanbacteria (Stocker-Wörgötter, 1995). The fungus provides the overall structure, including the naming fruitbody; in most species the photosynthetic cells are held in a subepidermal layer. In addition to generally benefiting both partners, the binary association also produces a number of secondary products that are not known to be formed by the respective bionts (Lawrey, 1995). This type of parabiotic relationship has arisen several times. Gargas *et al.* (1995) used small subunit ribosomal DNA to determine the relationship of 10 mycobionts to 65 other fungi and found there were at least five independent origins of the lichen habit: two distinct groups of ascomycetes (comprising seven species) and three of basidiomycetes (a single species each).

Evolutionary reduction in both ascomycetes and basidiomycetes has produced a diverse assortment of yeasts and yeast-like fungi that are probably variously derived from the mainline forms (de Hoog, Smith & Weijman, 1988; Kurtzman & Fell, 1996). The dichotomy between the two species sets is marked by distinct morphological and molecular biological characters (Moore, 1988*a*, 1996*a*). A number of these fungi are dimorphic (San-Blas & San-Blas, 1984) and can shift between a budding yeast phase (Mischke & Chant, 1995) and a mycelial sexual phase (Y $\leftrightarrow$ M). One or the other phase is often parasitic: in human pathogens such as *Blastomyces* and *Histoplasma*, the Y-phase requires the elevated temperature of the host (Kwon-Chung & Bennett, 1992) while, in plant pathogens such as bunts (*Tilletia*), smuts (*Ustilago*; Ruiz-Herrera *et al.*, 1995), and witches' broom (*Taphrina*), it is the M-phase that is parasitic and the Y-phase that is free-living; the anamorph of *Taphrina* has the separate name *Lalaria* (Moore, 1996*b*). Other examples are the exidiomycete orders (Table 2) Tremellales and Filobasidales (*Cryptococcus* Y-phase) and the Ustomycota (Table 2) in general. There are also dimorphic asexual yeasts, most notably a number of *Candida* spp., that alternate budding and pseudomycelial phases (Soll, 1986).

Parasitism is a recurring theme in biology, a polyphyletic trend from free-living to commensalism to more or less or total dependence on a host or hosts for nutrition. The reductive effect on a given parasite's morphology can range from minor to profound as structures no longer needed (body parts, synthetic pathways, fruiting structures) are cast to the evolutionary wind. In the course of becoming simpler, a parasite can concomitantly, in the span of time, become multifarious and bizarre as it exploits, adjusts to, and maximizes itself to its chosen niche. Fungi parasitize a wide range of plants, animals, and other fungi (Gee, 1995); major adaptive radiations occur in the Trichomycota (Zygomycetia) (Lichtwardt, 1986), the Laboulbeniales (pyrenomycetes) (Tavares; 1985), the Uredinales (Auriculariomycetes) (Savile, 1971; Scott & Chakravorty, 1982; Roy, 1993), and the Septobasidiales (Ustomycota) (Couch, 1938).

PARTICULAR EVOLUTIONARY TRENDS IN THE BASIDIOMYCETES

The independent sporothallus of the basidiomycetes has achieved functional diploidy while retaining haploid nuclei. This unique development, first seen in the ascomycetes, has been effected by splitting the two events of syngamy (fertilization) that run in sequence in plants and animals. Plasmogamy between monokaryotic gametothalli (primary mycelia) is followed by the close pairing of nuclei (Fig. 1) to establish the dikaryon (secondary mycelium) of indeterminate growth and indefinite life span (Smith *et al.*, 1992). Given the right environmental circumstances, fruiting bodies (tertiary mycelium) are produced containing basidia; within each basidium *karyogamy* establishes the zygote, the only diploid nucleus in the life cycle, followed quickly by meiosis and the production of four, exceptionally more, basidiospores. The close association of the paired, contrasting nuclei (Fig. 1) in the dikaryon is a virtual diploid and there is evidence that there is internuclear communication (Buller, 1931; Crowe, 1960; Gaber & Leonard, 1981). Unique to the basidiomycetes is the emergence of multi-allelic compatibility systems in which the mating type genes have many different alleles (Casselton & Economou, 1985; Petersen, 1995) with the consequence that somatogamy can occur between all gametothalli (monokaryons) except those with identical mating alleles.

Fruiting bodies are composed of organized, differentiated, gravitropic hyphae (Moore, 1991). In contrast to the relatively unexceptional secondary mycelium from which it arises, this structured tertiary mycelium has a large repertory of specialized hyphae and expresses a welter of morphological diversity that, among other things, is a strategy to enhance and prolong spore dispersal.

The subkingdom Basidiomycetia comprises two phyla (Table 2): the Basidiomycota and the Ustomycota. The phylum Basidiomycota, the basidiomycetes proper, is a homogeneous group characterized by 5S rRNA in Walker's Group 5 (Moore, 1988*b*), basidia producing basidiospores, and, except for the rusts and *Eocronartium*, the dolipore/parenthesome septal pore complex (Fig. 4; Table 4). The phylum Ustomycota (ustomycetes) however, has other 5S rRNA, ustidia producing sporidia, and nanoscopic pores with no paraseptal cytoplasmic differentiation (Oberwinkler & Bauer, 1989; Moore, 1988*a,b*; 1996*a,b,c*); most species are also dimorphic.

Fruit bodies of the Basidiomycota are large with extensive hymenia producing quantities of air borne basidiospores, often over extended periods of time because of basidiocarp modifications (Moore, 1971). There are many formal schemes for classifying these fungi; the easiest to comprehend, however, is the vernacular assortment into four general types plus the rusts: mushrooms (agarics, amanitas, and boletes); puffballs and their relatives, earth stars, stinkhorns, and other gasteromycetes; bracket, coral, dry rot and other aphyllophore fungi; and jelly fungi. These are polyphyletic

Table 4. *Types of basidia and septa in major taxa (Moore, 1996d)*

Taxon	Basidia[a]			Septa
	○	[=]	⊗	
Hymenomycetes	√			O_1P_1
Gasteromycetes	√			O_1P_1
Clavariomycetes	√			O_1P_2
Auriculariomycetes				
Auriculariales		√		O_1P_2
Eocronartiales		√		rust-type[b]
Uredinales		√		rust-type[b]
Exidiomycetes			√	O_1P_2
Tremellales			√	$O_1P_{3,4}$
Filobasidiales	√			$O_2P_{3\text{–}6}$

[a]Key to basidial morphology: ○ = non-septate; [=] = transverse septate; ⊗ = cruciate septate.
[b]A reduced septal morphology composed of simple pores with sharp edges and, frequently, an included pulley-shaped, electron-dense granule; the paraseptal cytoplasm is denser, organelle free, and delimited by domes of simple vesicles (see Moore, 1985*a*).

assemblages except, perhaps, the rusts with iterated patterns of development. Gasteromycetes, for example, are seen to be a consequence of the evolutionary fixation of an ontogeny in which certain mushrooms failed to open before spore discharge (Heim, 1971; Pegler & Young, 1979; Theirs, 1984; Bruns, Fogel & White, 1989). These developmental trends are accompanied by changes in basidiome structure, basidial morphology and septal ultrastructure. Mushrooms have evolved a number of developmental types (Watling, 1978) but, as a group, are characterized by ephemeral, synchronously formed fruit bodies that develop apace, produce spores, and soon perish. The basidia are undivided autobasidia and produce ballistospores. The allochronic polypores and other apyllophores have similar basidia but the basidiocarps have been enhanced by the addition of tough, thick-walled hyphae (Pegler, 1973). Such structures have a seasonal to many-years life span; a woody bracket, for example *Fomes*, can commence growth and spore production anytime during the year when conditions are favorable. In gasteromycetes the consequence of delayed basidiocarp opening is the internal wadding of the hymenium and the loss of the ballistic mechanism, reducing the basidia to apobasidia producing statismospores (Pegler & Young, 1979). In jelly fungi the fruiting body is gelatinous and can dry and revive repeatly; the basidia have become variously septate phragmobasidia and the spore attachment a needle-like spiculum that facilitates the tip of each basidial cell penetrating through the gelatinous matrix to the surface independant of its congeners; these fungi are also uniquely characterized by basidiospores that can become septate and form secondary spores.

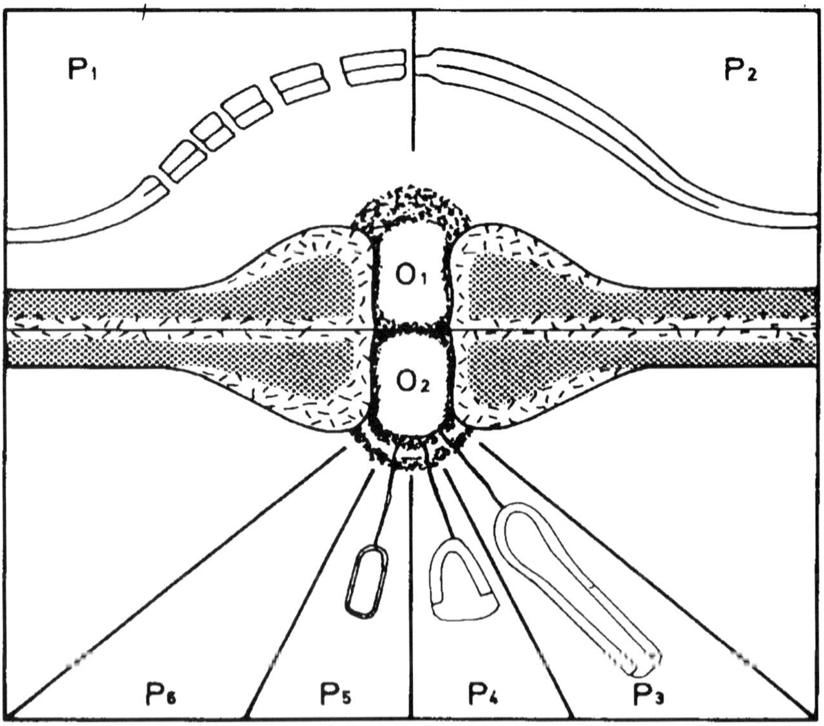

Fig. 4. Summary diagram of basic dolipore/parenthesome (d/p) septal morphologies (from Moore, 1985*a*). The diagram is divided across the page between the two types of occlusions found in the dolipore opening: in O1 septa it appears to be a granule while in O2 septa it appears to be a striated band. Associated with these septal types are several forms of parenthesome: P1 regularly perforate; P2 imperforate, P3–5 vesiculate, P6 absent. See Table 4 for major taxa characterized by combinations of these configurations.

The dolipore/parenthesome (d/p) septal complex of these fungi (Moore, 1985*a*) comprises three elements (Fig. 4): the *dolipore* (the inflated margin of the central pore which has the same general morphology in all species); the *occlusions* (base-to-base champagne-cork-like electron dense material in the pore channel and whose 'tops' in the pore orifice appear, in section, like either a granule (O_1) or a striated band (O_2)); and the parenthesomes (paraseptal, double membrane organelles differentiated from the endoplasmic reticulum that are either regularly perforate (P_1), or imperforate, except occasionally for a small central pore (P_2), or vesiculate (P_{3-5}), or absent (P_6)). The distribution of these variations is as follows (Moore, 1996*d*): O_1 occlusions occur in all groups except the Tremellales (*sens. str.*) which has the O_2 type; P_1 parenthesomes occur in mushrooms, gasteromycetes, and polypores; the P_2 type occurs in some aphyllophores (Moore, 1994) and all jelly fungi except the Tremellales (*sens. str.*) which has the P_{3-6} type. These variations can be phylogenetically interpreted as follows: O_2/P_{3-6} d/p septa are advanced, particularly as the Tremellales (*sens. str.*) is the only dimorphic order in the phylum; O_1/P_1 d/p septa are the basic type; and P_2 parenthesomes are intermediate. Regretably, however, little is known about the cell biology of this unique cytological complex, other than its morphology and that it forms very rapidly (Orlovich & Ashford, 1994). Parenthetically, *Pleurotus cystidiosus* Miller would be an excellent species for in depth studies of the d/p complex *if* its conidia could be induced to germinate: it has tetrapolar sexuality (Moore, 1985*b*), and its monokaryons and dikaryons form numerous coremia that produce apical conidia *ad libitum*.

Basidial and septal structure are fundamental characters of the Basidiomycota and are phylogenetically conservative. Used together (Table 4) they define the classes of the phylum (Moore, 1996*d*).

CONCLUSIONS

The Fungi share basic characters with the other crown taxa (mode of cell division with the prokaryotes, walled cells with the plants, heterotrophy and centrioles with the animals) yet have a fugue of unique attributes that demarcates them as a separate kingdom. They may have origins prior to the Cretaceous (Dennis, 1969; 1976; Smith & Crane, 1979; Hibbett, Grimaldi & Donoghue, 1995) but it was not until the rise of the flowering plants some 65 million years ago that they were to acquire a significant ecological role, particularly that of wood decay (Elliott *et al.*, 1979; Prillinger & Molitoris, 1979; Redhead & Ginns, 1985; Boddy & Watkinson, 1995; Wyatt & Broda, 1995), and become the abundant saprotrophic and parasitic organisms we know today.

Molecular biology has become a major force in microbial biosystematics (O'Donnell *et al.*, 1995; Doolittle *et al.*, 1996). It plays a major role in bacterial classification (Goodfellow & O'Donnell, 1994). Within the Fungi there are the aforementioned circumscription of the kingdom and the dichotomy of the basidiomycetes. This source of new information is also important in yeast systematics (Kurtzman & Fell, 1966) and is being exploited in analysing relationships in higher fungi (e.g. Berbee & Taylor, 1992; Boekhout *et al.*, 1992; Berres *et al.*, 1995; Gargas *et al.*, 1995; Luzoni & Vilgalys, 1995; Masclaux *et al.*, 1995).

REFERENCES

Ahmadjian, V. & Hale, M. E. (eds.) (1973). *The Lichens*. Academic Press, New York/London.

Allen, T. L. & Douglas, M. G. (1990). Organization of the nuclear pore complex in *Saccharomyces cerevisiae*. *Journal of Ultrastructure and Molecular Structure Research*, **102**, 95–108.

Andrews, J. H. (1995). Fungi and the evolution of growth form. *Canadian Journal of Botany*, **73** (Suppl. 1), S1206–12.

Bandoni, R. J., Bisalputra, A. A. & Bisalputra, T. (1967). Ascospore development in *Hansenula anomala. Canadian Journal of Botany*, **45**, 361–6.

Barr, D. J. S. & Désaulniers, N. L. (1989). The flagellar apparatus of the Oomycetes and Hyphochytriomycetes. In *The Chromophyte Algae*. Green, J. C., Leadbeater, B. S. C. & Diver, W. L., eds., pp. 343–55. Clarendon Press, Oxford.

Barrett, J. T. (1986). *Contemporary Classics in Plant, Animal, and Environmental Sciences*, p. 228, ISI Press, Philadelphia.

Bartnicki-Garcia, S. (1987). The cell wall: a crucial structure in fungal evolution. In *Evolutionary Biology of the Fungi*, Rayner, A. D. M., Brasier, C. M. & Moore, D., eds., pp. 389–403. Cambridge University Press, Cambridge.

Bartnicki-Garcia, S., Hergert, F. & Gierz, G. (1989). Computer simulation of fungal morphogenesis and the mathematical basis for hyphal (tip) growth. *Protoplasma,* **153**, 46–57.

Bartnicki-Garcia, S. & Lippman, E. (1989). Fungal cell wall composition. In *Practical Handbook of Microbiology*, O'Leary, W. M., ed., pp. 381–404, CRC Press, Boca Raton (Florida).

Beakes, G. W. (1987). Oomycete phylogeny: ultrastructural perspectives. In *Evolutionary Biology of the Fungi*, Rayner, A. D. M., Brasier, C. M. & Moore, D., eds., pp. 405–21. Cambridge University Press, Cambridge.

Beakes, G. W. (1989). Oomycete fungi: their phylogeny and relationship to chromophyte algae. In *The Chromophyte Algae*. Green, J. C., Leadbeater, B. S. C. & Diver, W. L., eds. pp. 325–42. Clarendon Press, Oxford.

Berbee, M. L. & Taylor, J. W. (1992). Convergence in ascospore discharge mechanism among pyrenomycete fungi based on 18S ribosomal RNA gene sequence. *Molecular Phylogenetics and Evolution*, **1**, 59–71.

Berbee, M. L. & Taylor, J. W. (1995). From 18S ribosomal sequence data to evolution of morphology among the fungi. *Canadian Journal of Botany*, **73** (suppl. 1), S677–83.

Berres, M. E., Szabo, L. J. & McLaughlin, D. J. (1995). Phylogenetic relationships in auriculariaceous basidiomycetes based on 25S ribosomal DNA sequences. *Mycologia*, **87**, 821–40.

Boddy, L. & Watkinson, S. C. (1995). Wood decomposition, higher fungi, and their role in nutrient redistribution. *Canadian Journal of Botany*, **73** (suppl. 1), S1377–83.

Boekhout, T. & Linnemans, W. A. M. (1982). Ultrastructure of mitosis in *Rhodosporidium toruloides. Studies in Mycology*, **22**, 23–38.

Boekhout, T., Yamada, Y., Weijman, C. J. M., Roeymans, H. J. & Batenburg-van der Vegte, W. H. (1992). The significance of coenzyme Q, carbohydrate composition and septal ultrastructure for the taxonomy of ballistoconidia-forming yeasts and fungi. *Systematic and Applied Microbiology*, **15**, 1–10.

Bourett, T. M. & McLaughlin, D. J. (1986). Mitosis and septum formation in the basidiomycete *Helicobasidium mompa. Canadian Journal of Botany*, **64**, 130–45.

Bruns, T. D., Fogel, R. & White, T. J. (1989). Accelerated evolution of a false-truffle from a mushroom ancestor. *Nature*, **339**, 140–2.

Bruns, T. D., White, T. J. & Taylor, J. W. (1991). Fungal molecular systematics. *Annual Review of Ecology and Systematics*, **22**, 525–64.

Buller, A. H. R. (1931). The diploidization [dikaryotization] of a haploid [monokaryotic] mycelium by a theoretically incompatible diploid [dikaryotic] mycelium. *Researches on Fungi*, **4**, 245–63.

Buller, A. H. R. (1933). Woronin bodies and their movements. *Researches on Fungi*, **5**, 127–30.

Burns, R. (1988). Chromosome movement *in vitro*. *Nature*, **331**, 479.

Carson, D. B. & Cooney, J. J. (1990). Microbodies in fungi, a review. *Journal of Industrial Microbiology*, **6**, 1–18.

Carter, B. L. A. (1978). The yeast nucleus. *Advances in Microbial Physiology*, **17**, 243–302.

Casselton, L. A. & Economou, A. (1985). Dikaryon formation. In *Developmental Biology of Higher Fungi*. Moore, D., Casselton, L. A. Wood, D. A. & Frankland, J. A., eds., pp. 213–29, Cambridge University Press, Cambridge.

Corner, E. J. H. (1964). *The Life of Plants*. The World Publishing Company Cleveland and New York.

Couch, J. N. (1938). *The Genus Septobasidium*. University of North Carolina KPress, Chapel Hill.

Crowe, L. K. (1960). The exchange of genes between nuclei of a dikaryon. *Heredity*, **15**, 397–405.

Cullings, K. W., Szaro, T. M. & Bruns, T. D. (1996). Evolution of extreme specialization within a lineage of ectomycorrhizal epiparasites. *Nature*, **379**, 63–6.

Dahlberg, A. & Stenlid, J. (1995). Spatiotemporal patterns in ectomycorrhizal populations. *Canadian Journal of Botany*, **73** (suppl. 1), S1222–30.

Dennis, R. L. (1969). Fossil mycelium with clamp connections from the Middle Pennsylvanian. *Science*, **163**, 670–1.

Dennis, R. L. (1976). *Paleosclerotium*, a Pennsylvanian age fungus combining features of modern ascomycetes and basidiomycetes. *Science*, **192**, 66–8.

Dick, M. W. (1995). Sexual reproduction in the *Peronosporomycetes* (chromistan fungi). *Canadian Journal of Botany*, **73** (suppl. 1), S712–24.

Doolittle, R. F., Feng, Da-Fei, Tsang, S., Cho, G. & Little, E. (1996). Determining divergence times of the major kingdoms of living organisms with a protein clock. *Science*, **271**, 470–6.

Durán, R. & Gray, P. M. (1989). Nuclear DNA, an adjunct to morphology in fungal taxonomy. *Mycotaxon*, **36**, 205–19.

Egashira, T., Tokunaga, J. & Tokunaga, M. (1972). Electron microscopy of nuclear membranes during somatic separation in *Neurospora crassa*. *The Journal of General Microbiology*, **71**, 203–6.

Elliott, C. G., Abou-Heilah, A. N., Leake, D. L. & Hutchinson, S. A. (1979). Analysis of wood-decaying ability of monokaryons and dikaryons of *Serpula lacrymans*. *Transactions of the British Mycological Society*, **73**, 127–33.

Erhard, M., Barker, D. & Green, J. (1988). Is chromosome condensation a phylogenetic marker? In *The Expanding Realm of Yeast-like Fungi*. de Hoog, G. S., Smith, M. Th. & Weijman, A. C. M., eds., pp. 267–77, Elsevier, Amsterdam.

Farkaŝ, V. (1990). Fungal cell walls, Their structure, biosynthesis and biotechnological aspects. *Acta Biotechnologica*, **10**, 225–38.

Gaber, R. F. & Leonard, T. J. (1981). Unilateral internuclear gene transfer and cell differentiation in *Schizophyllum*. *Nature*, **291**, 342–44.

Gargas, A., DePriest, P. T., Grube, M. & Tehler, A. (1995). Multiple origins of lichen symbioses in fungi suggested by SSU rDNA phylogeny. *Science*, **268**, 1492–95.

Gee, H. (1995). Mycological mystery tour. *Nature*, **375**, 276.

Ghabrial, S. A. (1994). New developments in fungal virology. *Advances in Virus Research*, **43**, 303–88.

Girbart, M. (1969). Die Ultrastruktur der Apikalregion von Pilzhyphen. *Protoplasma*, **67**, 413–41.

Girbart, M. (1979). A microfilamentous septal belt (FSB) during induction of cytokinesis in *Trametes versicolor* (L. ex Fr.). *Experimental Mycology*, **3**, 215–28.

Gooday, G. W. & Schofield, D. A. (1995). Regulation of chitin synthesis during growth of fungal hyphae: the possible participation of membrane stress. *Canadian Journal of Botany*, **73** (suppl. 1), S114–21.

Goodfellow, M. & O'Donnell, A. G. (eds). (1994). *Handbook of New Bacterial Systematics*. Academic Press, New York/London.

Grove, S. N. & Bracker, C. E. (1970). Protoplasmic organization of hyphal tips among fungi: vesicles and Spitzenkörper. *Journal of Bacteriology*, **104**, 989–1009.

Gunderson, J. H., Elwood, H., Ingold, A., Kindle, K. & Sogin, M. L. (1987). Phylogenetic relationships between chlorophytes, chrysophytes, and oomycetes. *Proceedings of the National Academy of Science USA*, **84**, 5823–27.

Harley, J. L. & Smith, S. E. (1983). *Mycorrhizal Symbiosis*. Academic Press, New York/London.

Heath, I. B. (1981). Nucleus-associated organelles in fungi. *International Review of Cytology*, **69**, 191–221.

Heath, I. B. (1995). Integration and regulation of hyphal tip growth. *Canadian Journal of Botany*, **73** (suppl. 1), S131–39.

Heath, I. B., Ashton, M.-L., Rethoret, K. & Heath, M. C. (1982). Mitosis and the phylogeny of *Taphrina*. *Canadian Journal of Botany*, **60**, 1696–725.

Heim, R. (1971). The interrelationships between the Agaricales and Gasteromycetes. In *Evolution in the Higher Basidiomycetes*. Petersen, R. H., ed., pp. 505–34. University of Tennessee Press, Knoxville.

Hibbett, D. S., Grimaldi, D. & Donoghue, M. J. (1995). Cretaceous mushrooms in amber. *Nature*, 377, 487.

Hoch, H. C. & Howard, R. J. (1980). Ultrastructure of freeze-substituted hyphae of the basidiomycete *Laetisaria arvalis*. *Protoplasma*, **103**, 281–97.

Hoog, G. S. de, Smith, M. Th. & Weijman, A. C. M., eds. (1988). *The Expanding Realm of Yeast-like Fungi*. Elsevier, Amsterdam.

Howard, K. L. & Moore, R. T. (1970). Ultrastructure of oosporogenesis in *Saprolegnia terrestris*. *Botanical Gazette*, **131**, 311–36.

Kilmartin, J. V. & Adams, A. E. M. (1984). Structural rearrangements of tubulin and actin during the cell cycle of the yeast *Saccharomyces*. *The Journal of Cell Biology*, **98**, 922–33.

King, S. M. & Hyams, J. S. (1982). The mitotic spindle of *Saccharomyces cerevisiae*, assembly, structure and function. *Micron*, **13**, 93–117.

Kole, A. P. (1965). Flagella. In *The Fungi, An Advanced Treatise*. Ainsworth G. C. & Sussman, A. S., eds., vol. 1, pp. 77–93. Academic Press, New York, London.

Kotov, V. & Reshetnikov, S. V. (1990). A stochastic model for early mycelial growth. *Mycological Research*, **94**, 577–86.

Kurtzman, C. & Fell, J. (eds.) (1966). *The Yeasts, A Taxonomic Study*. 4th edn. Elsevier, Amsterdam.

Kwon-Chung, K. J. & Bennett, J. E. (1992). *Medical Mycology*. Lee & Febiger, Philadelphia/London.

Lawrey, J. D. (1995). The chemical ecology of lichen mycoparasites: a review. *Canadian Journal of Botany*, **73** (suppl. 1), S603–8.

Lewis, D. H. (1987). Evolutionary aspects of mutualistic associations between fungi and phytosynthetic organisms. In *Evolutionary Biology of the Fungi*, Rayner, A. D. M., Brasier, C. M. & Moore, D., eds., pp. 161–78. Cambridge University Press, Cambridge.

Lichtwardt, R. W. (1986). *The Trichomycetes, Fungal Associates of Arthropods*. Springer-Verlag, New York.

Lu, B. C. (1978). Meiosis in *Coprinus*. VIII. A time-course study of the fusion and division of the spindle pole body during meiosis. *Journal of Cell Biology*, **76**, 761–66.

Luzoni, F. & Vilgalys, R. (1995). Integration of morphological and molecular data sets in estimating fungal phylogenies. *Canadian Journal of Botany*, **73**, (suppl. 1), S649–59.

Marchant, R. & Moore, R. T. (1973). Lomasomes and plasmalemmasomes in fungi. *Protoplasma*, **76**, 235–47.

Markham, P. & Collinge, A. J. (1987). Woronin bodies of filamentous fungi. *FEMS Microbiology Reviews*, **46**, 1–11.

Masclaux, F., Gueho, E., Hoog, G. S. de & Christen R. (1995). Phylogenetic relationships of human-pathogenic *Cladosporium* (*Xylohypha*) species inferred from partial LS rRNA sequences. *Journal of Medical and Veterinary Mycology*, **33**, 327–38.

Mayr, E. (1982). *The Growth of Biological Thought: Diversity, Evolution, and Inheritance*. Belknap Harvard, Cambridge.

Mischke, M. D. & Chant, J. (1995). The shape of things to come: morphogenesis in yeast and related patterns in other systems. *Canadian Journal of Botany*, **73** (suppl. 1), S234–42.

Money, N. P. (1995). Turgor pressure and the mechanics of fungal penetration. *Canadian Journal of Botany*, **73** (suppl. 1), S96–S102.

Moore, D. (1991). Perception and response to gravity in higher fungi. *New Phytologist*, **117**, 3–23.

Moore, R. T. (1964). Fine structure of mycota. 12. Karyochorisis – somatic nuclear division – in *Cordyceps militaris. Zeitschrift für Zellforschung*, **63**, 921–37.

Moore, R. T. (1965). The ultrastructure of fungal cells. In *The Fungi, an Advanced Treatise*, vol. 1. Ainsworth, G. C. & Sussman, A. S., eds., pp. 95–118, Academic Press, New York, London.

Moore, R. T. (1968). Fine structure of mycota. 13. Zoospore and nuclear cap formation in *Allomyces. Journal of the Elisha Mitchell Scientific Society*, **84**, 147–65.

Moore, R. T. (1971). An alternative concept of the fungi based on their ultrastructure. In *Recent Advances in Microbiology*, Perez-Miravete, A. & Pelaez, D., ed., pp. 49–64, Libreria International, Mexico City.

Moore, R. T. (1985*a*). The challenge of the dolipore/parenthesome septum. In *Developmental Biology of Higher Fungi*. ed. Moore, D., Casselton, L. A., Wood, D. A. & Frankland, J. A., eds., pp. 175–212, Cambridge University Press, Cambridge.

Moore, R. T. (1985*b*). Mating type factors in *Pleurotus cystidiosus. Transactions of the British Mycological Society*, **85**, 354–8.

Moore, R. T. (1988*a*). Micromorphology of yeasts and yeast-like fungi and its taxonomic implications. In *The Expanding Realm of Yeast-like Fungi*. de Hoog, G. S., Smith, M. Th. & Weijman, A. C. M., eds., pp. 203–26. Elsevier, Amsterdam.

Moore, R. T. (1988*b*). A reconnaissance of the division Ustomycota (Basidiomycotera). In *Taxonomy–Putting Plants and Animals in their Place*. Moriarty, C., ed., pp. 63–91. Royal Irish Academy, Dublin.

Moore, R. T. (1989). Alicean taxonomy – small characters made large. *Botanical Journal of the Linnean Society*, **99**, 59–79.

Moore, R. T. (1994). Third order morphology, TEM in the service of taxonomy. In *Identification and Characterization of Pest Organisms*. Hawksworth, D. L., ed., pp. 249–59, CAB International, Wallingford.

Moore, R. T. (1996*a*). Cytology and ultrastructure. In *The Yeasts, a Taxonomic Study*, 4th edn. [in press]. Kurtzman C. P. & Fell, J. eds. Elsevier, Amsterdam.

Moore, R. T. (1996*b*). *Lalaria*. In *The Yeasts, a Taxonomic Study*, 4th edn. [in press]. Kurtzman, C. P. & Fell, J., eds. Amsterdam, Elsevier.

Moore, R. T. (1996*c*). An inventory of the phylum Ustomycota. *Mycotaxon*, **59**, 1–31.

Moore, R. T. (1996*d*). The dolipore/parenthesome septum in modern taxonomy. In *Rhizoctonia Species*, Sney, B., Jabali-Hare, S., Neate, S. & Dijst, G. eds., pp. 13–34. Kluwer, Dordrecht.

Moore, R. T. & McAlear, J. H. (1961). Fine structure of mycota. 5. Lomasomes – previously uncharacterized hyphal structures. *Mycologia*, **53**, 194–200.

Moore, R. T., Cookson, J., Dykes, F. & Holman, J. G. (1991). Structure of fungal nuclei as visualized by computer reconstruction of permanganate fixed serial sections. *Royal Microscopical Society, 4th International Botanical Microscopy Meeting*, abs. PIx-2, p. 25. Durham.

Morris, N. R. (1990). Lower eukaryotic cell cycle: perspectives on mitosis from the fungi. *Current Opinion in Cell Biology*, **2**, 252–7.

Morton, J. P., Bentivenga, S. P. & Bever, J. D. (1995). Discovery, measurement, and interpretation of diversity in arbuscular endomycorrhizal fungi (Glomales, Zygomycetes). *Canadian Journal of Botany*, **73** (suppl. 1), S25–32.

Motta, J. J. (1969). Somatic nuclear division in *Armillaria mellea*. *Mycologia*, **61**, 873–86.

Nehrbass, U., Kern, H., Mutvei, A., Horstmann, H., Marshallsay, B. & Hurt, E. C. (1990). NSP1. A yeast nuclear pore protein localized at the nuclear pores exerts its essential function by its carboxy-terminal domain. *Cell*, **61**, 979–89.

Oberwinkler, F. & Bauer, R. (1989). The systematics of gastroid, auricularioid heterobasidiomycetes. *Sydowia*, **41**, 224–56.

O'Donnell, K. L. (1992). Ultrastructure of meiosis and the spindle pole body cycle in freeze-substituted basidia of the smut fungi *Ustilago maydis* and *Ustilago avenae*. *Canadian Journal of Botany*, **70**, 629–38.

O'Donnell, A. G., Goodfellow, M. & Hawksworth, D. L. (1995). Theoretical and practical aspects of the quantification of biodiversity among microorganisms. In *Biodiversity – Measurement and Estimation*. Hawksworth, D. L., ed., pp. 65–73, Chapman & Hall, London.

O'Donnell, K. L. & McLaughlin, D. J. (1984). Ultrastructure of meiosis in *Ustilago maydis*. *Mycologia*, **76**, 468–85.

Oakley, B. R. (1995). A nice ring to the centrosome. *Nature*, **378**, 555–6.

Olson, L. W. & Fuller, M. S. (1968). Ultrastructural evidence for the biflagellate origin of the uniflagellate fungal zoospore. *Archiv für Mikrobiologie*, **62**, 237–50.

Orlovich, D. A. & Ashford, A. E. (1994). Structure and development of the dolipore septum in *Pisolithus tinctorius*. *Protoplasma*, **178**, 66–80.

Paquin, R., Roewer, I., Wang, Z. & Lang, B. F. (1995). A robust fungal phylogeny using the mitochondrially encoded NAD5 protein sesquence. *Canadian Journal of Botany*, **73** (suppl. 1), S180–5.

Pegler, D. N. (1973). *The Polypores. The British Mycological Society Bulletin*, **7** (suppl. 1), 1–43.

Pegler, D. N. & Young, T. W. K. (1979). The gasteroid Russulales. *Transactions of the British Mycological Society*, **72**, 353–81.

Perotto, S., Perotto, R., Faccio, A., Schubert, A., Varma, A. & Bonfante, P. (1995). Ericoid mycorrhizal fungi: cellular and molecular bases of their interactions with the host plant. *Canadian Journal of Botany*, **73** (suppl. 1), S557–68.

Petersen, R. H. (1995). Contributions of mating studies to mushroom systematics. *Canadian Journal of Botany*, **73** (suppl. 1), S831–42.

Prillinger, H. & Molitoris, H. P. (1979). Genetic analysis in wood-decaying fungi. I. Genetic variation and evidence for allopatric speciation in *Pleurotus ostreatus* using phenoloxidase zymograms and morphological criteria. *Physiologia Plantarum*, **46**, 265–77.

Redhead, S. A. & Ginns, J. H. (1985). A reappraisal of agaric genera associated with brown rots of wood. *Transactions of the Mycological Society of Japan*, **26**, 349–81.

Rout, M. P. & Kilmartin, J. V. (1990). Components of the yeast spindle and spindle pole body. *The Journal of Cell Biology*, **111**, 1913–27.

Roy, B. A. (1993). Floral mimicry by a plant pathogen. *Nature*, **362**, 56–8.

Ruhle, J. L. & Marx, D. H. (1979). Fiber, food, fuel, and fungal symbionts. *Science*, **206**, 419–22.

Ruiz-Herrera, J., León, C. G., Guevara-Olvera, L. & Cárabez-Trejo, A. (1995). Yeast-mycelial dimorphism of haploid and diploid strains of *Ustilago maydis*. *Microbiology*, **141**, 695–703.

San-Blas, G. & San-Blas, F. (1984). Molecular aspects of fungal dimorphism. *CRC Critical Reviews in Microbiology*, **11**, 101–27.

Savile, D. B. O. (1971). Coevolution of the rust fungi and their hosts. *Quarterly Review of Biology*, **46**, 211–18.

Scott, K. J. & Chakravorty, A. K. (1982). *The Rust Fungi*. Academic Press, New York/London.

Setliff, E. C., Hoch, H. C. & Patton, R. F. (1974). Studies on nuclear division in basidia of *Poria latemarginata*. *Canadian Journal of Botany*, **52**, 2323–33.

Sietsma, J. H., Wösten, H. A. B. & Wessels, J. G. H. (1995). Cell wall growth and protein secretion in fungi. *Canadian Journal of Botany*, **73** (suppl. 1), S388–95.

Smith, M. L., Bruhn, J. N. & Anderson, J. B. (1992). The fungus *Armillaria bulbosa* is among the largest and oldest living organisms. *Nature*, **356**, 428–31.

Smith, P. H. & Crane, P. R. (1979). Fungal spores of the genus *Pesavis* from the Lower Tertiary of Britain. *Botanical Journal of the Linnean Society*, **79**, 243–8.

Soll, D. R. (1986). The regulation of cellular differentiation in the dimorphic yeast *Candida albicans*. *BioEssays*, **5**, 5–11.

Sparrow, A. H., Price, H. J. & Underbrink, A. G. (1972). A survey of DNA content per cell and per chromosome of prokaryotic and eukaryotic organisms, some evolutionary considerations. *Brookhaven Symposia in Biology*, pp. 451–94.

Stocker-Wörgötter, E. (1995). Experimental cultivation of lichens and lichen symbionts. *Canadian Journal of Botany*, **73** (suppl. 1), S579–89.

Tanaka, K. & Kanbe, T. (1986). Mitosis in the fission yeast *Schizosaccharomyces pombe* as revealed by freeze-substitution electron microscopy. *Journal of Cell Science*, **180**, 253–68.

Tavares, I. I. (1985). *Laboulbeniales (Fungi, Ascomycetes)*. J. Cramer, Braunschweig.

Theirs, H. D. (1984). The secotioid syndrome. *Mycologia*, **76**, 1–8.

Todd, N. K. & Rayner, A. D. M. (1980). Fungal individualism. *Science Progress Oxford*, **66**, 331–4.

Veenhuis, M. & Harder, W. (1988). Microbodies in yeasts, structure, function and biogenesis. *Microbiological Sciences*, **5**, 347–51.

Watling, R. (1978). From infancy to adolescence: Advances in the study of higher fungi. *Botanical Society of Edinburgh Transactions* **42** (suppl.), 61–73.

Wells, K. (1977). Meiotic and mitotic division in the Basidiomycotina. In *Mechanisms and Control of Cell Division*. Rost, T. L. & Gifford, E. M. Jr. eds., pp. 337–74. Dowden, Hutchinson & Ross, Stroudsberg, USA.

Wessels, J. G. H. (1986). Cell wall synthesis in apical hyphal growth. *International Review of Cytology*, **104**, 37–79.

Wessels, J. G. H. (1990). Role of cell wall architecture in fungal tip growth generation. In *Tip Growth in Plant and Fungal Cells*. Heath, I. B., ed., pp. 1–29, Academic Press, New York, London.

Wyatt, A. M. & Broda, P. (1995). Informed strain improvement for lignin degradation by *Phanerochaete chrysosporium*. *Microbiology*, **141**, 2811–22.

Zuck, R. K. (1953). Alternation of generations and the mode of nutrition. *Drew University Bulletin*, **41** (suppl.), 3–19.

MICROORGANISMS IN SYMBIOSIS: ADAPTATION AND SPECIALIZATION

A. E. DOUGLAS

Department of Biology, University of York, PO Box 373, York YO1 5YW, UK

INTRODUCTION

Microbial symbioses and the adaptive diversification of eukaryotes

The great majority of eukaryotes are the product of a symbiosis. It is now generally accepted that, early in their evolutionary history, eukaryotes acquired aerobic bacteria (probably α-Proteobacteria), which have evolved into mitochondria (Margulis, 1981; Cavalier-Smith, 1992). By forming this symbiosis, eukaryotes gained access to aerobic respiration. Nearly all extant eukaryotes are derived from the symbiosis with mitochondria; the lineages that diverged prior to acquisition of mitochondria are today represented by several relatively minor groups of protists, collectively known as the Archezoa (Cavalier-Smith, 1987; Schlegel, 1994).

The acquisition of aerobic respiration by symbiosis is undoubtedly a key factor contributing to the evolutionary success of eukaryotes in our predominantly aerobic world (see Fenchel, this volume). This evolutionary event was not, however, unique. Through symbiosis, eukaryotes have acquired a range of metabolic capabilities, including photosynthesis, nitrogen fixation and cellulose degradation; and the combination of the metabolic novelty of the microbial partner and the morphological complexity of the eukaryote has triggered various adaptive radiations (Margulis & Fester, 1991; Douglas, 1992). As examples, all the algae, plants and lichens are based on the acquisition of photosynthesis, ultimately from cyanobacteria; herbivory in mammals is founded on the capacity of gut microbiota in these animals to degrade cellulose; and the radiation of the Leguminosae, one of most speciose and diverse families of plants, is probably linked to the capacity of most legumes to accommodate nitrogen-fixing bacteria, called rhizobia (which include the genera *Rhizobium* and *Bradyrhizobium*).

Our understanding of these symbioses is, however, grossly unbalanced. As the opening paragraphs of this article illustrate, the importance of symbiosis as one of a few key factors shaping the evolutionary history and diversity of eukaryotes is generally accepted (Maynard-Smith & Szathmary, 1995). By comparison, the evolutionary consequences of symbiosis for microorganisms which provide the metabolic capabilities have been neglected.

The scope of this chapter

The primary focus of this article is the microorganisms acquired by eukaryotes. The chapter concerns associations in which the microbial donors of the metabolic capability (also known as symbionts) are located within the body of the eukaryote (or host). As the survey of these associations in Table 1 shows, the symbionts include Eubacteria (e.g. the nitrogen-fixing symbionts), Archaea (the methanogens in various protists) and eukaryotic microbes (i.e. protists, including various algae); and, for the hosts, all the major groups of Eukaryotes (protists, fungi, plants and animals) are represented.

For all these associations included in Table 1, the eukaryotic host can be viewed as a habitat occupied by the microbial symbionts. This chapter addresses two questions: first, what are the conditions within this habitat, and do symbiotic microorganisms exhibit a suite of metabolic and biochemical characteristics that can be interpreted as adapatations to their symbiotic lifestyle, and, second, do these adaptations for symbiosis reduce the capac-

Table 1. *A survey of microorganisms in symbiosis with eukaryotes*[a]

Metabolic capability	The microbial symbionts	The eukaryotic hosts
Photosynthesis	Algae (e.g. *Chlorella*, *Trebouxia* and *Symbiodinium*) and cyanobacteria (e.g. *Nostoc*)	Various protists, aquatic invertebrates, lichenized fungi
Nitrogen fixation	Rhizobia (*Rhizobium* and *Bradyrhizobium*)	Leguminous plants
	Frankia (an actinomycete)	Various dicot plants
	Cyanobacteria (e.g. Nostoc)	Various plants, fungi, diatoms
Essential nutrients (vitamins and essential amino acids)	Various bacteria (incl. β- and γ- Proteobacteria, flavibacteria) yeasts	Insects, protists
Cellulose degradation	Bacteria (e.g. *Ruminococcus*) and chytrid 'fungi'	Herbivorous vertebrates
	hypermastigote and trichomonad protists	Certain termites
Luminescence	*Photobacterium*, *Vibrio* and allied bacteria	Various fish and squid
Consumption of hydrogen	Methanogenic bacteria	Anaerobic protists

[a] See Douglas (1994) for further information.

ity of symbiotic microorganisms to utilize the environment, such that the microbes are specialized for symbiosis? The final section of the chapter considers the most specialized of all symbionts, those that have evolved into organelles.

ADAPTATIONS OF SYMBIOTIC MICROORGANISMS

The eukaryotic host as a habitat

There are two broad types of habitat occupied by microorganisms living in eukaryotes: extracellular and intracellular. The extracellular symbionts may reside between closely apposed host cells or, more frequently, within cavities or invaginations of the host body. They include the photosynthetic symbionts in lichens, the luminescent bacteria in the tubules of fish light organs, nitrogen-fixing cyanobacteria between corticel cells of cycad roots, and the gut microbiota of animals. Most of the intracellular symbionts are located in the host cell cytoplasm. They are invariably acquired by phagocytosis, and are usually retained within the host membrane derived from the phagosome. This novel structure 'housing' the symbionts can usefully be described as an organelle, the symbiosome (Roth, Jeon & Stacey, 1988). Intracellular symbionts are widespread in wall-less protists and invertebrate animals with phagocytotically active cell membranes (e.g. algae in ciliates, bacteria in insects), but rare in fungi and plants, in which phagocytosis is limited by cell walls. They are also apparently absent in vertebrates (the reasons are obscure, but could be related to the sophisticated immune system of these animals).

The supply of nutrients to many symbionts is probably severely restricted. Intracellular symbionts, in particular, derive their entire nutritional requirements from the surrounding host cell, and their nutrient acquisition may be controlled primarily by the transport properties of the surrounding symbiosome membrane. In the association between the legume soybean and *Bradyrhizobium*, the host symbiosome membrane represents a major permeability barrier. The supply of the dicarboxylate malate (the main carbon source of rhizobia in symbiosis) is limited by the low affinity of the dicarboxylate transporter on the symbiosome membrane (Udvardi *et al.*, 1989), and the host pools of glutamate (an amino acid readily utilised by rhizobia) are unavailable to the rhizobia in both soybeans and phaseolus beans because the symbiosome membrane is essentially impermeable to this amino acid (Udvardi, Salom & Day, 1988; Herrada, Puppo & Rigaud, 1989). There is also biochemical and cytochemical evidence that *Symbiodinium*, the dinoflagellate symbiont in corals and related cnidarians, cannot utilise the hosts' substantial cytoplasmic pool of inorganic phosphate, presumably because the symbiosome membrane lacks a phosphate transporter; and *Symbiodinium* is dependent on organic phosphorus compounds for its phosphorus nutrition (Jackson & Yellowlees, 1990; Rands, Loughman & Douglas, 1993). The condition of these symbionts contrasts with the generally favourable nutritional status of intracellular parasites. For example, Schwab, Beckers & Joiner (1994) describe the membrane surrounding *Toxoplasma gondii* as a molecular 'sieve', permitting the free exchange of molecules up to 1.3–1.9 kDa between the host cell cytoplasm and parasite.

Several authors (e.g. Haygood, 1993; McAuley, 1987) have linked the restricted supply of nutrients in symbiosis to the characteristically low rates of proliferation of microbial symbionts. For example, the population doubling time of many culturable symbionts is one or more orders of magnitude lower in symbiosis than in laboratory culture. Detailed data are available for symbiotic *Chlorella*, which have a *maximal* doubling time of 10 days in symbiosis and 10 hours in batch culture (Douglas & Huss, 1986). Nutrient-limitation is not, however, the basis of the low growth rates of some symbiotic microorganisms. Several unculturable bacterial symbionts may be genetically incapable of rapid growth. The intracellular bacteria in at least two insect groups, the aphids and tsetse flies, and the luminescent bacteria in the flashlight fish *Kryptophanaron alfredi* have just one copy of rRNA genes (most bacteria have several), and this may limit the rate of ribosome production, protein synthesis and, ultimately, growth (Untermann, Baumann & McLean, 1989; Wolfe & Haygood, 1993; Aksoy, 1995).

A further likely feature of the symbiotic habitat is the defensive response of the host to colonization by foreign organisms. Symbiotic microorganisms may generally evade or suppress these reactions, but host-mediated destruction of symbionts can occur. As an example, the hypersensitive reaction of plants, which is exhibited in response to various pathogens, is rarely triggered by symbiotic microorganisms; but it has been observed in alfalfa plants challenged with excessive numbers of rhizobia (Vasse, de Billy & Truchet, 1993). The intracellular 'habitat' may be a privileged site protected from many of the host defences. Consistent with this view, the intracellular bacterial symbiont of aphids, *Buchnera*, is rapidly lysed if they are released from the host cells into the body cavity of the insect (Hinde, 1971).

The ways in which microorganisms may overcome or evade host defensive responses have been studied in associations with intracellular parasites. For example, a variety of parasites, including *Chlamydia*, *Legionella* and *Toxoplasma*, modify the surrounding host membrane so that it is not competent to fuse with lysosomes, while *Coxiella* and *Leishmania* 'permit' lysosomal fusion and persist in the phagolysosome, and *Mycobacterium* causes the selective exclusion of H^+/ATPases from the surrounding host membrane (Russell, 1995; Sturgill-Koszycki *et al.*, 1994; Clemens & Horwitz, 1995). Intracellular symbionts are probably not generally subject to lysosomal attack, although the failure of the symbiosomal membrane to fuse with lysosomes has been demonstrated only in the hydra–*Chlorella* symbiosis (Hohman, McNeil & Muscatine, 1982). Intracellular rhizobia are subject to acidification via H^+/ATPase activity on the symbiosome membrane, and the symbiosomal pH may be as low as 5.5–6 units (Blumwald *et al.*, 1985). The symbiosome pH of sea anemones bearing *Symbodinium* and hydra with *Chlorella* is greater than 5.5 units and possibly close to neutrality (Rands *et al.*, 1992, 1993).

The restricted supply of nutrients, low growth rates and hazards of triggering host defensive responses together suggest that the host habitat, and particularly host cells, may generally be unfavourable or 'stressful' for symbiotic microorganisms (Moulder, 1985; Douglas, 1995). (Biological stress has been defined in a variety of different ways; in this chapter, it is described as any environmental factor that results in potentially injurious changes to organisms (see Hoffmann & Parsons, 1991).) In broad terms, this interpretation is further supported by the demonstration that intracellular bacteria have high levels of the chaparonin protein GroEL. This protein is present in all bacteria, and it is often described as a stress protein because its production is greatly increased when bacteria are exposed to a variety of stresses (e.g. high temperature). Elevated levels of GroEL have been described in the symbiotic bacteria in aphids and tsetse flies (Kakeda & Ishikawa, 1991; Aksoy, 1995) and *Rhizobium* in the root nodules of legumes (Choi, Ahn & Jeon, 1991), and also in some intracellular parasites (e.g. Shinnick, 1987; Vodkin & Williams, 1988). However, detailed interpretation of this biochemical indicator of stress is not feasible until the nature of the putative stress in eukaryotic cells has been established.

Symbiotic microorganisms as overproducers

The hosts in many associations derive substantial amounts of one or a few compounds from their symbiotic microorganisms. For example, legumes derive virtually all of the nitrogen fixed by their rhizobia, exclusively as ammonia, and the hosts of algae obtain photosynthesis-derived sugars (e.g. maltose from *Chlorella*), polyhydric alcohols (e.g. ribitol from *Trebouxia*) or glycerol (from *Symbiodinium*). The compounds translocated from symbionts to host cells are known as 'mobile compounds', and their sustained release represents a core adaptation of symbiotic microorganisms. The performance (survival, growth, reproduction etc.) of the host depends on this interaction, and linked to this, microorganisms which do not release these compounds are usually rejected by the host. For example, legume nodules abort prematurely if the rhizobia are unable to fix nitrogen (Brewin, 1991) and strains of *Chlorella* that release little photosynthate are lysed or expelled from the cells of hydra (McAuley & Smith, 1982).

At the metabolic level, this adaptation involves the overproduction of the mobile compounds by the symbionts and their efficient efflux from the symbiont cells. Recent studies on three symbiotic systems, the nitrogen-fixing rhizobia, the bacteria *Buchnera* in aphids and the dinoflagellate *Symbiodinium* in corals have revealed novel features in the regulation of metabolism, and a diversity of genetic and biochemical processes underlying nutrient release.

The rhizobia in legumes

In the rhizobium–legume symbiosis, the transport of ammonia from rhizobial cells to the surrounding plant cell cytoplasm is promoted by a very steep concentration gradient. For example, in soybean nodules, the ammonia concentration is 12 mM in the rhizobia and $<10\,\mu$M in the surrounding plant cell (Streeter, 1989). This condition is sustained primarily by the different activities of the ammonia assimilating enzyme, glutamine synthetase, in the rhizobia and plant cell (Fig. 1). The bacterial genes for glutamine synthetase are repressed, and rhizobia have virtually no capacity to assimilate the ammonia that they fix, but the host cell has high glutamine synthetase activity and assimilates rhizobial-derived ammonia very efficiently. Until recently, it was accepted that ammonia is translocated across the rhizobial cell membrane and legume symbiosome membrane by passive diffusion, but recent patch clamp analyses have identified an ammonium channel in the symbiosome membrane of soybean (Tyerman *et al.*, submitted).

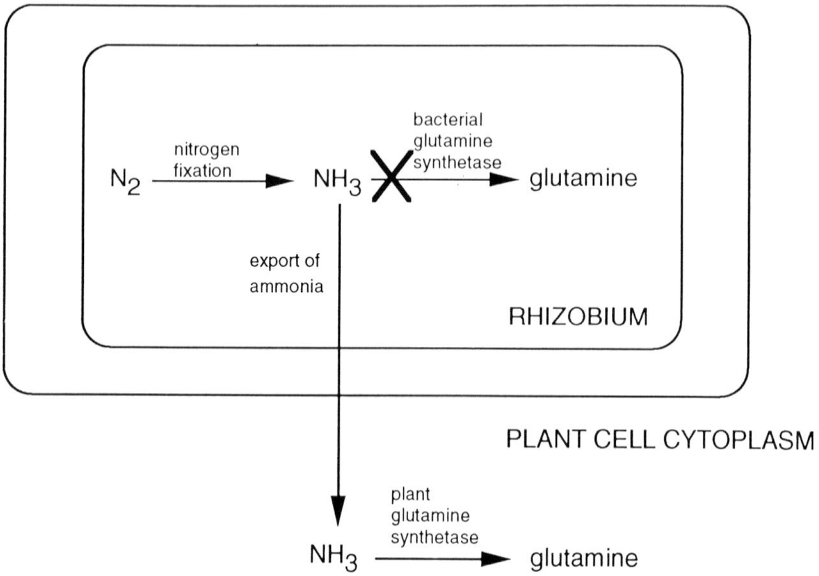

Fig. 1. Ammonia release from rhizobia and glutamine synthetase activity in rhizobia and the surrounding plant cell cytoplasm. Ammonia assimilation by the rhizobia is undetectable because glutamine synthetase production is repressed (see text for details).

From our understanding of the control of nitrogen fixation in nonsymbiotic bacteria such as *Klebsiella*, we might expect the high ammonia concentration in rhizobial cells to repress *nif* genes (via the *ntr* system), resulting in depressed nitrogen fixation rates. This does not occur because of a fundamental difference between the regulation of *nif* in rhizobia and other bacteria. In rhizobia, *nif* genes are expressed under microaerophilic conditions (as occur in root nodules), independent of the nitrogen supply. As demonstrated by de Philip, Batut & Bristard (1990), the oxygen sensor is a membrane-bound haemoprotein (FixL). Under low oxygen conditions, FixL phosphorylates a regulator protein (FixJ) resulting in the activation of *nif* genes.

The intracellular bacteria Buchnera *in aphids*

The nutrients acquired by aphids from their symbiotic bacteria *Buchnera* are essential amino acids. This has been demonstrated by a combination of dietary and metabolic experiments. Aphids with their bacteria are independent of a dietary supply of many or all essential amino acids, while aphids experimentally deprived of their bacteria have a dietary requirement for all essential amino acids (Mittler, 1971). The synthesis of essential amino acids and their translocation to the aphid tissues has been demonstrated by radiotracer studies (e.g. Douglas, 1988; Febvay *et al.*, 1995). However, nutrient release from *Buchnera* apparently ceases as soon as the bacteria are isolated from the aphids, and, as a result, we are entirely ignorant of the transporters in *Buchnera* and even whether the mobile compounds are amino acids or peptides.

The results of recent molecular analyses of the *Buchnera* genome suggest the basis for the overproduction of essential amino acids by *Buchnera*. Bracho *et al.* (1995) have demonstrated that *leuA-D*, the genes coding for leucine synthesis, are amplified and borne on a bacterial plasmid. If these plasmid-borne genes are expressed (this has not yet been demonstrated), *Buchnera* is potentially capable of synthesizing leucine at rates substantially higher than other bacteria with a single, chromosomal *leu* operon per genome. At present, it is not known whether the genes coding for the enzymes in the essential amino acid synthesis are generally amplified, but the data of Lai, Baumann & Baumann (1994) on the *trp* genes (coding for enzymes in tryptophan synthesis) suggest that the genetic organization of these genes is not uniform. In several aphid species, *trpEG*, which codes for anthranilate synthetase (the first enzyme in tryptophan biosynthesis) is amplified on a plasmid (distinct from the *leu*-plasmid), and the other *trp* genes (*trpDC(F)BA*) are chromosomal. The activity of anthranilate synthetase in most bacteria is inhibited by tryptophan, and as yet there is no indication that *Buchnera* enzyme is unusual. Lai *et al.* (1994) suggest that *Buchnera* cells overproduce anthranilate synthetase (reflecting the multiple gene copies of *trpEG*), resulting in a residual activity of the enzyme in the presence of tryptophan. In other words, amplification of *trpEG* reduces the sensitivity of tryptophan synthesis to feedback inhibition by tryptophan.

The dinoflagellate alga Symbiodinium *in corals and related animals*

Current understanding of photosynthate release by symbiotic algae is based entirely on metabolic studies, mostly involving ^{14}C-radioisotope. This contrasts with rhizobia and *Buchnera*, in which genetic and molecular analyses have been valuable tools. Photosynthate release in a variety of algae have three features in common: first, carbon is available for release for a very limited period, usually less than a few minutes, after fixation; and, secondly, no intracellular pool of the mobile compounds is detectable, suggesting that the synthesis and release of the mobile compound are closely linked; and third, release is (usually) abolished on separation from the host (e.g. Cernichiari, Muscatine & Smith, 1969; Sutton and Hoegh-Guldberg, 1990).

Recent studies have been conducted with *Symbiodinium*. They have focused primarily on the capacity of a homogenate of host tissue to induce some strains of this alga to continue to release photosynthate after isolation from the symbiosis. It has been suggested that photosynthate release is triggered by a single compound (the putative 'host factor') in the host homogenate. Hinde and coworkers (unpublished results) have indications that the factor may be a protein, while Gates *et al*. (1995) have demonstrated that simple mixtures of amino acids can mimic the effect of host homogenate. It is not known how host homogenate promotes photosynthate release. Ritchie, Eltringham & Hinde (1993) have shown that it does not influence the permeability of the *Symbiodinium* membrane to glycerol. Presumably, therefore, the host factor acts to promote the biosynthesis of glycerol, in preference to alternative metabolic fates of newly fixed photosynthate.

In summary, recent research has revealed a diversity of mechanisms underlying the overproduction of specific compounds by symbiotic microorganisms. They include:

(a) novel regulation of gene expression: the *nif* genes of rhizobia are controlled by oxygen tension and not (as in other bacteria) by the availability of combined nitrogen;

(b) novel genetic organization: certain genes coding for enzymes involved in the synthesis of essential amino acids are amplified and plasmid-borne in the bacterial symbionts of aphids;

(c) novel regulation of major metabolic pathways: the hosts of symbiotic algae can promote the metabolism of recently fixed photosynthate into the mobile compound, and the hosts of nitrogen-fixing bacteria suppress bacterial assimilation of the primary nitrogen-fixation product, ammonia.

These features can be considered as genetic and metabolic adaptations of microorganisms for symbiosis.

SPECIALIZATION OF SYMBIOTIC MICROORGANISMS

On the apparent rarity of symbiotic microorganisms in the environment

It is widely accepted, in the symbiosis literature, that most symbiotic microorganisms are ecologically restricted to their hosts, and have a very limited capacity to persist in the environment. The rhizobia are exceptional, in that they are readily isolated from soils that either support legumes or have borne legumes in recent decades (Sprent & Sprent, 1990), indicating that these symbiotic bacteria maintain freeliving populations independent of the host for extended periods. Recent studies with *Rhizobium leguminosarum*, however, indicate that a substantial fraction of the soil population may be unable to infect the legume hosts. For example, 83 (98%) of 85 isolates of *R. leguminosarum* bv. phaseoli obtained by Segovia *et al.* (1991) from a single field in Mexico, and 8 (62%) of 13 isolates of *R. leguminosarum* obtained by Laguerre, Bardin & Amarger (1993) from a field in France were not infective. It has been known for some years that, in *Rhizobium*, the genes conferring the ability to infect legumes are located on a single plasmid (the Sym plasmid). The non-infective isolates of Segovia *et al.* (1991) and Laguerre *et al.* (1993) became infective when complemented with the appropriate Sym plasmid, suggesting that they probably lacked the Sym plasmid or bore a non-functional plasmid. It appears that individual lineages of *Rhizobium* in the soil can switch between infective and noninfective states by the acquisition and loss, respectively, of the Sym plasmid (although the frequency of these events may be relatively low and vary between different chromosomal × plasmid combinations) (Schofield *et al.*, 1987). It is very likely that *Rhizobium* bearing the Sym plasmid are at a selective advantage in the presence of plants, but at a selective disadvantage in the soil (where the plasmid is functionless but costly to maintain); and, consequently, the relative abundance of infective over non-infective *Rhizobium* in the soil would be expected to decline progressively with time and distance from suitable legume hosts. At an ecological level, we can consider the Sym plasmid, but not the bacterium, to be specialized for symbiosis. This distinction between the bacterium and plasmid, however, does not apply to all rhizobia because, in *Bradyrhizobium*, the symbiosis-genes coding for symbiosis-specific functions are borne on the chromosome. Indeed, the situation in *Rhizobium* may be unique among symbiotic microorganisms.

The view that most symbiotic microorganisms are restricted to their host environment has been developed by Moulder (1979, 1985), who likened hosts, especially host cells, to 'extreme' environments, such as deserts, hot springs, salt pans. The organisms in the abiotic extreme environments (deserts etc.) are ecologically specialized; for example, the bacteria which inhabit hot springs are not also abundant in temperate waters, presumably because the physiological adaptations and life history traits that 'fit' these organisms for high temperatures preclude their utilisation of other environments. The proposed parallel between symbionts and the microorganisms in extreme environments is strengthened by the physiological and biochemical evidence that microorganisms in symbiosis are stressed (as discussed earlier in this chapter).

The principal evidence cited in the literature to support the notion that symbiotic microorganisms are specialized is that most are not routinely identified in samples of water, soil etc., and many have never been recovered from the environment. This evidence is, however, flawed. Most traditional sampling methods involve cultivation of microorganisms on routine microbiological media, and fail to detect any microorganisms that do not grow well *in vitro*. Recent analyses of the microbial diversity have used PCR and other DNA methods independent of cultivation, and revealed that the unculturable microorganisms may comprise more than 90% of the phylogenetic diversity and 30% of the microbial biomass in many habitats (e.g. Fuhrman, McCallum & Davis, 1993; De Long *et al.*, 1994; Amann, Ludwig & Scheifer, 1995).

The relevance of the limitations of traditional methods of microbial enumeration to symbiotic systems is illustrated by the recent study of Lee & Ruby (1995) on the association between the squid *Euprymna scolopes* and luminescent bacteria *Vibrio fischeri*. *V. fischeri* is generally culturable. By incubation of sea water samples on appropriate media, Lee & Ruby (1995) estimated the density of *V. fischeri* in the water column as 2–4 CFU ml^{-1}; but they obtained 100-fold higher estimates of the bacterial density both by direct hybridization of samples with a DNA probe specific to symbiotic *V. fischeri* and by quantification of the rates of infection of bacteria-free squid. Lee & Ruby (1995) concluded that up to 99% of the viable and infective *V. fischeri* in the water column are unculturable. In other words, the traditional method to enumerate luminescent bacteria underestimated the abundance of symbiotic *V. fischeri* by two orders of magnitude.

Methods independent of cultivation *in vitro* have not been used to assess symbiotic microorganisms in the environment. We simply lack the information to decide whether most symbiotic microorganisms are rare apart from their hosts. However, most vertically transmitted symbionts, such as the intracellular bacteria in insects, are unlikely to occur widely apart from their hosts. Many may not have ready access to the external environment, and they are presumably under no selection pressure to retain functions specifically required for a freeliving existence. Arguably, the most extreme expression of the specialization of vertically transmitted symbionts is their transformation into organelles, as is considered below.

Symbiont-derived organelles as the 'ultimate' specialisation of symbiotic microorganisms

Intracellular symbiotic microorganisms can be considered to have evolved into organelles of eukaryotic cells if genes essential to their functioning have been transferred to the host nucleus (Douglas, 1992). Symbiont-derived organelles are, consequently, specialized specifically to the host lineage bearing 'their' genes.

The specialization of symbiont-derived organelles can be illustrated by the chloroplast symbioses in sacoglossan molluscs. Various sacoglossans acquire plastids from the multicellular algae on which they feed. The plastids are photosynthetically active in the animal tissues, providing sugars to the animal host, but they cannot persist indefinitely in the animal (Gallop, Bartrop & Smith, 1980) because the animal genome lacks the plastid-specific genes. For example, ribulose bisphosphate carboxylase, the enzyme responsible for photosynthetic carbon dioxide fixation, cannot be synthesized in the mollusc–plastid association because the small subunit of this protein is coded in the algal nucleus.

Although the transfer of DNA to the host nucleus defines the transformation of symbiotic microorganisms into organelles, the evolutionary origins of the organelles can only be established with confidence if the organelles retain DNA. For example, it is impossible to test the idea of Cavalier-Smith (1990) that peroxisomes are symbiont-derived organelles that have lost all their DNA.

The value of the 'residual' DNA in symbiont-derived organelles in establishing the evolutionary history of organelles is indicated by recent research on the 'complex' plastids in various algae. It has been known for many years that the plastids in various algae have either three or four bounding membranes (and not the two membranes observed in chlorophytes, terrestrial plants, and rhodophytes). These multiple membranes have been interpreted as evidence for the lateral transfer of photosynthetic symbionts among protists (Gibbs, 1978, 1981). It is argued that certain nonphotosynthetic protists have gained photosynthesis by acquiring algae (i.e. eukaryotes with plastids) as symbionts. As illustrated in Fig. 2, the inner two membranes of the resultant complex plastid are those of the plastid, the third membrane is the cell membrane of the algal symbiont, and the fourth (outermost) membrane of the plastid is the symbiosome membrane of the host.

For most algae with complex plastids, the scenario outlined in Fig. 2 is untestable because only the genomes of the host (the nucleus) and the plastid are present. Two groups of algae, the cryptomonads and chlorarachniophytes, have complex plastids with a third genome, located between the second and third membranes (Fig. 2), the predicted location of the nucleocytoplasm of the algal symbiont in the hypothesis of Gibbs (1978). This

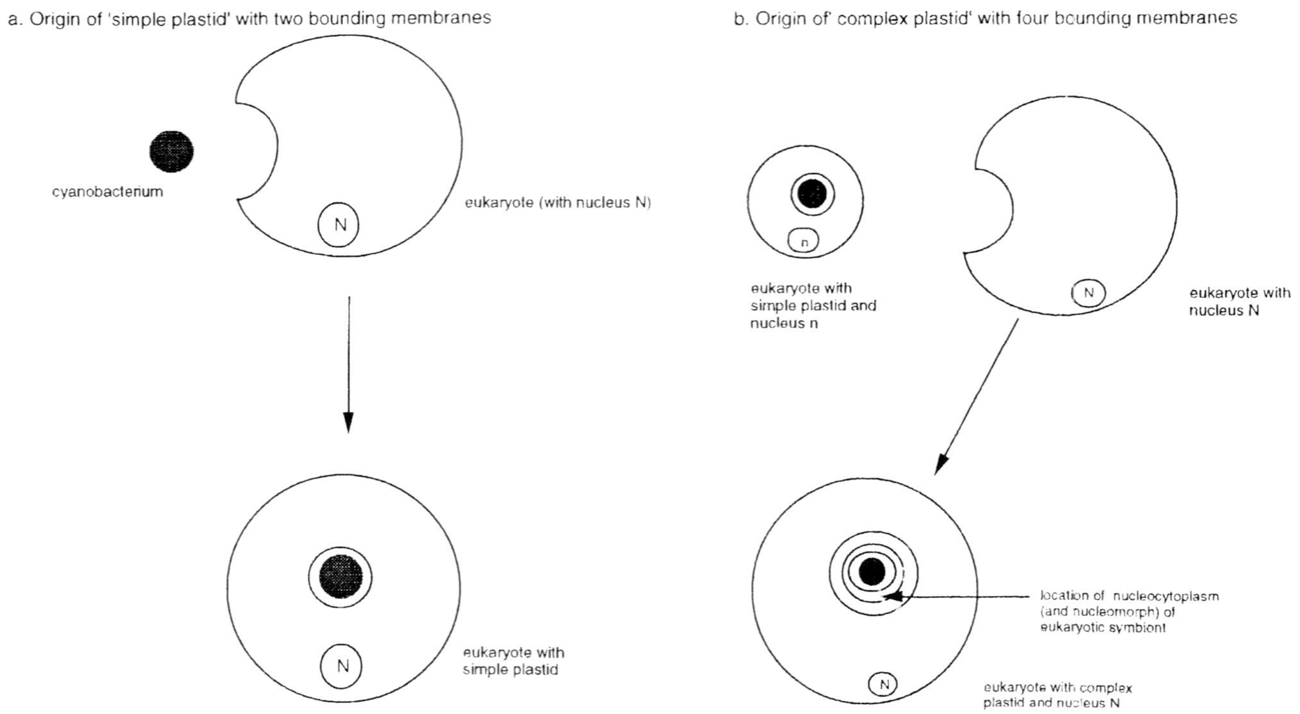

Fig. 2. The origin of plastids in algae and plants: (*a*) Simple plastids are derived from a cyanobacterium acquired by a protist. (*b*) Complex plastids are generated when a eukaryotic algae (as in (*a*)) is acquired by another protist. The nucleus of the eukaryotic alga may be lost, or retained as a nucleomorph.

genome was first recognized in ultrastructural studies, as a small, nucleus-like structure, which was called a nucleomorph (Greenwood, Griffiths & Santore, 1977). Recent molecular analyses have demonstrated that, in both cryptomonads and chlorarachniophytes, the nucleomorph has a haploid genome size of just 550–660 kb (it is the smallest known eukaryotic genome). The DNA is organized into three chromosomes, all of which bear rRNA genes (Eschbach *et al.*, 1991; McFadden *et al.*, 1994) and one of which has the gene for the heat-shock protein Hsp70 (Hofmann *et al.*, 1994; Rensing *et al.*, 1994). Sequence analysis of the rRNA genes have revealed that the nucleomorph of cryptomonads is probably allied with rhodophytes (Douglas *et al.*, 1991), with the implication that the ancestors of cryptomonads acquired unicellular rhodophytes as symbionts. However, the nucleomorph of chlorarachniophyte *Chlorarachnion* cannot, as yet, be allied with any extant algal group, on the basis of its rRNA sequence (McFadden *et al.*, 1994).

At present, we do not understand why some, but apparently not all, genomes are retained in symbiont-derived organelles. Organellar DNA is certainly not required for division of the symbiont-derived organelles, as is indicated by the persistence of mitochondria lacking DNA in the *petite* mutants of yeasts. McFadden and Gilson (1995) have suggested that the eukaryotic nucleus in complex plastids is retained (as the nucleomorph) only if it contains plastid-specific genes. This suggestion, however, begs the question why genes are more likely to be transferred to the host nucleus from a cyanobacterial genome than from nucleomorph. The retention of the mitochondrial genome may be linked to various discrepancies between the genetic code in these organelles and the 'universal code' in the eukaryotic nucleocytoplasm. For example, CUA codes for leucine in the nucleocytoplasm, and threonine in yeast mitochondria, while UGA is a stop codon in the universal code and a codon for tryptophan in animal mitochondria. This explanation is not, however, relevant to the retention of plastid genome, which does not depart from the universal code. A final possible reason arises from the fact that organelle-specific proteins that are coded in the nucleus are synthesized in the cytoplasm and subsequently imported to the organelles. Heijne (1986) has suggested that certain genes may be retained in the organelles because the structural conformation of the protein prevents their effective translocation from the cytoplasm into the organelles.

The genomes of all symbiont-derived organelles are small, and contain relatively few protein-coding genes. For example, the human mitochondrial genome bears just 13 protein-coding genes, whose expression requires a translation machinery of rRNA and tRNA molecules, coded by 24 mitochondrial genes (Gray, 1989). The resources for this apparently inefficient system are provided by the nucleocytoplasm, and the maintenance of these functional organelle genomes is, presumably, costly to eukaryotic cells.

CONCLUSIONS

Much of the literature on symbiosis treats the microorganisms in symbiosis as a coherent group, with the underlying assumption that they have many common characteristics, which adapt them to the symbiotic environment. Various studies, as reviewed in this article, provide partial support for this view. In particular, all intracellular microorganisms examined to date have high levels of the GroEL protein, suggesting that the habitat in eukaryotic cells may be stressful. There are also clear indications that the microorganisms have restricted access to various nutrients. In other respects, the adaptations of symbiotic microorganisms are very diverse. The metabolic and genetic basis of nutrient release by symbionts varies between different systems, and the little information available suggests that host defences may be avoided by a variety of mechanisms. Furthermore, symbiotic microorganisms may vary widely in the degree of specialization for symbiosis. Although we certainly need more information on this issue, it is already clear that some microorganisms, such as the rhizobia, sustain substantial freeliving populations, while the existence of others, especially vertically-transmitted symbionts and the symbiont-derived organelles, is inextricably linked to that of the eukaryotes which house them. These findings suggest that we should be cautious in extrapolating conclusions from the study of one type of association to other systems. Generalizations about symbiotic microorganisms should be based on research on a range of different symbioses.

REFERENCES

Aksoy, S. (1995). Molecular analysis of the endosymbionts of tsetse flies: 16S rDNA locus and over-expression of a chaparonin. *Insect Molecular Biology*, **4**, 23–9.

Amann, R. I., Ludwig, W. & Scheifer, K-H. (1995). Phylogenetic identification and in situ detection of individual microbial cells without cultivation. *Microbiological Reviews*, **59**, 143–69.

Blumwald, E., Fortin, M. G., Rea, P. A., Verma, D. P. S. & Poole, R. J. (1985). Presence of the host-plasma membrane type H^+-ATPase in the membrane envelope enclosing the bacteroids in soybean root nodules. *Plant Physiology*, **78**, 665–72.

Bracho, A. M., Martinez-Torres, D., Moya, A. & Latorre, A. (1995). Discovery and molecular characterisation of a plasmid localised in *Buchnera* sp. bacterial endosymbiont of the aphid *Rhopalosiphum padi*. *Journal of Molecular Evolution*, **41**, 67–73.

Brewin, N. J. (1991). Development of the legume root nodule. *Annual Reviews of Cell Biology*, **7**, 191–226.

Cavalier-Smith, T. (1987). The origin of cells: a symbiosis between genes, catalysts and membranes. *Cold Spring Harbor Symposia on Quantitative Biology*, **52**, 805–24.

Cavalier-Smith, T. (1990). Symbiotic origin of peroxisomes. *Endocytobiology IV*, Nardon, P. Gianinazzi-Pearson, V., Grenier, A. M., Margulis, L. and Smith, D. C., eds. pp. 515–521. Institut National de la Recherche Agronomique, Paris.

Cavalier-Smith, T. (1992). The number of symbiotic origins of organelles. *BioSystems*, **28**, 91–106.

Cernichiari, E., Muscatine, L. & Smith, D. C. (1969). Maltose excretion by the symbiotic algae of *Hydra viridis*. *Proceedings of the Royal Society of London B*, **173**, 557–76.

Choi, E. Y., Ahn, T. S. & Jeon, K. W. (1991). Elevated levels of stress proteins associated with bacterial symbiosis in *Amoeba proteus* and soybean root nodule cells. *BioSystems*, **25**, 205–12.

Clemens, D. L. & Horwitz, M. A. (1995). Characterisation of the *Mycobacterium tuberculosis* phagosome and evidence that phagosomal maturation is inhibited. *Journal of Experimental Medicine*, **181**, 257–70.

De Long, E. F., Wuk-Y., Prezelli, B. B. & Jovine, R. V. M. (1994). High abundance of Archaea in Antarctic marine picoplankton. *Nature*, **371**, 695–7.

de Philip, P., Batut, J. & Bristard, P. (1990). *Rhizobium meliloti* FixL is an oxygen sensor and regulates *R. meliloti nifA* and *fix K* genes differently in *Escherichia coli*. *Journal of Bacteriology*, **172**, 4255–62.

Douglas, A. E. (1988). Sulphate utilisation in an aphid symbiosis. *Insect Biochemistry*, **18**, 599–605.

Douglas, A. E. (1992). Symbiosis in evolution. *Oxford Surveys in Environmental Biology*, **8**, 347–82.

Douglas, A. E. (1994). *Symbiotic Interactions*. Oxford University Press, Oxford.

Douglas, A. E. (1995). The ecology of symbiotic microorganisms. *Advances in Ecological Research*, **26**, 69–103.

Douglas A. E. & Huss, V. A. R. (1986). On the characteristics and taxonomic position of symbiotic *Chlorella*. *Archives for Microbiology*, **145**, 80–4.

Douglas, S. E., Murphy, C. A., Spencer, D. F. & Gray, M. W. (1991). Cryptomonad algae are evolutionary chimaeras of two phylogenetically-distinct unicellular eukaryotes. *Nature*, **350**, 148–51.

Eschbach, S., Hofmann, C. J. B., Maier, U-G., Sitte, P. & Hansmann, P. (1991). A eukaryotic genome of 660 kb: karyotype of nucleomorph and cell nucleus of the cryptomonad alga, *Pyrenomonas salina*. *Nucleic Acids Research*, **19**, 1779–81.

Febvay, G., Liadouze, I., Guillard, J. & Bonnot, G. (1995). Analysis of energetic amino acid metabolism in *Acyrthosiphon pisum*: a multidimensional approach to amino acid metabolism in aphids. *Archives of Insect Biochemistry and Physiology*, **29**, 45–69.

Fuhrman, J. A., McCallum, K. & Davis, A. A. (1993). Phylogenetic diversity of subsurface marine microbial communities from the Atlantic and Pacific Oceans. *Applied and Environmental Microbiology*, **59**, 1294–302.

Gallop, A., Bartrop, J. & Smith, D. C. (1980). The biology of chloroplast acquisition by *Elysia viridis*. *Proceedings of the Royal Society of London B*, **207**, 335–49.

Gates, R. D., Hoegh-Guldberg, O., McFall-Ngai, M. J., Bil', K. Y. & Muscatine, L. (1995). Free amino acids exhibit anthozoan 'host factor' activity: they induce the release of photosynthate from symbiotic dinoflagellates *in vitro*. *Proceedings of the National Academy of Sciences USA*, **92**, 7430–2.

Gibbs, S. P. (1978). The chloroplasts of *Euglena* may have evolved from symbiotic green algae. *Canadian Journal of Botany*, **56**, 2883–9.

Gibbs, S. P. (1981). The chloroplasts of some algal groups may have evolved from eukaryotic algae. *Annals of the New York Academy of Science*, **361**, 193–208.

Gray, M. W. (1989). Origin and evolution of mitochondrial DNA. *Annual Reviews in Cell Biology*, **5**, 25–50.

Greenwood, A. D., Griffiths, H. B. and Santore, U. J. (1977). *British Phycological Journal*, **12**, 119.

Haygood, M. G. (1993). Light organ symbioses in fishes. *Critical Reviews in Microbiology*, **19**, 191–216.

Heijne, G. (1986). Why mitochondria need a genome. *FEBS Letters*, **198**, 1–4.

Herrada, G., Puppo, A. & Rigaud, J. (1989). Uptake of metabolites by bacteroid-containing vesicles and by free bacteroids from french bean nodules. *Journal for General Microbiology*, **135**, 3165–71.

Hinde, R. (1971). The control of the mycetome symbiotes of the aphids *Brevicoryne brassicae, Myzus persicae* and *Macrosiphum rosae. Journal of Insect Physiology,* **17**, 1791–800.

Hoffmann, A. A. & Parsons, P. A. (1991). *Evolutionary Genetics and Environmental Stress*. Oxford University Press, Oxford, UK.

Hofmann, C. J. B., Rensing, S. A., Hauber, M. M., Martin, W. F., Muller, S. B., Couch, J., McFadden, G. I., Igloi, G. L. & Maier, U-G. (1994). The smallest known eukaryotic genomes encode a protein gene: towards an understanding of nucleomorph functions. *Molecular and General Genetics*, **243**, 600–4.

Hohman, T. C., McNeil, P. L. & Muscatine, L. (1982). Phagosome-lysosome fusion inhibited by algal symbiont of *Hydra viridis. Journal of Cell Biology*, **94**, 56–63.

Jackson, A. E. & Yellowlees, D. J. (1990). Phosphate uptake by zooxanthellae isolated from corals. *Proceedings of the Royal Society of London B*, **242**, 201–4.

Kakeda, K. & Ishikawa, H. (1991). Molecular chaparone produced by an intracellular symbiont. *Journal of Biochemistry*, **110**, 583–7.

Laguerre, G., Bardin, M. & Amarger, N. (1993). Isolation from soil of symbiotic and nonsymbiotic *Rhizobium leguminosarum* by DNA hybridization. *Canadian Journal of Microbiology*, **39**, 1142–9.

Lai, C. Y., Baumann, L. and Baumann, P. (1994). Amplification of *trpEG*: adaptation of *Buchnera aphidicola* to an endosymbiotic association with aphids. *Proceedings of the National Academy of Sciences USA*, **91**, 3819–23.

Lee, K-H. and Ruby, E. G. (1995). Symbiotic role of the viable but nonculturable state of *Vibrio fischeri* in Hawaiian coastal seawater. *Applied and Environmental Microbiology*, **61**, 278–83.

McAuley, P. J. (1987). Nitrogen limitation and amino acid metabolism of *Chlorella* symbiotic with green algae. *Planta*, **171**, 532–8.

McAuley, P. J. & Smith, D. C. (1982). The green hydra symbiosis. V. Stages in the intracellular recognition of algal symbionts by digestive cells. *Proceedings of the Royal Society of London B*, **216**, 7–23.

McFadden, G. I. & Gilson, P. R. (1995). Something borrowed, something green: lateral transfer of chloroplasts by secondary endosymbiosis. *Trends in Ecology and Evolution*, **10**, 12–17.

McFadden, G. I., Gilson, P. R., Hofmann, C. J. B., Adcock, G. J. & Maier, U-G. (1994) Evidence that an amoeba acquired a chloroplast by retaining part of an engulfed eukaryotic alga. *Proceedings of the National Academy of Sciences*, **91**, 3690–4.

Margulis, L. (1981). *Symbiosis in Cell Evolution* (2nd edition). Freeman, San Fransisco, USA.

Margulis, L. & Fester, R. (1991). *Symbiosis as a Source of Evolutionary Innovation: Speciation and Morphogenesis*. The MIT Press, Cambridge, Massachusetts.

Maynard Smith, J. and Szathmary, E. (1995). *The Major Transitions in Evolution*. W. H. Freeman, New York.

Mittler, T. E. (1971). Dietary amino acid requirements of the aphid *Myzus persicae* affected by antibiotic uptake. *Journal of Insect Physiology*, **101**, 1023–8.

Moulder, J. W. (1979). The cell as an extreme environment. *Proceedings of the Royal Society of London B*, **204**, 199–210.

Moulder, J. W. (1985). Comparative biology of intracellular parasitism. *Microbiological Reviews*, **49**, 298–337.

Rands, M. L., Douglas, A. E., Loughman, B. C. & Hawes, C. R. (1992). The pH of the perisymbiont space in the green hydra–*Chlorella* symbiosis: an immunocytochemical investigation. *Protoplasma*, **170**, 90–3.

Rands, M. L., Loughman, B. C. & Douglas, A. E. (1993). The symbiotic interface in an alga-invertebrate symbiosis. *Proceedings of the Royal Society of London B*, **253**, 161–5.

Rensing, S. A., Goddemeier, M., Hofmann, C. J. B. & Maier, U-G. (1994). The presence of a nucleomorph HSP70 gene is a common feature of Crytophyceae and Chlorachniophyceae. *Current Genetics*, **26**, 451–5.

Ritchie, R. J., Eltringham, K. & Hinde, R. (1993). Glycerol uptake by zooxanthellae of the temperate hard coral *Plesiastrea versipora*. *Proceedings of the Royal Society of London B*, **253**, 189–95.

Roth, L. E., Jeon, K. & Stacey, G. (1988). Homology in endosymbiotic systems: the term 'symbiosome'. In *Molecular Genetics of Plant–Microbe Interactions*. Palacios, R. & Verma, D. P. S., pp. 220–5. APS Press, St Paul, Minnesota.

Russell, D. G. (1995). *Mycobacterium* and *Leishmania* – stowaways in the endosomal network. *Trends in Cell Biology*, **5**, 125–8.

Schlegel, M. (1994). Molecular phylogeny of eukaryotes. *Trends in Ecology and Evolution*, **9**, 330–5.

Schofield, P. R., Gibson, A. H., Dudman, W. F. & Watson, J. M. (1987). Evidence for genetic exchange and recombination of *Rhizobium* symbiotic plasmids in a soil population. *Applied and Environmental Microbiology*, **53**, 2942–7.

Schwab, J. C., Beckers, C. J. M. & Joiner, K. A. (1994). The parasitiphorous vacuole membrane surrounding intracellular *Toxoplasma gondii* functions as a molecular sieve. *Proceedings of the National Academy of Sciences USA*, **91**, 509–13.

Segovia, L., Pinero, D., Palacios, R. and Martinez-Romero, E. (1991). Genetic structure of a soil population of nonsymbiotic *Rhizobium leguminosarum*. *Applied and Environmental Microbiology*, **57**, 426–33.

Shinnick, T. M. (1987). The 65-kilodalton antigen of *Mycobacterium tuberculosis*. *Journal of Bacteriology*, **69**, 1080–8.

Sprent, J. I. & Sprent, P. (1990). *Nitrogen Fixing Organisms*. Chapman and Hall, England.

Streeter, J. G. (1989). Estimation of ammonium concentration in the cytosol of soybean nodules. *Plant Physiology*, **90**, 779–82.

Sturgill-Koszycki, S., Schlesinger, P. H., Chakraborty, P., Haddix, P. L., Collins, H. L., Fok, A. K., Allen, R. D., Gluck, S. L., Heuser, J. & Russell, D. G. (1994). Lack of acidification of *Mycobacterium* phagosomes produced by exclusion of the vesicular proton-ATPase. *Science*, **263**, 678–81.

Sutton, D. C. & Hoegh-Guldberg, O. (1990). Host–zooxanthella interactions in four temperate marine invertebrate symbioses: assessment of effect of host extracts on symbionts. *Biological Bulletin*, **175**, 178–86.

Udvardi, M. K., Price, G. C., Gresshoff, P. M. & Day, D. A. (1989). A dicarboxylate transporter on the peribacteroid membrane of soybean nodules. *FEBS Letters*, **231**, 36–40.

Udvardi, M. K., Salom, C. L. & Day, D. A. (1988). Transport of L-glutamate across the bacteroid but not the peribacteroid membrane of soybean nodules. *Molecular Plant–Microbe Interactions*, **1**, 250–4.

Untermann, B. M., Baumann, P. & McLean, D. (1989). Pea aphid symbiont relationships established by analysis of 16S rRNAs. *Journal of Bacteriology*, **171**, 2970–4.

Vasse, J., de Billy, F. & Truchet, G. (1993). Abortion of infection during the *Rhizobium meliloti*–alfalfa symbiotic interaction is accompanied by a hypersensitive reaction. *The Plant Journal*, **4**, 555–66.

Vodkin, M. H. & Williams, C. (1988). A heat shock protein operon in *Coxiella burnetti* produces a major antigen homologous to a protein in both *Mycobacteria* and *Escherichia coli*. *Journal of Bacteriology*, **170**, 1227–34.

Wolfe, C. J. & Haygood, M. G. (1993). Bioluminescent symbionts of the Caribbean flashlight fish (*Kryptophanaron alfredi*) have a single rRNA operon. *Molecular Marine Biology and Biotechnology*, **2**, 189–97.

EVOLUTION OF METABOLIC PATHWAYS

L. DIJKHUIZEN

Department of Microbiology, Groningen Biomolecular Sciences and Biotechnology Institute (GBB), University of Groningen, Kerklaan 30, 9751 NN Haren, the Netherlands

INTRODUCTION

The analysis of the evolution of metabolic pathways is especially rewarding with micro-organisms: (i) they can be grown in large population sizes in the laboratory under strictly controlled and strongly selective conditions; (ii) only a relatively short period of time is required to obtain many successive generations; (iii) they only have relatively small genome sizes. Over the years a wealth of information has become available about a large variety of metabolic pathways in micro-organisms and the enzymes involved. Most of these pathways are involved in catabolic and/or biosynthetic functions; they are backed up by sets of proteins functioning in signal reception, transduction, and regulation, allowing an accurate global control of cellular metabolism carefully balanced with the environmental conditions. These metabolic pathways are of varying complexity and may consist of either a few or many enzyme steps.

With the rapid emergence of ever more sophisticated techniques for the analysis of protein structures and enzyme mechanisms, nucleotide sequencing, data analysis, and *in vitro* protein engineering, the study of structure/function relationships of these proteins has increasingly become possible. This has also provided further insights into the evolution of enzyme function in the origin of metabolic pathways and how the latter may have arisen.

In bacteria incubated under strongly selective conditions, genotypic adaptation may occur, allowing *de novo* utilization of new growth substrates present or resulting in the appearance of mutants resistant to toxic compounds. The study of these phenomena allows an analysis of evolutionary mechanisms under laboratory conditions as well as the construction of strains of micro-organisms possessing particular desirable traits, and is sometimes referred to as *in vivo* protein engineering.

Current knowledge on the evolution of metabolic pathways is reviewed in this chapter. Emphasis is given to information that has become available from recent studies of the evolution of enzyme function and of genotypic adaptation mechanisms.

EVOLUTION OF ENZYME FUNCTION

What do protein sequences tell us?

The classical view of the origin of metabolic pathways is that of retro-evolution, with the last reaction in the pathway being the first to evolve, and the first enzyme being the last one to appear (Horowitz, 1945). This comes from the notion that the development of isolated new enzyme activities would lack adaptive significance unless the enzymes catalysing the final steps in a biosynthetic pathway were also available. Consecutive enzymes in biochemical pathways, which bind the same substrate and product molecules, may have evolved by a series of gene duplication events, followed by divergence. As the multiple genes for a pathway are generally organized in an operon under a common control mechanism, this prompted the suggestion that all structural genes of an operon originated from a common ancestral gene (Horowitz, 1965). Consequently, all the member proteins of a particular pathway would be related to each other. Alternatively, the multiple genes for a pathway may have originated from different ancestral genes in different regions of the chromosome, and were transposed subsequently to their present positions. Jensen (1976) suggested that genes encoding enzymes catalysing analogous reactions would be more likely to be homologous than would genes encoding sequential enzyme steps.

With the accumulation of nucleotide sequence data for complete biochemical pathways, evidence is increasing that the genes encoding the consecutive enzymes in a metabolic pathway arose from different antecedent genes. To date very few convincing similarities at the sequence level have been found between the different enzymes in metabolic pathways, except for short structural motifs which are involved in binding of common nucleotides, cofactors, substrates or products. Therefore assembly of new biosynthetic pathways most likely proceeded by the recruitment of already existing enzymes. It is conceivable that early micro-organisms possessed the ability to make only a limited number of different active enzymes. However, those enzymes might have possessed very broad substrate specificities (Jensen, 1976; Lazcano, Fox & Oró, 1992). Evolution could have led to enzyme specialization with duplicate genes diverging to code for enzymes with an improved catalytic activity and specificity for a particular substrate, the substrate specialization being further reinforced by the development of regulatory mechanisms (Jensen, 1976). This process finally resulted in the establishment of the contemporary enzyme families catalysing specific classes of chemical reactions.

A large number of translated protein sequences are now available for *Escherichia coli* together with a wealth of data about their functions. A systematic search for amino acid sequence similarities of each *E. coli* protein to all others available revealed that extended sequence similarities coupled with functional similarities are widespread among the proteins of *E. coli* whose sequences are known (Labedan & Riley, 1995). Thus in many cases a single ancestral protein may have been the evolutionary source of a range of enzymes catalysing similar reactions with different substrates. Genome sequencing in a larger variety of micro-organisms undoubtedly will provide further interesting data on such paralogous gene sequences (homologous genes that diverged after duplication events), allowing a further analysis of metabolic evolution. Also a detailed analysis of the evolution of the mandelate pathway showed that for each step an apparently more widespread enzyme, that catalyzes a similar chemical reaction, had been used as the source of the new activity (Petsko *et al.*, 1993). Further support has also been derived from data now available for enzymes of many other pathways, for instance those of the β-ketoadipate pathway (catechol, 1,2-dioxygenase, muconate cycloisomerase, muconate lactonizing enzyme) (Petsko *et al.*, 1993; Schlömann, 1994; Elsemore & Ornston, 1995).

What do protein structures tell us?

The rapidly expanding knowledge in this field has led to the recognition that the total number of protein structural types is much smaller than the total number of proteins. This most likely reflects the evolutionary history of protein diversity. A striking example is provided by proteins with a $(\beta/\alpha)_8$-barrel super-secondary structure, the so-called TIM-barrel fold. It consists of a core of eight twisted parallel β-strands arranged close together, like staves, into a barrel. The α-helices that connect the parallel β-strands are all on the outside of this barrel. Many enzymes have now been found to have this fold as a significant part of their structures. It is still debated whether the TIM-barrels themselves have undergone convergent or divergent evolution, or both (Farber, 1993; Doolittle, 1994). The large diversity of functions of these proteins could reflect a general convergence to the same structure which has arisen independently several times during evolution, either as a result of the intrinsic stability of these barrels or their ease of formation (Branden & Tooze, 1991; Doolittle, 1994). Sequence similarities between proteins that contain this structure, however, are only observed in the case of enzymes that perform analogous functions in different micro-organisms. Thus only for these proteins is descent from a common ancestor clearly discernable. Two examples are discussed in the following sections.

The α-amylase family

Glycosyl hydrolase enzymes have been classified into more than 45 families on the basis of amino acid sequence similarities. Enzymes with different substrate specifications are sometimes found in the same family, indicating an evolutionary divergence, leading to acquisition of new specificities (Davies & Henrissat, 1995). The α-amylase family (Janeček, 1994; Jespersen *et al.*, 1991, 1993; Svensson, 1994) consists of starch hydrolases and related enzymes, for instance α-amylase, cyclodextrin glycosyltransferase (CGTase), pullulanase, and other enzymes which catalyse cleavage (or formation) or α-1, 4- and α-1, 6-glucosidic bonds in malto-oligosaccharides or polysaccharides such as amylose and amylopectin. At least ten crystal structures are currently available of amylolytic enzymes. These three-dimensional structures, and the known primary structures of many other amylolytic enzymes which show clear sequence similarities, indicate that the α-amylase family is indeed descended from a common ancestor. These enzymes possess a number of common structural domains (Fig. 1), of which the A domain of around 400 amino acids folds into the catalytic TIM-barrel structure. The functions of the other domains are generally not known, except for the E domain, which is a starch granule-binding domain and may allow enzymes such as CGTase to use new raw starch as a substrate. Common catalytic and binding events in the α-amylase family have been reviewed (Svensson, 1994). The general characteristics of the active sites of the α-amylase family proteins are very similar in that they carry the catalytic triad Asp328, Asp229 and Glu257 (numbering for the *Bacillus circulans* strain 251 CGTase; Lawson *et al.*, 1994) at the same position in the A domain (Fig. 2). These three carboxylates are strongly conserved in members of the α-amylase family and, where studied, their modification via site-directed mutagenesis results in virtually complete loss of activity. Members of the α-amylase family thus possess similar active sites and $(\beta/\alpha)_8$-barrel structures but represent 18 different enzymes specificities (Svensson, 1994). This large diversity in enzyme specificity may arise by variations in substrate

Fig. 1. Domain-level architecture in starch-degrading and related enzymes. Adapted from Jesperson *et al.* (1991). White areas indicate segments in which no similarity to other sequences has been found. ттттттт indicates heavily glycosylated 'hinge' regions in the glucoamylases. See Jespersen *et al.* (1991) for definitions of domains A–J and references. TAA, Taka-amylase A, *Aspergillus oryzae*; G4α, maltotetrao-hydrolase, *Pseudomonas saccharophila*; αSli, α-amylase, *Streptomyces limosus*; G2α, maltogenic α-amylase, *Bacillus stearothermophilus*; CGT, cyclodextrin glycosyltransferase, *Bacillus circulans*; BE, branching enzyme, *Escherichia coli*; PuKa, pullulanase, *Klebsiella aerogenes*; αPu, α-amylase-pullulanase, *Clostridium thermohydrosulfuricum*; G5α, maltopentaose-producing amylase, alkaliphilic Gram-positive bacterium; βSb, β-amylase, soybean; βCt, β-amylase, *Clostridium thermosulfurogenes*; Mal, maltase, *Saccharomyces cerevisiae*; 1,6G, oligo-1,6-glucosidase, *Bacillus cereus*; DxG, dextran glucosidase, *Streptococcus mutans*; Iso, isoamylase, *Pseudomonas amyloderamosa*; PuBs, pullulanase, *B. stearothermophilus*; NPu, neopullulanase, *B. stearothermophilus*; βBp, β-amylase, *Bacillus polymyxa*; GAn, glucoamylase, *Aspergillus niger*; GRh, glucoamylase, *Rhizopus oryzae*.

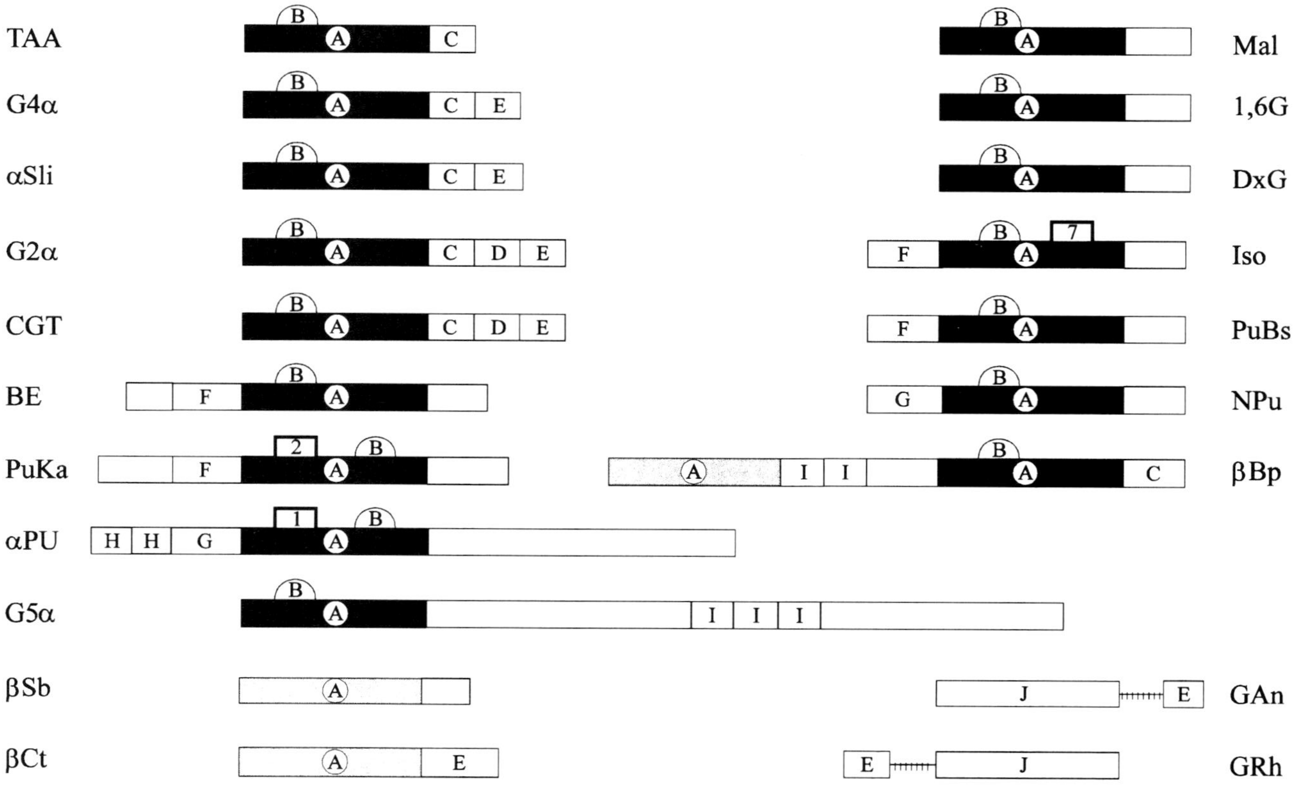
TAA
G4α
αSli
G2α
CGT
BE
PuKa
αPU
G5α
βSb
βCt
Mal
1,6G
DxG
Iso
PuBs
NPu
βBp
GAn
GRh
A
B
C
D
E
F
G
H
I
J
1
2
7

Fig. 1.

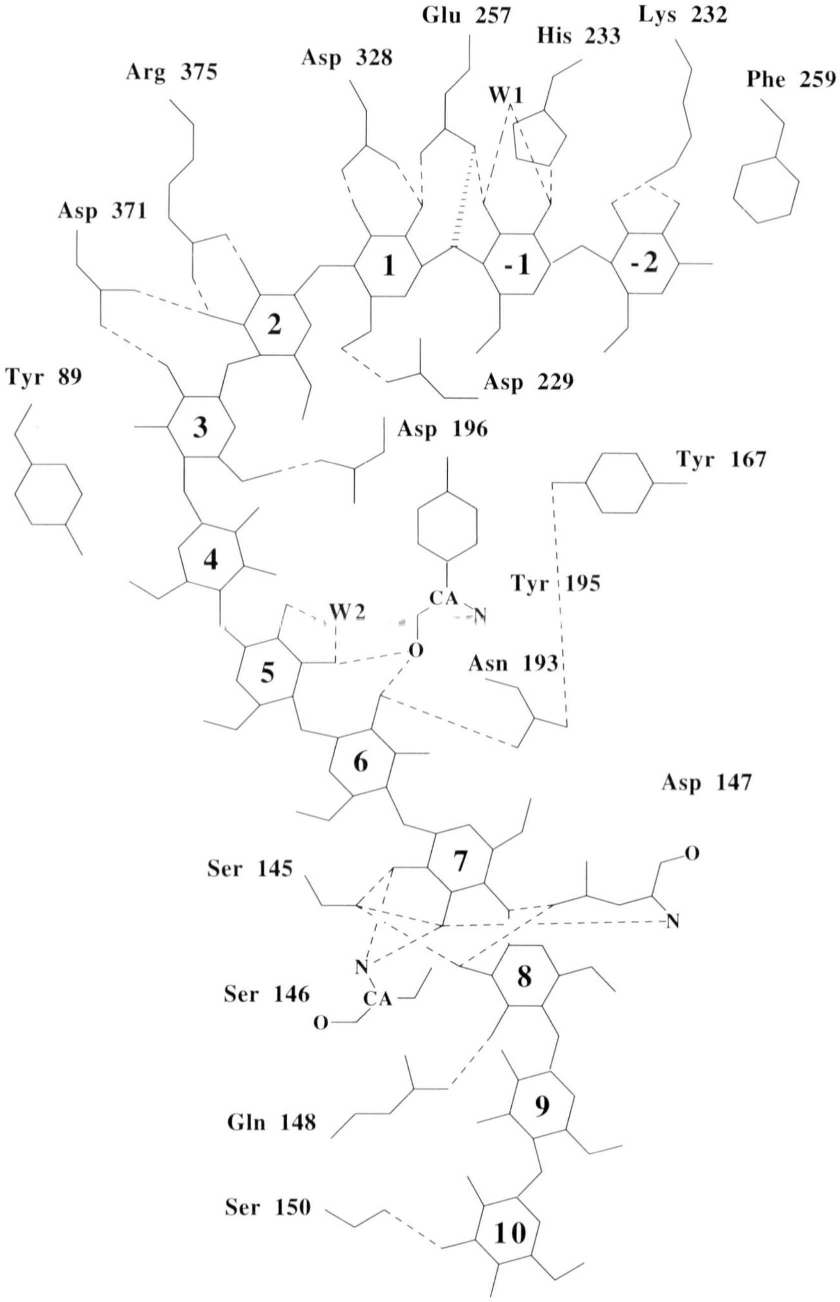

Fig. 2. Summary of hydrogen bonds between *Bacillus circulans* strain 251 CGTase and the maltononaose inhibitor bound at the active site. Adapted from Strokopytov *et al.* (1996).

Table 1. *Cyclization activity (units/mg), starch conversion and product specificity of* Bacillus circulans *strain 251 wild type and mutant CGTases*[a]

Mutants	Cyclization activity	Conversion of starch into cyclodextrins (%)	Product ratio (%)			Conversion (%) of starch into G1–G4 oligosaccharides
			α	β	γ	
Y195	280	39.3	13	64	23	0
Y195F	175	38.8	15	64	20	0
Y195W	74	33.3	18	63	19	2–4
Y195L	143	24.4	0	86	14	6–10
Y195G	22	24.8	19	64	17	16–20

[a]Results of 45 h incubations of CGTase proteins (0.1 unit of β-cyclodextrin-forming activity per ml) with 10% jet-cooked starch.
Adapted from Penninga *et al.* (1995).

binding at the loops between the β-strands and α-helices (Janeček, 1994; Svensson, 1994).

The three-dimensional structures of the *B. circulans* strain 8 (Klein & Schulz, 1991), strain 251 (Lawson *et al.*, 1994), and *Bacillus stearothermophilus* (Kubota *et al.*, 1991) CGTase proteins have been determined, revealing strong similarities. Site-directed mutagenesis made clear that the active site residue Tyr195 plays an important role in the CGTase cyclization and coupling reactions (Table 1; Fig. 2) (Fujiwara *et al.*, 1992; Nakamura, Haga & Yamane, 1994; Penninga *et al.*, 1995; Sin *et al.*, 1994). This aromatic residue occupies a dominant position in the active site and is strongly conserved in CGTase proteins (as Tyr or Phe). In contrast, α-amylase proteins possess a Gly, Ser or Val residue at this position. Soaking experiments with *B. circulans* strain 251 CGTase protein crystals, the G4 inhibitor acarbose and a G6 malto-oligosaccharide, resulted in binding of a G9 inhibitor in the active site. The three-dimensional structure of this protein-inhibitor complex was solved at 2.6 Å resolution, allowing identification of the interactions between sugar molecules and amino acid residues at these nine sugar-binding subsites (Fig. 2) and elucidation of the cyclization mechanism of CGTase (Strokopytov *et al.*, 1996). The loop forming sugar-binding subsites 6, 7 and 8 (amino acid residues 145–148) appears to be very important for the specificity of CGTase. Tyr195 and these sugar-binding subsites thus may constitute important evolutionary adaptations in the CGTase protein, allowing production of cyclic cyclodextrin compounds from starch (Lawson *et al.*, 1994; Penninga *et al.*, 1995; Strokopytov *et al.*, 1996). Organisms employing CGTase enzymes for conversion of starch into cyclodextrins also possess cell-bound cyclomalto-dextrinase enzymes, converting cyclodextrins into intracellular glucose, maltose and maltotriose. Use of the latter set of enzymes for starch utilization may confer an additional competitive advantage over organisms using an extracellular

α-amylase producing linear (G1–G6) malto-oligosaccharides outside the cell.

The α/β hydrolase family

The family of α/β hydrolase fold enzymes (Ollis *et al.*, 1992) comprises a wide variety of hydrolytic enzymes of different phylogenetic origin and catalytic function, that have practically no sequence similarity. Nevertheless, these enzymes share a similar three-dimensional fold and arrangement of catalytic residues. In view of the similarity in structure, overall topology, catalytic triad residues, and conserved loops at the active site, Ollis *et al.* (1992) concluded that the α/β hydrolase fold enzymes diverged from a common ancestor. The enzyme haloalkane dehalogenase belongs to this family and provides a very interesting example of the evolution of enzyme function (Dijkstra & Verschueren, 1996). This enzyme cleaves carbon–halogen bonds and plays a crucial role in the degradation of halogenated hydrocarbons. The structure of the enzyme from *Xanthobacter autotrophicus* GJ10, catalysing the first step in 1,2-dichloroethane metabolism, was solved at 2.0 Å resolution and its reaction mechanism has been unravelled by X-ray crystallography, site-directed mutagenesis and oxygen isotope incorporation studies (Franken *et al.*, 1991; Pries *et al.*, 1994*a*; Verschueren *et al.*, 1993*a*,*b*). The enzyme consists of two domains: a main domain composed of a β-sheet of eight central β-strands connected by α-helices, and a cap domain composed of five helices lying on top of the main domain. The folding of the main domain is identical to that of other members of the α/β hydrolase fold family (Ollis *et al.*, 1992). The enzymes of this family share a common mechanism of hydrolysis for a variety of substrates. *In vivo* selection experiments designed to grow *X. autotrophicus* strain GJ10 on longer chain halogenated compounds such as 1-chlorohexane, resulted in isolation of a number of mutant dehalogenases (Pries *et al.*, 1994*b*). Their analysis revealed that the cap domain itself, and the region linking it to the main domain, play an important role in determining substrate specificity. Most likely these mutations have an effect on the size or shape of the substrate binding cavity or on the relative positions of the two domains and thereby affect the binding site between the two domains (Dijkstra & Verschueren, 1996).

The evolutionary origin of the haloalkane dehalogenase is unknown. It is likely that this is a 'retooled' enzyme that became specifically adapted to use halogenated compounds relatively recently. It remains to be determined from which primeval ancestor α/β hydrolase protein this haloalkane dehalogenase evolved and which mutations took place during its adaptation. Major differences between haloalkane dehalogenase and other α/β hydrolase fold family members concern the presence of the cap domain, furnishing the substrate binding pocket, and of the Trp125 and Trp175 residues that are involved in binding the halogen atom and stabilizing the reaction transition state (Verschueren *et al.*, 1993*b*; Janssen, van der Ploeg & Pries, 1994; Pries *et al.*, 1994*a*; Dijkstra & Verschueren, 1996).

Thus the general conclusion is that the evolution of catalytic functions and new metabolic pathways involved the recruitment of similar enzymatic activities from pre-existing pathways. In the words of Petsko *et al.* (1993) 'getting the chemistry right, it would seem, is the hard part; specificity is relatively easy to deal with later'. The latter has been demonstrated in detail in laboratory experiments on the substrate specificity of various pentose and pentitol converting enzymes (Mortlock & Gallo, 1992), and acetamidase in *Pseudomonas aeruginosa* (Clarke & Drew, 1988). Duplication and divergence of genes could be the most frequently used mechanism by which a primitive genome was enlarged and biochemical complexity was introduced (Jensen, 1976; Labedan & Riley, 1995; Lazcano *et al.*, 1992).

Two further examples from pathways studied in greatest detail, glycolysis and aromatic amino acid biosynthesis, will be presented in the following sections.

THE GLYCOLYTIC PATHWAY

Glycolysis is a very ancient, central metabolic, pathway that is found (at least in part) in all organisms. This is one of the best studied metabolic pathways, with a wealth of data available on nucleotide sequences and protein crystal structures for enzymes from many different organisms. On the basis of the known distribution of glycolytic enzymes in Prokaryota and Eukaryota, Fothergill-Gilmore & Michels (1993) proposed a convincing scenario for how the pathway may have been assembled, namely from the bottom up, essentially as a reversal of gluconeogenesis. This evolutionary mechanism would have given rise to glucose catabolic pathways with greater and greater net yields of ATP molecules. Glycolysis appears to have arisen by a chance assembly of independently evolving enzymes, since there are very few convincing similarities at the sequence level between different enzymes in the glycolytic pathway. A comparison of the sequences of a particular enzyme from phylogenetically distant organisms, on the other hand, showed clear similarities in most cases, indicating that they have diverged from a common ancestor. Also the core structures and catalytic mechanisms of these homologous enzymes are the same (Fothergill-Gilmore & Michels, 1993). Only the evolution of phosphofructokinase (PFK), considered as a critical and relatively late step in enabling the complete present-day glycolytic sequence, will be dealt with in more detail here.

The most commonly occurring PFK enzyme uses ATP as phospho donor. This enzyme is found in all major kingdoms, with the probable exceptions of Archaea, and certain protists and Bacteria. Recently, evidence was presented for the presence of an ADP–PFK enzyme in the hyperthermophile *Pyrococcus furiosus*, belonging to the Archaea, but no sequence data have yet become available for this protein (Kengen *et al.*, 1994). An alternative relatively rare form of PFK that uses inorganic pyrophosphate (PPi) as phospho donor is also found, for instance in the potato plant, in protozoa such as *Entamoeba histolytica, Giardia lamblia* and *Naeglaria fowleri*, and in *Propionibacterium freudenreichii* (for review see Mertens, 1991). In contrast to ATP–PFK, the PPi–PFK enzyme catalyzes the reversible phosphorylation of fructose-6-phosphate. PPi–PFK enzymes are present in phylogenetically distant groups and show strong variations in subunit size and quaternary structure. This has led to the suggestion that PPi–PFK proteins arose on various occasions from ATP–PFK enzymes in response to anaerobic growth conditions (Mertens, 1991). Characterization of a PPi–PFK enzyme from the aerobic actinomycete *Amycolatopsis methanolica*, with biochemical characteristics intermediate between those of ATP- and PPi-dependent PFK enzymes, and a study of its phylogeny recently has provided further insights (see Fig. 3) (Alves *et al.*, 1994, 1996). The deduced amino acid sequence of this PPi–PFK showed extensive similarities to both ATP- and PPi–PFK enzymes. Construction of a phylogenetic tree using the neighbour-joining method (Fig. 3) showed clearly that the PPi–PFK enzymes form a monophyletic group. This contradicts the hypothesis that PPi dependent enzymes evolved many times from ATP-dependent enzymes (Mertens, 1991) and supports the view that both types of PFK enzymes evolved from a common ancestor. Interestingly, the simplest PFK proteins, PPi-dependent homodimeric enzymes lacking allosteric control, are encountered in *P. freudenreichii* and *E. histolytica*, which have an obligate anaerobic metabolism (Mertens, 1991). Alves *et al.* (1996) proposed that the ancestral PFK resembled the enzymes encountered in the latter organisms from which the allosteric (hetero)tetrameric PPi-dependent enzymes and the tetrameric and hetero-octameric ATP-dependent enzymes encountered in plants, bacteria, insects, yeasts and mammals were derived (Fig. 3).

AROMATIC AMINO ACID BIOSYNTHESIS

The biosynthesis of aromatic amino acids proceeds initially via a common pathway of seven enzymes, the shikimate pathway, starting with the condensation of erythrose-4-phosphate and phosphoenolpyruvate by DAHP synthase, leading to the branchpoint intermediate chorismate (Fig. 4). From chorismate there are three major branches leading to L-Phe, L-Tyr and L-Trp. The sequence of enzymatic steps and the intermediates in the shikimate pathway and the L-Trp specific branch are found consistently throughout nature. The pathways to L-Phe and L-Tyr, however, show some variations. L-Phe is synthesized via either phenylpyruvate and/or arogenate while L-Tyr is synthesized via 4-hydroxyphenylpyruvate and/or arogenate.

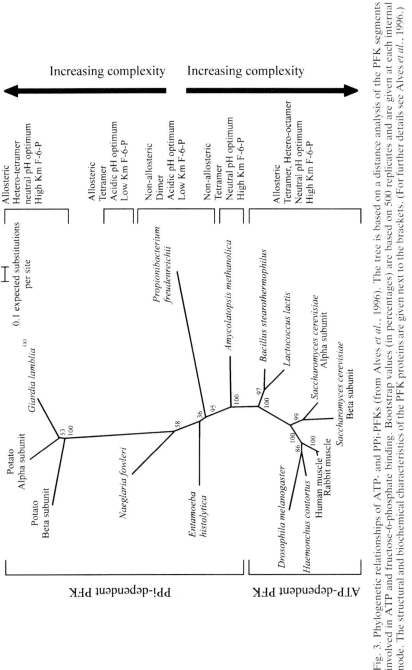

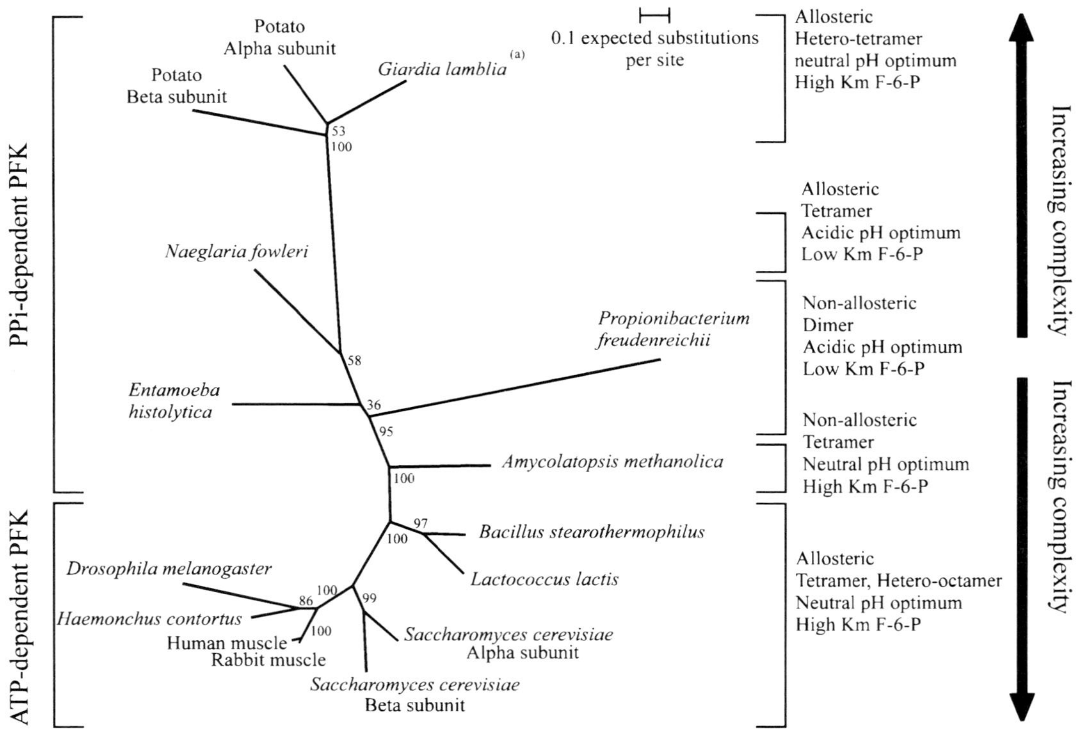

Fig. 3. Phylogenetic relationships of ATP- and PPi-PFKs (from Alves *et al.*, 1996). The tree is based on a distance analysis of the PFK segments involved in ATP and fructose-6-phosphate binding. Bootstrap values (in percentages) are based on 500 replicates and are given at each internal node. The structural and biochemical characteristics of the PFK proteins are given next to the brackets. (For further details see Alves *et al.*, 1996.)

In some organisms both pathways leading to L-Phe and L-Tyr are present. The biochemistry and evolution of these pathways has been extensively reviewed over the years (Bentley, 1990; Jensen, 1985, 1992). Only two different aspects will be covered here.

In yeast and fungi steps two to six of the shikimate pathway (converting 3-dehydroquinate into EPSP; Fig. 4) are localized on a pentafunctional enzyme. This AROM protein consists of five domains, each showing homology to the individual enzymes of *E. coli*, for instance. The gene encoding the AROM protein appears to have evolved via multiple gene

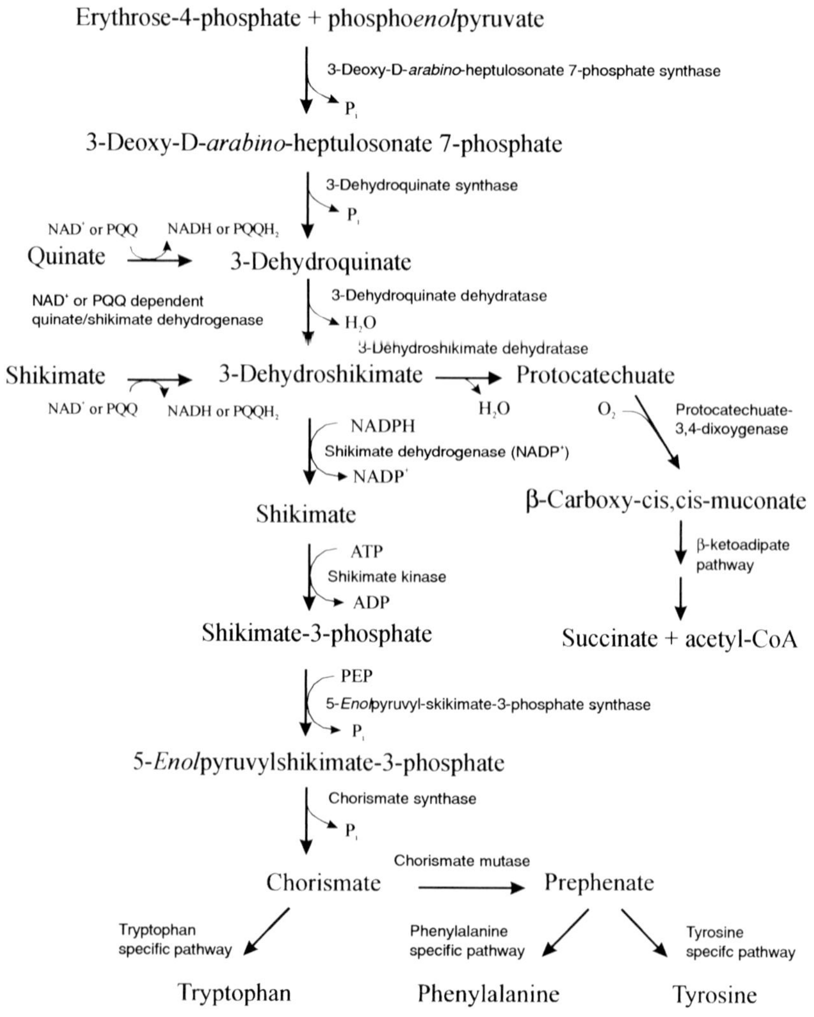

Fig. 4. Schematic representation of the biosynthesis of aromatic amino acids via the shikimate pathway and the catabolism of quinate and shikimate. (From Euverink *et al.*, 1992.)

Table 2. *Properties and function of 3-dehydroquinate dehydratases from various organisms*

		K_m (μM)	Subunit M_r ($\times$1000)	Structure	Function
Type I	*Escherichia coli*	18	29	Dimer	Biosynthetic
	Acinetobacter calcoaceticus	–	–	–	Catabolic
	Neurospora crassa	5	165*	Dimer	Biosynthetic
	Aspergillus nidulans	–	171*	Dimer	Biosynthetic
	Plants	27	59†	Monomer	Biosynthetic
Type II	*Neurospora crassa*	70	18.5	Dodecamer	Catabolic
	Aspergillus nidulans	150	16.5	Dodecamer	Catabolic
	Streptomyces coelicolor	650	16	Dodecamer	Biosynthetic
	Mycobacterium tuberculosis	–	14‡	Dodecamer	Biosynthetic
	Amycolatopsis methanolica	121	12	Dodecamer	Biosynthetic/ Catabolic
	Acinetobacter calcoaceticus	–	–	–	Biosynthetic

*, Part of the AROM complex; †, Part of the 3-dehydroquinate dehydratase/NADP-dependent shikimate dehydrogenase bifunctional enzyme; ‡, personal communication from A.R. Hawkins.
Adapted from Euverink *et al.* (1992).

fusions (Duncan, Edwards & Coggins, 1987). Initially, channelling of substrates to the next active site was thought to be the physiological role of the AROM protein. In view of the substantial leakage of shikimate pathway intermediates from the protein, this appears less likely (Hawkins & Lamb, 1995). The function of this protein is possibly in the coordinate expression of a single gene instead of five different genes, allowing a higher degree of regulatory control (Duncan, Edwards & Coggins, 1988).

The enzyme 3-dehydroquinate dehydratase and 3-dehydroshikimate dehydratase catalyze consecutive reactions, allowing growth of micro-organisms on quinate by its conversion to protocatechuate and subsequent metabolism by the β-ketoadipate pathway (Fig. 4). 3-Dehydroquinate dehydratase also plays an essential role in the shikimate pathway. The evolutionary history of 3-dehydroquinate dehydratase proteins appears particularly complex. Two forms of 3-dehydroquinate dehydratase are known: the type I enzyme, a generally thermolabile, dimeric protein, and the type II enzyme, a thermostable, dodecameric protein, with relatively small subunits (Table 2). In yeast and fungi the type I enzymes are present in the AROM protein and function in the shikimate biosynthetic pathway, whereas the type II enzymes are involved in quinate catabolism (Kleanthous *et al.*, 1992). Both types of 3-dehydroquinate dehydratase enzymes are not only present in fungi but also in bacteria. In some bacteria, e.g. *E. coli* unable to grow on quinate, only the type I enzyme with a biosynthetic function is present. In contrast, it is the type II enzyme that functions in the shikimate pathway of *Streptomyces coelicolor, Mycobacterium tuberculosis*, and *A. methanolica*. Of these, only the last organism is able to grow on quinate. Mutant and enzyme evidence showed that in *A. methanolica* one and the same type II enzyme served the dual biosynthetic and catabolic functions (Euverink *et al.*, 1992). Finally, *Acinetobacter calcoaceticus* was shown to employ a type I enzyme in quinate catabolism and circumstantial evidence suggests that a thermostable (type II) enzyme functions in the shikimate pathway (Elsemore & Ornston, 1995). The available sequence data combined with the information about the biochemical properties and quaternary structures of dehydroquinate dehydratase (Table 2), indicate that the type I and type II enzymes evolved from different ancestral proteins. Also the reaction mechanisms of these two types of dehydroquinate dehydratase enzymes appear to be different (opposite stereochemistry) (Harris *et al.*, 1993). Thus, dehydroquinate dehydratase provides an example of an enzyme activity that has evolved from at least two different ancestral genes. Also during the divergence of different bacterial cell lines from the cell lines of eukaryotes, swapping of genes for catabolic and biosynthetic dehydroquinate dehydratases must have taken place (Elsemore & Ornston, 1995).

MECHANISMS OF GENOTYPIC ADAPTATION

Stress and spontaneous mutagenesis

Recent years have seen a strong fundamental interest in the mechanisms employed by bacteria for phenotypic and genotypic adaptation under various stress conditions. The adaptive phenomena and strategies associated with growth under nutrient limitation, or general stress and starvation conditions, are most clearly visible in bacteria on the borderline between growth and nongrowth. The latter situation may frequently occur in many natural environments with a low supply of nutrients. As an example of phenotypic adaptation, carbon starvation induces the emergence of *E. coli* and *Vibrio* cells resistant to a variety of stresses (heat shock, oxidative stress, osmotic stress), via a complex turn-on/turn-off pattern of protein synthesis. Evidence now exists for a general starvation induced programme for survival in bacteria, involving special regulatory mechanisms, for instance a starvation-inducible sigma factor encoded by katF (*rpoS*) gene in *E. coli* (Matin, 1991; Siegele & Kolter, 1992; Kjelleberg *et al.*, 1993). Under strongly selective stress conditions, phenomena of genotypic adaptation become apparent as well. Spontaneous mutations, the source of evolutionary change, may be lethal, neutral or silent, disadvantageous or, more rarely, advantageous. The latter 'useful' mutations are of primary evolutionary relevance. There is evidence to suggest that under stress conditions non-growing cells are able to increase their spontaneous mutation rates and that environmental pressure actually may direct the type of mutation that occurs to improve growth conditions (Cairns, Overbaugh & Miller, 1988; Hall, 1990; Foster, 1993, 1995). However, the biochemical and molecular mechanisms involved in what has become termed directed, or selection induced, or adaptive mutations, remain to be elucidated.

Arber (1993) describes four different categories of mechanisms for spontaneous mutagenesis, each of which have their particular roles in metabolic evolution. These are: (i) replication infidelities, and (ii) effects of external/internal environmental mutagens; both these mechanisms can result in base mispairing, which may be corrected by repair systems involving several enzymes, but those remaining unrepaired may give rise to base substitutions (Friedberg, Walker & Siede, 1995). (iii) DNA rearrangements, involving mobile genetic elements such as insertion sequence (IS) elements and transposons, which may cause gene disruption, deletion, inversion, amplification, and replicon fusion. These rearrangements are not limited to the exponential growth phase but also may occur in the stationary phase. (iv) Acquisition of genetic information from other living organisms. Transposition of an IS element as well as of a composite transposon can occur to natural gene vectors, such as transferable plasmids. This can give rise to horizontal gene transfer to other bacterial strains via plasmid conjugation. Such processes have played an important role in recent times in the dispersion of drug-resistant characters (Bennett, 1995) and biodegradative activities (Sayler *et al.*, 1990) to a wide variety of bacterial strains. Horizontal gene transfer is discussed in more detail in the chapter by P. Gogarten in this volume.

Selection in continuous cultures

Spontaneous mutations in micro-organisms may be selected in continuous flow cultivation systems (Novick & Szilard, 1950*a*,*b*). Growth under nutrient limitation in continuous cultures provides strongly selective stress conditions. Micro-organisms generally respond to this initially with major phenotypical adaptation (Harder & Dijkhuizen, 1983; Harder, Dijkhuizen & Veldkamp, 1984), but prolonged incubation often results in genotypical adaptation as well (Dykhuizen & Hartl, 1983). The latter involves the appearance of spontaneous mutants that may rapidly become dominant since they grow faster than the average population at a given, constant dilution rate, either because they exhibit an increased μ_{max}, and/or possess an increased affinity for the limiting nutrient (Dykhuizen & Hartl, 1983; Dijkhuizen & Harder, 1992; Dykhuizen, 1990). Growth in nutrient-limited continuous cultures can thus produce fitter strains with either higher levels of substrate-capturing or growth-limiting enzymes, or alternatively, enzymes with improved catalytic properties, such as greater substrate affinity, or shifts in pH and temperature optima. The available data show that this may either be based on spontaneous mutations in regulatory systems (constitutive expression of inducible enzymes), gene duplication, mutations in structural genes that alter the substrate affinity or specificity of an enzyme, or activation of cryptic or silent gene systems (Beacham, 1987; Mortlock & Gallo, 1992). Recent examples include the hyperproduction of the binding-protein-dependent sugar transport systems in *Agrobacterium* species (Cornish, Greenwood & Jones, 1989), amidase in *Methylophilus methylotrophus* (Silman, Carver & Jones, 1989), the evolution of phenylacetamidase from acetamidase in *P. aeruginosa* (Clarke & Drew, 1988), xylitol dehydrogenase from ribitol dehydrogenase in *Klebsiella aerogenes* (Mortlock & Gallo, 1992), and the isolation of mutants of the fungus *Fusarium graminearum* A3/5 with improved glucose uptake or improved growth characteristics (Wiebe *et al.*, 1994*a*,*b*). Where studied in detail, data show that the substrate specificity of an enzyme can be altered by a single-site mutation and that a succession of amino acid substitutions can produce a family of enzymes with novel activities (Clarke & Drew, 1988). Selection of mutants with altered substrate specificity of an enzyme not necessarily requires employment of continuous cultures, but also may be carried out by way of selection on solid media (Clarke & Drew, 1988). Selective pressure in continuous cultures, however, may be considerably higher because of the employment of cell cultures of relatively high density under specific nutrient-limiting conditions over prolonged periods of time (Dijkhuizen & Harder, 1992; Dykhuizen & Hartl, 1983).

Adaptation to xenobiotic compounds

Most information about the evolutionary events resulting in adaptation to xenobiotic compounds can be derived from the comparison of related catabolic enzymes and homologies between catabolic genes. Only a single example will be discussed here. Several other examples of the evolution of catabolic pathways for degradation of halogenated compounds are discussed elsewhere (van der Meer *et al.*, 1992; Janssen, 1994).

For many chloroaromatic compounds, the degradative pathways converge at chlorosubstituted catechols as central intermediates. In such cases, dehalogenation takes place after ring cleavage has been accomplished. Bacteria degrading chloroaromatic compounds via *ortho*-cleavage of chlorocatechols seem, in general, to have two sets of enzymes: one set for catechol and a separate set for chlorocatechol catabolism. The genes for the first of these are located on the chromosome, the genes for the second generally on a catabolic plasmid. Only the (chloro)catechol 1,2-dioxygenase and (chloro)muconate cycloisomerase steps are common to both pathways and show high sequence similarities. The enzymes of the catechol catabolic pathway, however, are in general unable to catalyse an efficient conversion of chlorocatechols, or metabolites derived. Those bacteria utilizing chloroaromatic compounds as sole source of carbon and energy, metabolizing them via chlorocatechols as intermediates, have developed mechanisms to overcome these problems. The available evidence suggests that the evolution of what has been named the modified ortho-cleavage pathway involved all three mechanisms for genotypic adaptation outlined above (van der Meer *et al.*, 1992; Schlömann, 1994). Spontaneous mutations conferred to the enzymes the required altered substrate specificity for chlorocatechols and chloromuconates; horizontal gene transfer allowed acquisition of genes for dienelactone hydrolase and maleylacetate reductase; extensive gene rearrangements resulted in coupling of chlorocatechol 1,2-dioxygenase and chloromuconate cycloisomerase genes, also fusing these with the newly acquired genes encoding dienelactone hydrolase and maleylacetate reductase. The chlorocatechol catabolic operons encoded by the plasmids pAC27 (*clc* operon; *Pseudomonas putida* AC866), pJP4 (*tfd* operon; *Alcaligenes eutrophus* JMP134) and pP51 (*tcb* operon; *Pseudomonas* sp. P51) share similar structures and organization, and were obtained from strains that were isolated on different continents. The available evidence indicates that the structural and the regulatory genes for chlorocatechol degradation coevolved, giving rise to the *clc, tfd, tcb* operon versions (van der Meer *et al.*, 1992; Coco *et al.*, 1993; Schlömann, 1994). This raised the question whether the chlorocatechol pathway evolved recently, in response to the appearance of manmade pollutants, or whether chlorocatechol degradation evolved long ago, its original function being the catabolism of naturally occurring halogenated compounds. Chlorosubstituted and bromosubstituted aromatics have been shown to exist in nature in considerable concentrations and this may have provided the selection pressure necessary to account for the evolution of an ancient chlorocatechol degradation operon (Schlömann, 1994). In an attempt to estimate the age of the chlorocatechol pathway, he calculated that, for instance, the *clc–tcb* genes diverged approximately 70 million years ago. However, there is virtually no information concerning the mutation rates in bacteria living in nature and the responses of such rates to the presence of potential (xenobiotic) substrates. Clearly, the phenomena of directed mutations (see above) require further investigation in this respect. It cannot for instance be ruled out that a diversity of stress factors, including chemical pollutants (Blom, Harder & Matin, 1992) stimulate error-prone DNA replication and hence accelerate DNA evolution (van der Meer *et al.*, 1992).

CONCLUDING REMARKS

Micro-organisms have the potential to adapt themselves functionally and structurally to changes in their environment. Moreover, the versatility of microbes is such that, when placed under appropriate selective conditions, they may develop novel properties, either via activation of cryptic and silent genes, or via the selection of mutants in regulatory or structural genes. This relationship between metabolic adaptation and environmental changes remains one of the biggest challenges for future fundamental and applied research (Harder & Dijkhuizen, 1990). Insights into the mechanisms of the evolution of metabolic pathways may further stimulate several applications. Appropriate stress conditions in laboratory experiments may be used to select for improved industrial biocatalysts, displaying better or new kinetic characteristics, or higher activity or stability at extremes of pH, temperature, or solvent concentrations. The latter approach is sometimes referred to as *in vivo* protein engineering (Hermes *et al.*, 1990). Another example is the microbial adaptation to utilization of xenobiotic compounds, novel and possibly toxic compounds, often introduced into the environment by mankind in recent years. Suitably adapted organisms may be applied in bioremediation processes. Conceivably, the strongly selective conditions in chemostat cultures provide the best approach for development of such strains. A third example concerns the mechanisms involved in the emergence and spread of antibiotic resistance factors (Bennett, 1995). Although the situation has become quite desperate in recent years, detailed knowledge of the processes involved might prevent the world from entering a new era without effective antibiotics against pathogenic bacteria spreading infectious diseases. A fourth example concerns the assessment of the risks of applying, or releasing, genetically engineered micro-organisms. Finally, using a combination of techniques from various disciplines (physiology, molecular biology, biochemistry), rational approaches for metabolic pathway engineering increasingly become possible (Sahm, 1993), resulting in construction of strains overproducing primary or secondary metabolites, or high enzyme protein levels, or suitable biocatalysts. Thus studies into the evolution of metabolic pathways are important for both fundamental and practical reasons.

ACKNOWLEDGEMENTS

Parts of this work were supported by the Netherlands Technology Foundation (STW), which is subsidised by the Netherlands Organization for the Advancement of Pure Research (NWO), and by the Netherlands Programme Committee for Biotechnology of the Ministry of Economic Affairs.

REFERENCES

Alves, A. M. C. R., Euverink, G. J. W., Hektor, H. J., Hessels, G. I., van der Vlag, J., Vrijbloed, J. W., Hondmann, D., Visser, J. & Dijkhuizen, L. (1994). Enzymes of glucose and methanol metabolism in the actinomycete *Amycolatopsis methanolica*. *Journal of Bacteriology*, **176**, 6827–35.

Alves, A. M. C. R., Meijer, W. G., Vrijbloed, J. W. & Dijkhuizen, L. (1996).

Characterization and phylogeny of the *pfp* gene of *Amycolatopsis methanolica* encoding PPI-dependent phosphofructokinase. *Journal of Bacteriology*, **78**, 149–55.

Arber, W. (1993). Evolution of prokaryotic genomes. *Gene*, **135**, 49–56.

Beacham, I. R. (1987). Silent genes in prokaryotes. *FEMS Microbiology Reviews*, **46**, 409–17.

Bennett, P. M. (1995). The spread of drug resistance. In *Population Genetics of Bacteria*, Baumberg, S., Young, J. P. W., Wellington, E. M. H. & Saunders, J. R., eds., pp. 317–44. Cambridge University Press, Cambridge.

Bentley, R. (1990). The shikimate pathway. A metabolic tree with many branches. *Critical Reviews in Biochemistry and Molecular Biology*, **25**, 307–84.

Blom, A., Harder, W. & Matin, A. (1992). Unique and overlapping pollutant stress proteins of *Escherichia coli*. *Applied and Environmental Microbiology*, **58**, 331–4.

Branden, C. & Tooze, J. (1991). *Introduction to Protein Structure*. Garland Publishing, Inc., New York.

Cairns, J., Overbaugh, J. & Miller, S. (1988). The origin of mutants. *Nature*, **335**, 142–5.

Clarke, P. H. & Drew, R. (1988). An experiment in enzyme evolution. Studies with *Pseudomonas aeruginosa* amidase. *Bioscience Reports*, **8**, 103–20.

Coco, W. M., Rothmel, R. K., Henikoff, S. & Chakrabarty, A. M. (1993). Nucleotide sequence and initial functional characterization of the *clcR* gene encoding a LysR family activator of the *clcABD* chlorocatechol in *Pseudomonas putida*. *Journal of Bacteriology*, **175**, 417–27.

Cornish, A., Greenwood, J. A. & Jones, C. W. (1989). Binding-protein-dependent sugar transport by *Agrobacterium radiobacter* and *A. tumefaciens* grown in continuous cultures. *Journal of General Microbiology*, **135**, 3001–13.

Davies, G. & Henrissat, B. (1995). Structures and mechanisms of glycosyl hydrolases. *Structure*, **3**, 853–9.

Dijkhuizen, L. & Harder, W. (1992). Applications of prokaryotes. In *The Prokaryotes*, 2nd edn. Balows, C. A., Trüper, H. G., Dworkin, M., Harder, W. & Schleifer, K. H., eds.), pp. 197–206. Springer, New York.

Dijkstra, B. W. & Verschueren, K. H. G. (1996). Evolutionary aspects of catalysis by haloalkane dehalogenase. In press.

Doolittle, R. F. (1994). Convergent evolution: the need to be explicit. *Trends in Biochemical Sciences*, **19**, 15–18.

Duncan, K., Edwards, R. M. & Coggins, J. R. (1987). The penta-functional arom enzyme of *Saccharomyces cerevisiae* is a mosaic of monofunctional domains. *Biochemical Journal*, **246**, 375–86.

Duncan, K., Edwards, R. M. & Coggins, J. R. (1988). The *Saccharomyces cerevisiae ARO1* gene. An example of the co-ordinate regulation of five enzymes on a single biosynthetic pathway. *FEBS Letters*, **241**, 83–8.

Dykhuizen, D. E. (1990). Experimental studies of natural selection in bacteria. *Annual Review of Ecology and Systematics*, **21**, 373–98.

Dykhuizen, D. E. & Hartl, D. L. (1983). Selection in chemostats. *Microbiological Reviews*, **47**, 150–68.

Elsemore, D. A. & Ornston, L. N. (1995). Unusual ancestry of dehydratases associated with quinate catabolism in *Acinetobacter calcoaceticus*. *Journal of Bacteriology*, **177**, 5971–8.

Euverink, G. J. W., Hessels, G. I., Vrijbloed, J. W., Coggins, J. R. & Dijkhuizen, L. (1992). Purification and characterization of a dual function 3-dehydroquinate dehydratase from *Amycolatopsis methanolica*. *Journal of General Microbiology*, **138**, 2449–57.

Farber, G. K. (1993). An α/β-barrel full of evolutionary trouble. *Current Opinion Structural Biology*, **3**, 409–12.

Foster, P. L. (1993). Adaptive mutation: the uses of adversity. *Annual Review of Microbiology*, **47**, 467–504.

Foster, P. L. (1995). Adaptive mutation. In *Population Genetics of Bacteria*, Baumberg, S., Young, J. P. W., Wellington, E. M. H. & Saunders, J. R., eds), pp. 13–30. Cambridge University Press, Cambridge.

Fothergill-Gilmore, L. A. & Michels, P. A. M. (1993). Evolution of glycolysis. *Progress in Biophysics and Molecular Biology*, **59**, 105–236.

Franken, S. M., Rozeboom, H. J., Kalk, K. H. & Dijkstra, B. W. (1991). Crystal structure of haloalkane dehalogenase: an enzyme to detoxify halogenated alkanes. *EMBO Journal*, **10**, 1297–302.

Friedberg, E. C., Walker, G. C. & Siede, W. (1995). *DNA Repair and Mutagenesis*. ASM Press, Washington, DC.

Fujiwara, S., Kakihara, H., Sakaguchi, K. & Imanaka, T. (1992). Analysis of mutations in cyclodextrin glucanotransferase from *Bacillus stearothermophilus* which affect cyclization characteristics and thermostability. *Journal of Bacteriology*, **174**, 7478–81.

Hall, B. G. (1990). Spontaneous point mutations that occur more often when they are advantageous than when they are neutral. *Genetics*, **126**, 5–16.

Harder, W. & Dijkhuizen, L. (1983). Physiological responses to nutrient limitation. *Annual Review of Microbiology*, **37**, 1–23.

Harder, W. & Dijkhuizen, L. (1990). Microbial physiology and rec-DNA technology: the microbe, more than a bag of enzymes. In *Proceedings of the 5th European Congress on Biotechnology*, vol. 1, Christiansen, C., Munck, L. and Villadsen, J., eds., pp. 23–30. Munksgaard International Publisher, Copenhagen.

Harder, W., Dijkhuizen, L. & Veldkamp, H. (1984). Environmental regulation of microbial metabolism. In *The Microbe 1984. Part II, Prokaryotes and Eukaryotes*, Kelly, D. P. & Carr, N. G., eds., pp. 51–95.

Harris, J., Kleanthous, C., Coggins, J. R., Hawkins, A. R. & Abell, C. (1993). Different mechanistic and stereochemical courses for the reactions catalyzed by type I and type II dehydroquinases. *Journal of the Chemical Society*, **1993**, 1080–1.

Hawkins, A. R. & Lamb, H. K. (1995). The molecular biology of multidomain proteins. Selected examples. *European Journal of Biochemistry*, **232**, 7–18.

Hermes, H. F. M., Peeters, W. P. H., Peeters, P. J. H., Schepers, C. H. M., Kamphuis, J., Schoemaker, H. E. & Meijer, E. M. (1990). 'In vivo protein engineering': a goal-orientated approach for biocatalyst screening and optimization. In *Proceedings of the 5th European Congress on Biotechnology*, vol. 1, Christiansen, C., Munck, L. and Villadsen, J., eds., pp. 225–8. Munksgaard International Publisher, Copenhagen.

Horowitz, N. H. (1945). On the evolution of biochemical synthesis. *Proceedings of the National Academy of Sciences, USA*, **31**, 153–7.

Horowitz, N. H. (1965). *Evolving Genes and Proteins*, Bryson, V. & Vogel, H. J., eds., Academic Press, New York.

Janeček, S. (1994). Parallel β/α-barrels of α-amylase, cyclodextrin glycosyltransferase and oligo-1,6-glucosidase versus the barrel of β-amylase: evolutionary distance is a reflection of unrelated sequences. *FEBS Letters*, 353, 119–23.

Janssen, D. B. (ed.) (1994). Genetics of biodegradation of synthetic compounds. *Biodegradation*, **5**, 145–377 (Special issue).

Janssen, D. B., van der Ploeg, J. R. & Pries, F. (1994). Genetics and biochemistry of 1,2-dichloroethane degradation. *Biodegradation*, **5**, 249–57.

Jensen, R. A. (1976). Enzyme recruitment in evolution of new function. *Annual Review of Microbiology*, **30**, 409–25.
Jensen, R. A. (1985). Biochemical pathways in prokaryotes can be traced backward through evolutionary time. *Molecular Biology of Evolution*, **2**, 92–108.
Jensen, R. A. (1992). An emerging outline of the evolutionary history of aromatic amino acid biosynthesis. In *The Evolution of Metabolic Function*, Mortlock, R. P., ed., pp. 205–36. CRC Press, Boca Raton.
Jespersen, H. M., MacGregor, E. A., Sierks, M. R. & Svensson, B. (1991). Comparison of the domain-level organization of starch hydrolases and related enzymes. *Biochemical Journal*, **280**, 51–5.
Jespersen, H. M., MacGregor, E. A., Henrissat, B., Sierks, M. R. & Svensson, B. (1993). Starch- and glycogen-debranching and branching enzymes: prediction of structural features of the catalytic $(\beta/\alpha)_8$-barrel domain and evolutionary relationship to other amylolytic enzymes. *Journal of Protein Chemistry*, **12**, 791–805.
Kengen, S. V. M., de Bok, F. A. M., van Loo, N. D., Dijkema, C., Stams, A. J. M. & de Vos, W. M. (1994). Evidence for the operation of a novel Embden-Meyerhof pathway that involves ADP-dependent kinase during sugar fermentation by *Pyrococcus furiosus*. *Journal of Biological Chemistry*, **26**, 17537–41.
Kjelleberg, S., Albertson, N., Flärdh, K., Holmquist, L., Jouper-Jaan, Å, Marouga, R., Östling, J., Svenblad, B. & Weichart, D. (1993). How do non-differentiating bacteria adapt to starvation? *Antonie van Leeuwenhoek*, **63**, 333–41.
Kleanthous, C., Deka, R., Davis, K., Kelly, S. M., Cooper, A., Harding, S. E., Price, N. C., Hawkins, A. R. & Coggins, J. R. (1992). A comparison of the enzymological and biophysical properties of two distinct classes of dehydroquinase enzymes. *Biochemical Journal*, **282**, 687–95.
Klein, C. & Schulz, G. E. (1991). Structure of cyclodextrin glycosyltransferase refined at 2.0 Å resolution. *Journal of Molecular Biology*, **217**, 737–50.
Kubota, M., Matsuura, Y., Sakai, S. & Katsube, Y. (1991). Molecular structure of *B. stearothermophilus* cyclodextrin glucanotransferase and analysis of substrate binding site. *Denpun Kagaku*, **38**, 141–6.
Labedan, B. & Riley, M. (1995). Widespread protein sequence similarities: Origins of *Escherichia coli* genes. *Journal of Bacteriology*, **177**, 1585–8.
Lawson, C. L., van Montfort, R., Strokopytov, B., Rozeboom, H. J., Kalk, K. H., de Vries, G. E., Penninga, D., Dijkhuizen, L. & Dijkstra, B. W. (1994). Nucleotide sequence and X-ray structure of cyclodextrin glycosyltransferase from *Bacillus circulans* strain 251 in a maltose-dependent crystal form. *Journal of Molecular Biology*, **236**, 590–600.
Lazcano, A., Fox, G. E. & Oró, J. F. (1992). Life before DNA: The origin and evolution of early archean cells. In *The Evolution of Metabolic Function*, Mortlock, R. P., ed., pp. 237–95. CRC Press, Boca Raton.
Matin, A. (1991). The molecular basis of carbon-starvation-induced general resistance to *Escherichia coli*. *Molecular Microbiology*, **5**, 3–10.
Meer, J. R. van der, Vos, W. M., de Harayama, S. & Zehnder, A. J. B. (1992). Molecular mechanisms of genetic adaptation to xenobiotic compounds. *Microbiological Reviews*, **56**, 677–94.
Mertens, E. (1991). Pyrophosphate-dependent phosphofructo-kinase, an anaerobic glycolytic enzyme? *FEBS Letters*, **285**, 1–5.
Mortlock, R. P. & Gallo, M. A. (1992). Experiments in the evolution of catabolic pathways using modern bacteria. In *The Evolution of Metabolic Function*, Mortlock, R. P., ed., pp. 1–13. CRC Press, Boca Raton.
Nakamura, A., Haga, K. & Yamane, K. (1994). Four aromatic residues in the active center of cyclodextrin glucanotransferase from alkalophilic *Bacillus* sp. 1011:

effects of replacements on substrate binding and cyclization characteristics. *Biochemistry*, **33**, 9929–36.

Novick, A. & Szilard, L. (1950*a*). Description of the chemostat. *Science*, **112**, 715–16.

Novick, A. & Szilard, L. (1950*b*). Experiments with the chemostat on spontaneous mutation of bacteria. *Proceedings of the National Academy of Sciences, USA*, **36**, 708–19.

Ollis, D. L., Cheah, E., Cygler, M., Dijkstra, B. W., Frolow, F., Franken, S. M., Harel, M., Remington, S. J., Silman, I., Schrag, J., Sussman, J. L., Verschueren, K. H. G. & Goldman, A. (1992). The α/β hydrolase fold. *Protein Engineering*, **5**, 197–211.

Penninga, D., Strokopytov, B., Rozeboom, H. J., Lawson, C. L., Dijkstra, B. W., Bergsma, J. & Dijkhuizen, L. (1995). Site-directed mutations in Tyrosine 195 of cyclodextrin glycosyltransferase from *Bacillus circulans* strain 251 affect activity and product specificity. *Biochemistry*, **34**, 3368–76.

Petsko, G. A., Kenyon, G. L., Gerlt, J. A., Ringe, D. & Kozarich, J. W. (1993). On the origin of enzymatic species. *Trends in Biochemical Sciences*, **18**, 372–6.

Pries, F., Kingma, J., Pentenga, M., van Pouderoyen, G., Jeronimus-Stratingh, C. M., Bruins, A. P. & Janssen, D. B. (1994*a*). Site-directed mutagenesis and oxygen isotope incorporation studies of the nucleophilic aspartate of haloalkane dehalogenase. *Biochemistry*, **33**, 1242–7.

Pries, F., van den Wijngaard, A. J., Bos, R., Pentenga, M. & Janssen, D. B. (1994*b*). The role of spontaneous cap domain mutations in haloalkane dehalogenase specificity and evolution. *Journal of Biological Chemistry*, **269**, 17490–4.

Sahm, H. (1993). Metabolic design. In *Biotechnology*, 2nd edn., vol. 1, Rehm, H.-J., Reed, G., Pühler, A. & Stadler, P., eds., pp. 189–221. VCH-Verlag, Weinheim.

Sayler, G. S., Hooper, S. W., Layton, A. C. & King, J. M. H. (1990). Catabolic plasmids of environmental and ecological significance. *Microbial Ecology*, **19**, 1–20.

Schlömann, M. (1994). Evolution of chlorocatechol catabolic pathways. *Biodegradation*, **5**, 301–21.

Siegele, D. A. & Kolter, R. (1992). Life after log. *Journal of Bacteriology*, **174**, 345–8.

Silman, N. J., Carver, M. A. & Jones, C. W. (1989). Physiology of amidase production by *Methylophilus methylotrophus*: isolation of hyperactive strains using continuous culture. *Journal of General Microbiology*, **135**, 3153–64.

Sin, K. A., Nakamura, A., Masaki, H., Matsuura, Y. & Uozumi, T. (1994). Replacement of an amino acid residue of cyclodextrin glucanotransferase of *Bacillus ohbensis* doubles the production of γ-cyclodextrin. *Journal of Biotechnology*, **32**, 283–8.

Strokopytov, B., Knegtel, R. M. A., Penninga, D., Rozeboom, H. J., Kalk, K. H., Dijkhuisen, L. & Dijkstra, B. W. (1996). Structure of cyclodextrin glycosyltransferase complexed with a maltonoaose inhibitor at 2.6 Å resolution. Implications for product specificity. *Biochemistry*, **35**, 4241–9.

Svensson, B. (1994). Protein engineering in the α-amylase family: catalytic mechanism, substrate specificity, and stability. *Plant Molecular Biology*, **25**, 141–57.

Verschueren, K. H. G., Kingma, J., Rozeboom, H. J., Kalk, K. H., Janssen, D. B. & Dijkstra, B. W. (1993*a*). Crystallographic and fluorescence studies of the interaction of haloalkane dehalogenase with halide ions. Studies with halide compounds reveal a halide binding site in the active site. *Biochemistry*, **32**, 9031–7.

Verschueren, K. H. G., Seljée, F., Rozeboom, H. J., Kalk, K. H. & Dijkstra, B. W.

(1993*b*). Crystallographic analysis of the catalytic mechanism of haloalkane dehalogenase. *Nature*, **363**, 693–8.

Wiebe, M. G., Robson, G. D., Oliver, S. G. & Trinci, A. P. J. (1994*a*). Use of a series of chemostat cultures to isolate 'improved' variants of the Quorn[R] mycoprotein fungus, *Fusarium graminearum* A3/5. *Journal of General Microbiology*, **140**, 3015–21.

Wiebe, M. G., Robson, G. D., Oliver, S. G. & Trinci, A. P. J. (1994*b*). Evolution of *Fusarium graminearum* A3/5 grown in a glucose-limited chemostat culture at a slow dilution rate. *Journal of General Microbiology*, **140**, 3023–9.

GENE DUPLICATIONS AND HORIZONTAL GENE TRANSFER DURING EARLY EVOLUTION

J. PETER GOGARTEN, ELENA HILARIO AND LORRAINE OLENDZENSKI

Department of Molecular and Cell Biology, University of Connecticut, 75 North Eagleville Rd, Storrs, CT 06269-3044, USA

INTRODUCTION

The evolutionary history of organisms can be reconstructed using various information sources: the fossil and geological records, the comparative analysis of biochemical pathways, and the reconstruction of molecular phylogenies from macromolecules found in extant organisms. Analysis of early prokaryotic evolution is complicated because the availability of archaean microfossils is limited for geological reasons, and because different microbial groups have members with very similar morphology. A continuous representation of morphological forms in the fossil record is at present impossible (*cf.* Schopf, 1993 and this volume). To unravel the relationships between the major groups of prokaryotes, molecular markers have been used. The best studied of these is undoubtedly the small subunit ribosomal RNA (Olsen, Woese & Overbeek, 1994), but many other macromolecules, especially protein sequences have been useful in studying the relationships among prokaryotic kingdoms and domains. However, at best, the analysis of extant macromolecules can only yield information on the phylogeny of the molecules under study. Even if molecular phylogenies could be resolved without ambiguity, the step from molecular phylogeny to species phylogeny would remain complicated because genetic information has also been transferred horizontally between independent evolutionary lineages (Smith, Feng & Doolittle, 1992). The transfer of resistance genes among species of bacteria (Gray & Fitch, 1983), and the confirmation of the endosymbiotic origin of mitochondria and plastids and the subsequent transfer of organellar genes to the nucleus (for review see Margulis, 1993) provide ample evidence that genetic information is transferred horizontally. It is therefore not surprising that comparative analysis of different molecular phylogenies reveals a net-like structure of the species phylogeny (Gogarten, 1995; Hilario & Gogarten, 1993).

The focus of the following discussion is on the inference of species phylogeny from sometimes conflicting molecular phylogenies. The properties of the last common ancestor are discussed incorporating likely cases of horizontal gene transfer and lineage fusion.

CONFLICTING MOLECULAR PHYLOGENIES

The reconstruction of phylogenies from different molecular markers often results in phylogenetic trees with different topologies. Within the present discussion we will ignore that trees with identical topology can differ drastically with respect to branch lengths because of varying substitution rates for each molecule. Molecules remain recognizably similar over billions of years only because they are under enormous selection pressure. Without selection, sequences would be completely randomized over time due to substitutions. There is no reason to assume that the selection pressure on any given molecule is constant throughout evolution, nor is it reasonable to assume that the selection pressure varied in parallel for different molecular markers. For the purposes of this discussion, we will consider varying branch lengths only in the context of artefacts generated during the reconstruction process. There are three possible explanations that can reconcile molecular phylogenies with conflicting topologies.

Hypothesis 1

The phylogenies under consideration are not sufficiently resolved. The reasons for this could be severalfold. The data might not contain sufficient information, i.e. the molecule being analyzed may be too short, or regions within the molecule may be saturated with substitutions or not variable enough and thus would not contain useful phylogenetic information. The algorithms used to construct the phylogeny might not make use of all the information present in the data. Over time, additional substitutions can occur that obscure a phylogenetic signal and the algorithm used might be fooled by noise added during later stages of evolution. This would lead to incorrect resolution of deeper bifurcations. Some of these ill resolved branching patterns can be recognized using statistical analyses, for example maximum likelihood methods or analyses of bootstrapped data sets (*cf.* Felsenstein, 1988). However, artefacts that are due to a systematic bias in the data that are not appropriately considered in the analysis (for an example on nucleotide bias, see Lockhart *et al.*, 1992), or that are due to the attraction of long branches (e.g. Hillis, Huelsenbeck & Swofford, 1994) are not always associated with low bootstrap values.

Hypothesis 2

The phylogenies under consideration are comparing paralogous genes. An ancient gene duplication can give rise to two diverging gene copies per organism (gene A and B). If an analysis compares gene types A and B, the split between A and B reflects the gene duplication and not a speciation event; i.e., genes A and B are paralogous (Fitch, 1970) and not orthologous. Paralogous genes have been proven extremely useful in studying the early evolution of life. Some ancient gene duplications had already occurred before the time of the last common ancestor. Analysis of these duplicated genes can reach back to the time before the last common ancestor (e.g. Gogarten & Taiz, 1992). In particular, ancient duplicated genes allow the placement of the last common ancestor in the universal tree of life using one of the paralogous genes as an outgroup for the other (Gogarten *et al.*, 1989; Iwabe *et al.*, 1989). However, if paralogous genes are mistaken to be orthologous, the resulting phylogenies reflect not only speciation, but also the gene duplication event. When comparing two conflicting rooted phylogenies, the assumption of paralogous genes explains the deeper of the conflicting bifurcations as a gene duplication event. (See the discussion of the evolution of HSP70 homologues below for examples.)

Hypothesis 3

Genes were *transferred horizontally* between organisms. Scenarios involving the horizontal transfer of genetic material can be further subdivided according to the amounts of genetic material involved in the transfer.

3a) Xenology (Gray & Fitch, 1983)
One or a few genes are transferred from one line of descent into another. In this case the position of the xenologous sequence will reflect the phylogenetic position of the donor.

3b) Synology (Gogarten, 1994)
Formerly independent lineages fused to form a new line of descent. The transfer of single genes or operons leads to difficulties in finding the correct species tree; the occurrence of transfer events that included substantial portions of the respective genomes point toward major events in the history of life.

The two types of horizontal gene transfer can be distinguished by the comparison of phylogenies constructed from different molecular markers. Synology will give rise to two incompatible patterns each of which is strongly supported by many molecular markers. In contrast, xenology is revealed as a deviation of a single molecular marker from the majority consensus of other molecular markers. Both cases of horizontal gene transfer reconcile conflicting rooted phylogenies by explaining the more recent of the conflicting bifurcations as a case of horizontal transfer.

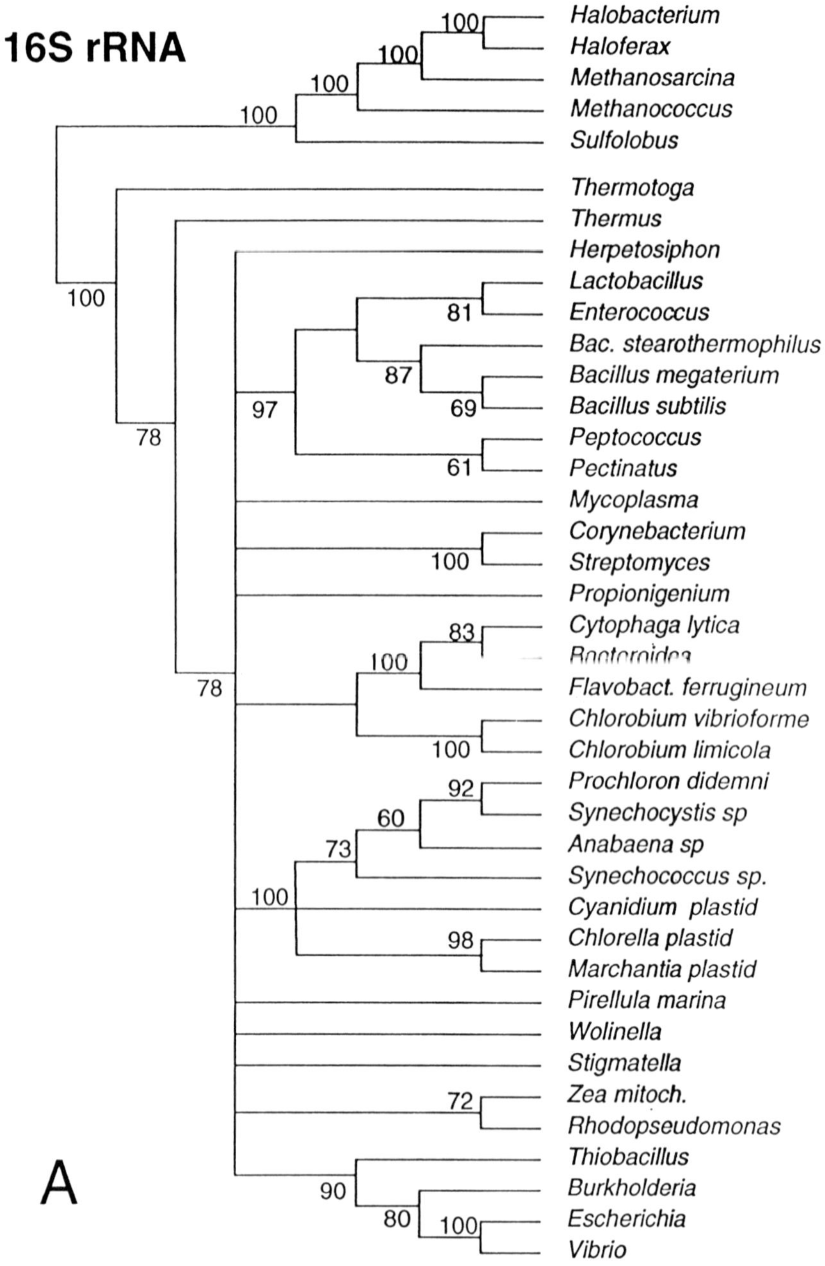

Fig. 1. Molecular phylogenies calculated using parsimony analyses for 16S rRNAs (A) and the catalytic H^+-ATPase subunits (B). Sequences and alignment used for A were retrieved from the Ribosomal Database Project (Larsen *et al.*, 1993). The aligned sequences were analyzed using parsimony analyses as implemented in Phylip 3.5c (Felsenstein, 1993). The *Thermotoga* ATPase subunit sequence was kindly provided by Dr. W. Ludwig, University of Munich.

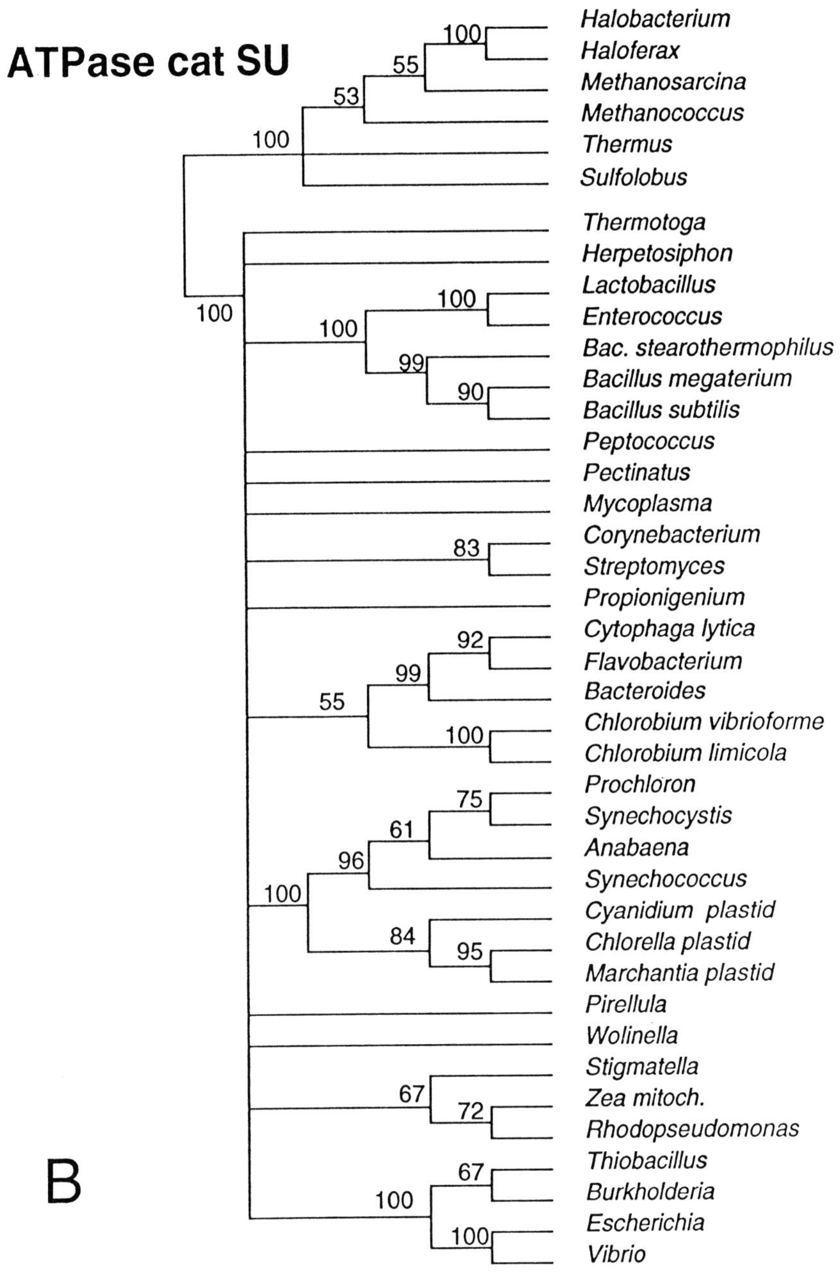

The other H^{+}-ATPase sequences were retrieved from GenBank. ATPase sequences were aligned using ClustalW (Thompson, Higgins & Gibson, 1994). All branches that were supported in less then 50% of the bootstrapped samples were collapsed. Numbers indicate the percentage of bootstrapped samples that support the branches.

16S rRNA VERSUS CATALYTIC H^+-ATPase SUBUNITS

To date, the best-studied molecular markers are the 16S-like ribosomal RNAs. At present more than 2600 different 16S rRNA sequences are available through the Ribosomal Database Project (Larsen *et al.*, 1993). An advantage of 16S rRNA is that ribosomes play a central role in cellular information processing. Since the ribosome interacts specifically with many cellular components and has coevolved over time with these components in each of the major cell lineages, it is unlikely that this molecule would have been transferred between distantly related organisms. Fig. 1A depicts parsimony analyses of samples bootstrapped from eubacterial and archaeal 16S rRNA sequences. All branches that were found in less than 50% of the bootstrapped samples are collapsed in the depicted tree. Most of the major eubacterial and archaeal clades are well defined by their 16S rRNA sequences as well as by additional biochemical and molecular characters. However, the radiation that gave rise to the major eubacterial groups cannot be resolved reliably using parsimony analysis.

With the discovery that the vacuolar type ATPase (V-ATPases) of the eukaryotic endomembrane system are homologous to the bacterial coupling factor ATPases (Zimniak *et al.*, 1988; for reviews on structure, function, and distribution of V-ATPases, see Harvey & Nelson, 1992) an additional marker became available that was present in all three cell lineages, i.e. in eubacteria, archaea and in the nucleocytoplasmic component of the eukaryotes. An advantage of ATPases as a marker for early evolution is that the ATP hydrolyzing headgroup of all three ATPase types (eubacterial or F-type, archaeal or A-type, and eukaryotic or V-type) consists of two homologous subunit types. Apparently, the gene duplication that gave rise to these two subunit types occurred before the three cellular lineages diverged from each other. Using this ancient gene duplication to root the universal tree of life, the last common ancestor was placed between eubacteria on one side and eukaryotes and archaea on the other (Gogarten *et al.*, 1989). This placement of the root has since been supported through the analysis of other ancient duplicated genes (elongation factors, dehydrogenases and t-RNAs: Iwabe *et al.*, 1989; aminoacyl t-RNA synthetases: Brown & Doolittle, 1995). The closer relationship between archaea and eukaryotes was also suggested by many other characters that had not undergone an ancient gene duplication, e.g. RNA polymerases (Puhler *et al.*, 1989), TATA binding proteins (Marsh *et al.*, 1994; Langer *et al.*, 1995). However, without the use of an outgroup, one cannot exclude the possibility that the characters shared between the archaea and eukaryotes reflect the state of the last common ancestor and that these characters were changed or abolished in the lineage leading to the eubacteria.

The ATPases clearly reflect the division of life into three distinct and separate cellular lineages or domains (Gogarten *et al.*, 1992). Within each of the three domains there is good agreement between the phylogenies calculated for the 16S rRNAs and ATPases subunits. Figure 2 shows a phylogenetic reconstruction of the evolution of the archaeal and vacuolar type ATPases (calculated using amino acid sequences). In agreement with the analyses of small subunit rRNA (Sogin *et al.*, 1989) and elongation factors (Hashimoto *et al.*, 1994) *Giardia lamblia* is placed at the base of the eukaryotic domain. In the past the phylogenetic placement of *Giardia* was hampered by the extremely high GC contents of its 16S-like rRNA. Although amino acids can be subject to convergent evolution (Gandbhir *et al.*, 1995), the influence of amino acid bias on phylogenetic reconstruction is far less pronounced then the effect of nucleotide bias. The grouping of red algae together with green algae and plants is also supported by the bootstrap analysis (but see Sogin, this volume). Taken together with the analyses of plastid encoded markers (see below, and Morden *et al.*, 1992) this finding leads support to the monophyletic group consisting of red and green algae.

Ludwig *et al.* (1993) compared eubacterial phylogenies calculated for 16S rRNA and F-ATPase catalytic subunits. The only noteworthy difference that these authors observed in distance matrix analyses was the placement of *Propionigenium modestum*: 16S rRNA suggests a closer association between *Propionigenium* and some Gram positive bacteria, whereas the F-ATPase subunits suggested an association between *Propionigenium* and cyanobacteria. Protein parsimony, which avoids the bias introduced through similar codon usage in unrelated species, also suggests the same association between *Propionigenium* and cyanobacteria. However, the bootstrap analyses depicted in Fig. 1B suggest that this finding might not be significant. Both analyses (16S rRNA and catalytic ATPase subunits) show *Propionigenium* as emerging as part of the major unresolved eubacterial radiation. This radiation is even less reliably resolved using ATPase subunits than with the 16S rRNAs. This is due mainly to the fact that in the case of the ATPases the outgroup (the archaeal sequences) share even less informative sites with the ingroup than in the case of the small subunit rRNAs. In other parts of the phylogeny the ATPases provide a more robust resolution, e.g., the ATPase subunit from the red algal plastid forms a monophyletic group with its homologues from the plastids of green algae and plants. This finding is not influenced by the high A/T content of the organellar genes, because the analysis was performed on the level of amino acids.

Considering the species included in Fig. 1, one well supported difference between the 16S rRNA and ATPase catalytic subunit phylogenies is the location of the sequences from *Thermus thermophilus* (also known as *T. aquaticus* HB8). Based on rRNAs and other biochemical characters (cell wall, lipids, antibiotic resistance) the genus *Thermus* is classified as a eubacterium, with closest similarities to the genus *Deinococcus* (Woese, 1987; Hensel *et al.*, 1986). However, *Thermus thermophilus* does not have an F-ATPase like other eubacteria, but an A/V-ATPase (Yokoyama,

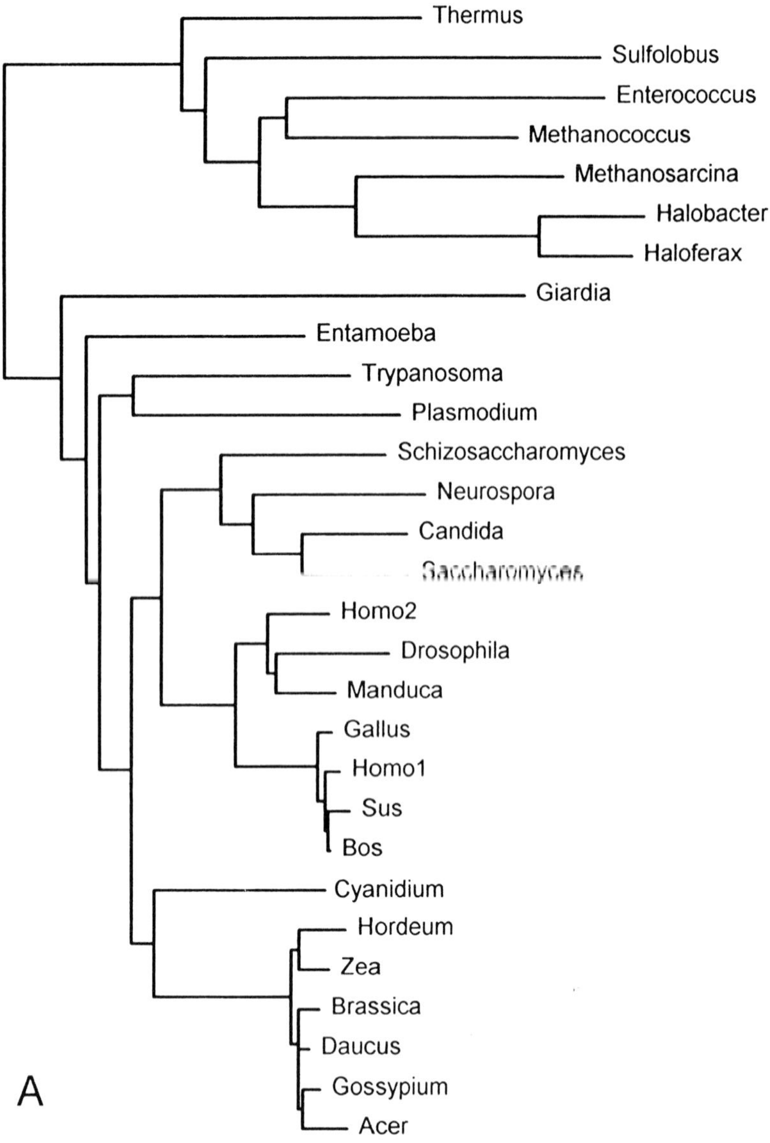

Fig. 2. Molecular phylogenies of the catalytic subunits of vacuolar and archaeal type H^+-ATPases calculated using distance matrix (A) and protein parsimony analyses (B) using the programs PROTDIST, FITCH and PROTPARS as implemented in Phylip 3.5c (Felsenstein, 1993). All branches in (B) that were supported in less than 50% of the bootstrapped samples were collapsed. Numbers in B give the percentage of bootstrapped samples that support the branch to the left of the number. Sequences were aligned as described in Fig. 1.

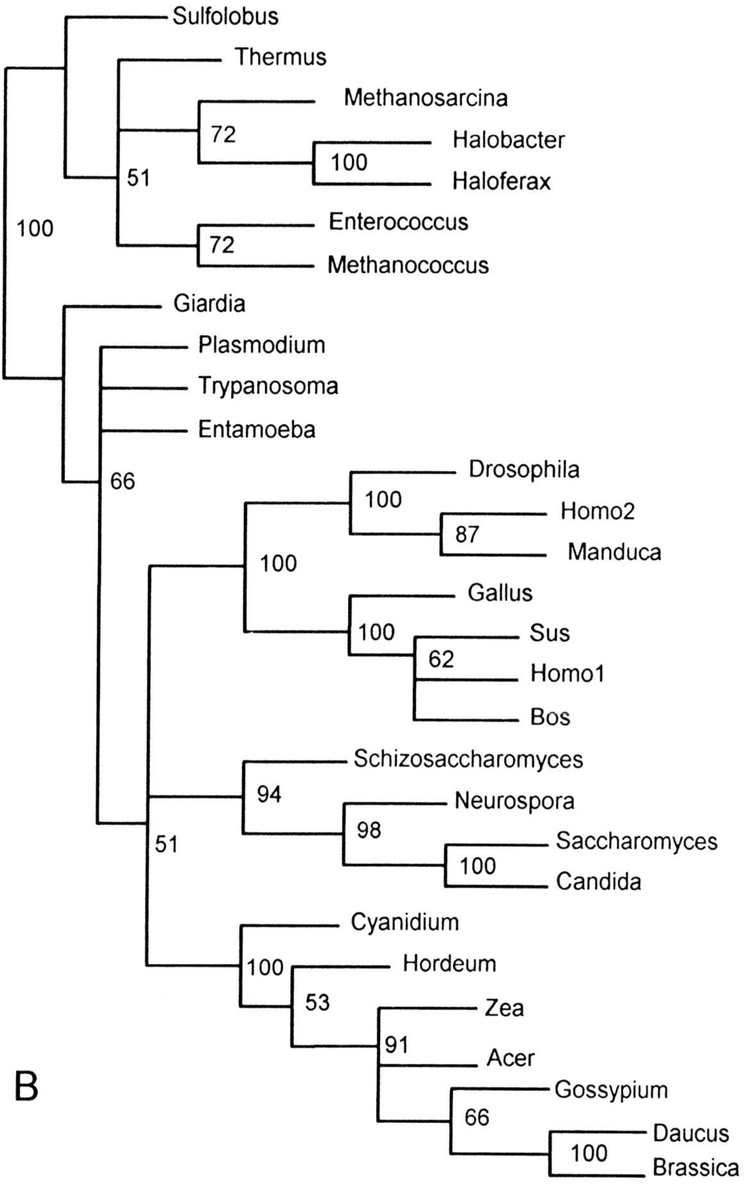
Sulfolobus
Thermus
Methanosarcina
72
Halobacter
100
Haloferax
51
100
Enterococcus
72
Methanococcus
Giardia
Plasmodium
Trypanosoma
Entamoeba
66
Drosophila
100
Homo2
87
Manduca
100
Gallus
100
Sus
62
Homo1
Bos
Schizosaccharomyces
94
Neurospora
98
Saccharomyces
51
100
Candida
Cyanidium
100
Hordeum
53
Zea
91
Acer
Gossypium
66
Daucus
100
Brassica

B

Oshima & Yoshida, 1990; Tsutsumi *et al.*, 1991). Both major subunits of the *Thermus* ATPase are clearly archaeal in character (Gogarten *et al.*, 1992). A similar result was obtained for a sodium pumping ATPase from another eubacterium, *Enterococcus hirae*. This bacterium has a normal F-ATPase that groups as expected with other low GC Gram positives (see Fig. 1B); in addition *E. hirae* has a sodium pumping ATPase that based on sequence, subunit composition and inhibitor sensitivity is classified as a vacuolar or archaeal type ATPase (Takase *et al.*, 1994) both of these archaeal type ATPases found in eubacteria group with the archaeal coupling factor ATPases (Hilario & Gogarten, 1993, see also Fig. 2). If these ATPases were acquired by horizontal gene transfer, their position in the phylogeny suggests that they were transferred from the archaeal lineage (in two separate events) to these eubacteria. Based only on the 16S rRNA and ATPase data it cannot be decided whether this conflict is due to unrecognized paralogy as suggested by Forterre *et al.* (1993) or due to horizontal gene transfer. However, the assumption of paralogy would lead one to conclude that the archaea and eubacteria do not form two distinct domains, but that the characters that suggest the clear separation between the two prokaryotic domains all had been duplicated in the last common ancestor and one set of these were selectively lost in the lineages leading to the eubacteria (with the exception of the V/A-ATPase in the lineages leading to *Enterococcus* and *Thermus*), whereas the complementary set was lost in the lineages leading to archaea and the nucleocytoplasmic component (Hilario & Gogarten, 1993, see also Fig. 5B).

TWO DISTINCT PROKARYOTIC DOMAINS?

As stated above, many characters support the division of prokaryotes into two distinct and separate groups, the archaea and the eubacteria (see also Table 1). In the case of unrooted phylogenies or considering non-polarized characters, it can be argued that the root of the universal tree of life might be located within one of the two prokaryotic domains. Analysis of ancient duplicated genes, in particular, H^+-ATPases, elongation factors and amino-acyl tRNA synthetases placed the last common ancestor between the two prokaryotic domains. These molecular markers suggested that all (if they are monophyletic) or a subgroup (if they are paraphyletic) of the archaea form the sistergroup to eukaryotic nucleocytoplasm. Recently however, a growing list of characters has been discovered that do not group the archaea as separate from the eubacteria. The most prominent of these are glutamine synthetases (Tiboni, Cammarano & Sanangelantoni, 1993; Kumada *et al.*, 1993; Brown *et al.*, 1994; Pesole *et al.*, 1995), homologues to the 70 kDa heat shock proteins (HSP70) (Gupta & Singh, 1992; Gupta & Golding, 1993), and glutamate dehydrogenases (Benachenhou-Lafha, Forterre & Labedan, 1993; Hilario & Gogarten, 1993). In all of these cases as well as in the case of

Table 1. *Characters supporting the division of prokaryotes into two distinct groups*

16S like rRNA
other ribosomal RNAs (23S and 5S)
ribosomal proteins
RNA polymerases
promoter organization
membrane lipids (ether lipids characterize archaebacteria) and fatty acid synthetase
cell wall structure (murein in eubacteria)
flagella (eubacterial or archaebacterial flagellin)
sensitivity towards antibiotics
ribosome binding motif on mRNAs
tRNAs* (Iwabe *et al.*, 1989)
dehydrogenases* (Iwabe *et al.*, 1989)
proton pumping ATPases* (Gogarten *et al.*, 1989)
elongation factors 1α/TU and 2/G* (Cammarano *et al.*, 1992)
aminoacyl tRNA synthetases* (Brown & Doolittle, 1995)

*These characters underwent an ancient gene duplication before the three domains diverged from each other. All of them support the archaea as the sister group to the eukaryotic nucleocytoplasm.
Modified from Zillig *et al.*, 1992

Table 2. *Ancient duplicated genes that support a close association between the archaea and the Gram positive eubacteria*

Homologues of the 70 kD heat shock proteins (HSP70) (Gupta & Golding, 1993)
Glutamine synthetases (Brown *et al.*, 1994; Kumada *et al.*, 1993; Tiboni *et al.*, 1993; Pesole *et al.*, 1995)
Glutamate dehydrogenases (Benachenhou-Lafha *et al.*, 1993)
Carbamoylphosphate synthetase (Lazcano, Puente, Gogarten, unpublished observations; Schofield, 1993)

carbamoylphosphate synthetase (Schofield, 1993; Lazcano, Puente & Gogarten, unpublished observations) the molecules have undergone ancient gene duplications, and in most of the cases the root appears to be placed between all of the prokaryotes on one side, and the eukaryotes on the other (for discussion of the HSP70 homologues see below). In the case of other molecular markers that allow a better resolution among the prokaryotes, it appears that the archaea group are among the Gram positive eubacteria, suggesting a possible genomic contribution from the Gram positive eubacteria to the archaea. The clade consisting of the Gram positives and archaea appears well supported by the respective data (e.g. Brown *et al.*, 1994; Gupta & Golding, 1993). Figure 3 compares unrooted trees for HSP70 homologues, catalytic ATPase subunits and 16S rRNA for similar sets of species. Using the tree comparison algorithm as implemented

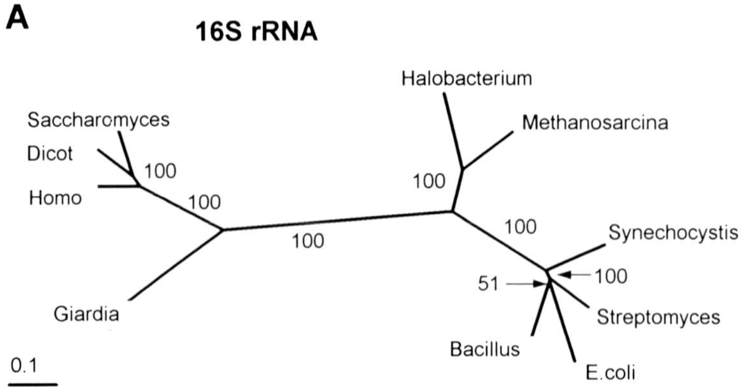

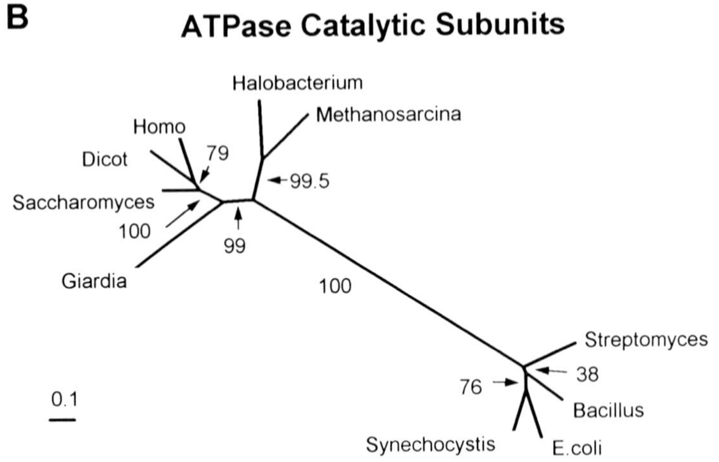

Tree Topology	Steps	Diff Steps	Its S.D.	Significantly worse?
16S tree	3026.0	26.0	14.0814	No
Conv. wisd.	3014.0	14.0	14.1524	No
ATPase tree	3000.0	<------ best		
HSP tree	4046.0	1046.0	63.3049	Yes

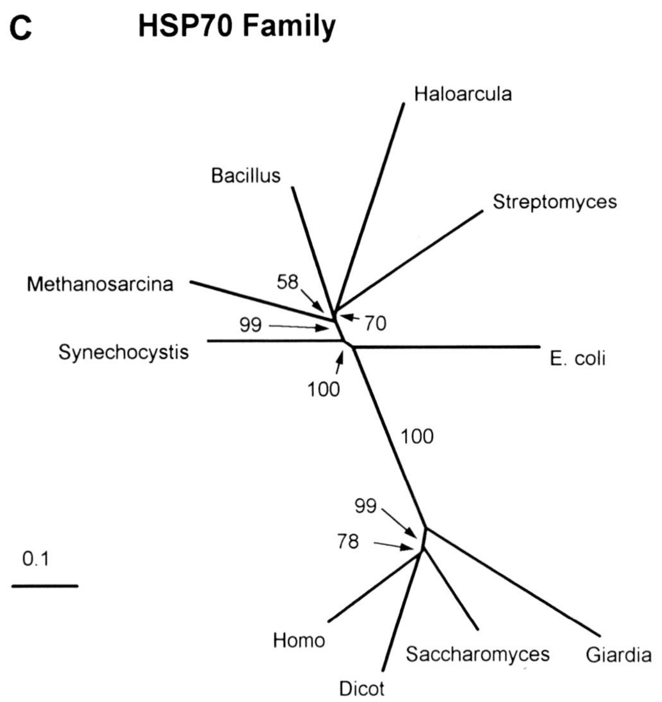

Tree Topology	Steps	Diff Steps	Its S.D.	Significantly worse?
16S tree	2692.0	104.0	21.7406	Yes
Conv. wisd.	2694.0	106.0	21.9240	Yes
ATPase tree	2675.0	87.0	20.6299	Yes
HSP tree	2588.0	<------ best		

Fig. 3. Unrooted molecular phylogenies calculated using nucleotide (A) and protein parsimony analyses. Tree topologies were compared using the user-tree option of the PROTPARS program (Felsenstein, 1993). The topology denoted as 'conventional wisdom' groups the two Gram positives together (*Streptomyces, Bacillus*), and it groups *Saccharomyces* together with the human sequence (*Homo*). Using parenthesis to denote the groups, this topology is as follows: (((*E. coli* (*Streptomyces, Bacillus*)), *Synechocystis*), (*Giardia* (Plant (*Homo, Saccharomyces*))), (*Methanosarcina, Halobacterium*). The aligned 16S rRNA sequences were obtained from the Ribosomal Database Project (Larsen *et al.*, 1993); the ATPase subunits and the HSP70 homologues were aligned using ClustalW (Thompson *et al.*, 1994). The depicted trees correspond to the most parsimonious trees calculated using DNAPARS and PROTPARS for the 16S rRNA and the amino acid sequences, respectively. Numbers give the percentage of bootstrapped samples that contained the group distal to this number. Branch lengths were calculated with PROTDIST, DNADIST and FITCH using the topology calculated with parsimony as user defined tree. Sequences from the following species were used: *Arabidopsis thaliana* (Dicot in 16S rRNAs); *Bacillus subtilis*; *Daucus carota* (Dicot in ATPases); *Escherichia coli*; *Giardia lamblia*; *Haloarcula marismortui*; *Halobacterium halobium*; *Homo sapiens*; *Methanosarcina mazei* (for HSP70s), *M. barkeri* (for ATPases and 16S rRNAs); *Petunia* sp. (Dicot in HSP70s) *Saccharomyces cerevisiae*; *Streptomyces griseus* for HSP70, *S. lividans* for 16S rRNAs and ATPases; *Synechocystis* sp. strain PCC 6803.

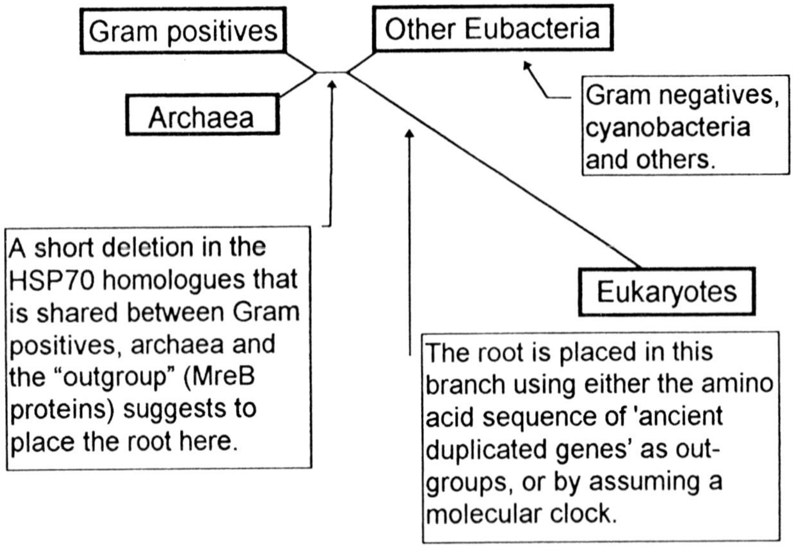

Fig. 4. Schematic diagram of the unrooted phylogenies that were calculated for glutamine synthetases and homologues of the 70kDa heat shock proteins (HSP70). The longest internal branch connects the prokaryotes to the eukaryotes. Within the prokaryotes these molecules do not reflect a clear separation into archaea and eubacteria. Rather, the archaea were found to group among the Gram positive eubacteria. Two of the internal branches have been suggested for the placement of the root. However, the arguments remain inconclusive in both cases. See text for discussion.

by Felsenstein (1993), the difference between the 16S rRNA and the ATPase trees appears to be not significant; in contrast, the topology obtained for the HSP70 sequences is incompatible with the ATPase data and vice versa (Fig. 3).

The amino terminal half of the HSP70s is homologous to the MreB proteins found in *Bacillus* and *E. coli*. Based on a gap that is shared between the MreB proteins and the HSP70s from Gram positives and archaea, Gupta and Golding (1993) placed the root of the HSP70 phylogeny between eukaryotes and Gram negatives (and other eubacteria) on one side and the Gram positive and the archaea on the other (see Fig. 4). However, the alignment between MreB proteins and HSP70s introduces several gaps throughout the MreB sequence. It therefore appears likely that the gap in the MreB proteins and the gap in some of the HSPs is due to convergence and does not necessarily reflect the ancestral state. If the amino acid sequence of the MreB proteins is used as an outgroup for the HSP70 phylogeny, the root is placed in the branch connecting pro and eukaryotes (Fig. 4). Depending on the chosen alignment, the bootstrap probabilities for this placement of the root are up to 100%; however, there are few (or no) informative positions present in the alignment between MreB and HSP70. Some positions are conserved in all sequences; these positions allow the detection of homology between the sequences. The z-values (Needleman & Wunsch, 1970; Lipman & Pearson, 1985) for pairwise comparisons between MreB and HSP70 amino acid sequences are between 4 and 10. However, the non-conserved positions appear saturated with substitutions. Thus the MreBs are either attracted to the longest internal branch (with parsimony analyses) or the centre of the phylogenetic tree (with distance matrix analyses).

Regardless of the question of the placement of the root in the HSP70 phylogeny, the difference between those molecular markers that reflect a close association between Gram positives and archaea and those that reflect two distinct prokaryotic domains does not appear to be due to ill resolved branching patterns (see Fig. 3; Gupta & Golding, 1993; Brown *et al.*, 1994). If one wants to reconcile the conflict assuming paralogous genes, one would have to postulate that the clear separation between eubacteria and archaea, which is found for many molecular markers (Table 1), is due to so far unrecognized ancient gene duplications (Fig. 5A). Using paralogy as an explanation, the true phylogeny would then be reflected by HSP70 homologues, glutamate dehydrogenases, and glutamine synthetases (i.e. Fig. 5B). This explanation is unlikely for two reasons: (i) H^+-ATPases, elongation factors, RNA polymerases, aminoacyl tRNA synthetases, and ribosomes still have comparable functions in all extant organisms. It seems unlikely that the last common ancestor would have had duplicate divergent copies for all of these activities. (ii) It seems extremely unlikely that nearly identical sets of these characters were lost several times in the lineages leading to the different eubacterial groups, whereas exactly the complementary set would have had to have been lost multiple times in the lineages leading to present day archaea and the eukaryotic nucleocytoplasm (see Fig. 5B).

The third and remaining possibility is that the genes that show the close association between the archaea and Gram positive eubacteria were obtained by the archaea via horizontal gene transfer (Fig. 5C). The same or at least a similar association is found for a variety of different molecular markers (Table 2). This suggests that a significant portion of the genome was involved in this transfer. Many of the genes that suggest a closer association between eubacteria and archaea are involved in biosynthetic pathways (e.g. glutamine synthetase, carbamoylphosphate synthetase, glutamate dehydrogenase). Some other well-studied enzymes, for example those involved in nitrogen fixation (Chien & Zinder, 1994; Souillard *et al.*, 1988) also suggest a close association between Gram positives and at least some archaea. Similar results were also reported for randomly obtained *Pyrococcus furiosus* sequence fragments (Frank T. Robb, University of Maryland, personal communication).

These findings suggest that many enzymes involved in biosynthetic pathways were obtained by the archaea through horizontal transfer from one or several Gram positive bacteria. If this interpretation is correct, the

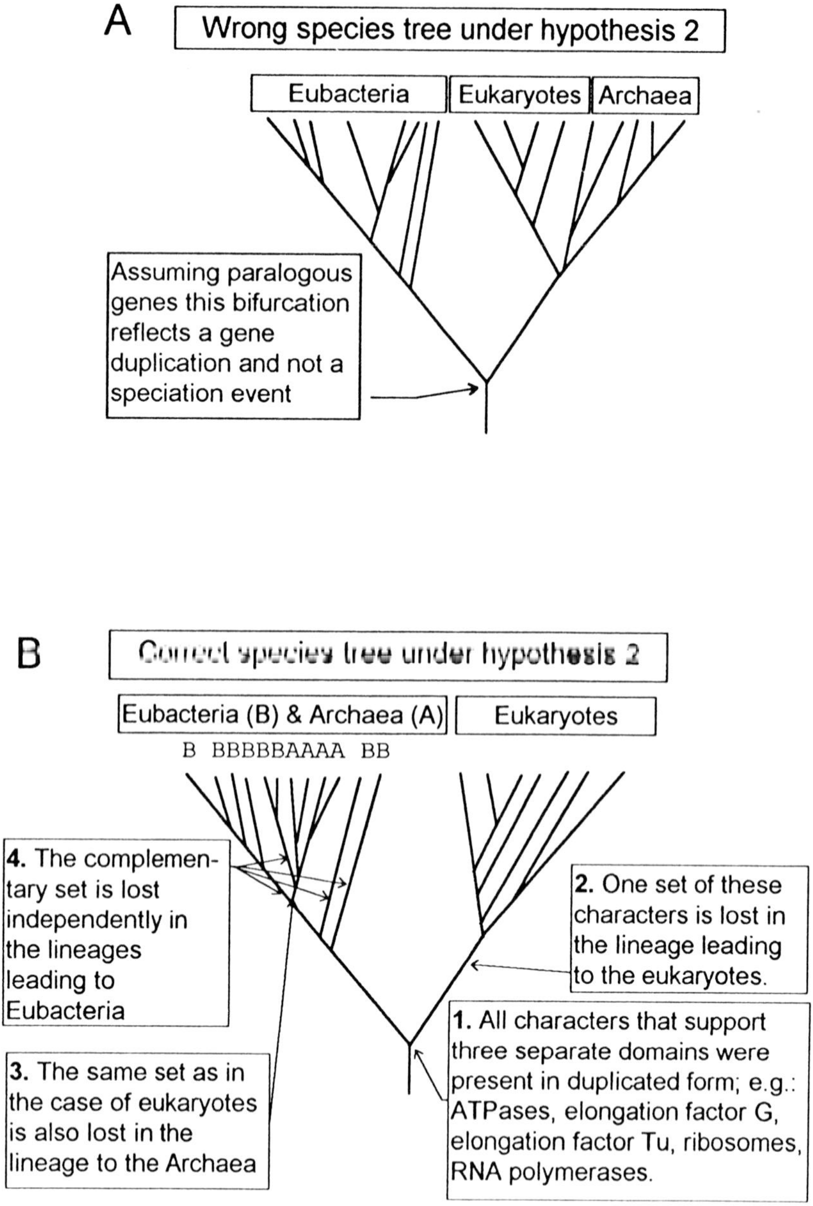
A
Wrong species tree under hypothesis 2
Eubacteria
Eukaryotes
Archaea
Assuming paralogous genes this bifurcation reflects a gene duplication and not a speciation event
B
Correct species tree under hypothesis 2
Eubacteria (B) & Archaea (A)
Eukaryotes
B BBBBBAAAA BB
4. The complementary set is lost independently in the lineages leading to Eubacteria
2. One set of these characters is lost in the lineage leading to the eukaryotes.
1. All characters that support three separate domains were present in duplicated form; e.g.: ATPases, elongation factor G, elongation factor Tu, ribosomes, RNA polymerases.
3. The same set as in the case of eukaryotes is also lost in the lineage to the Archaea

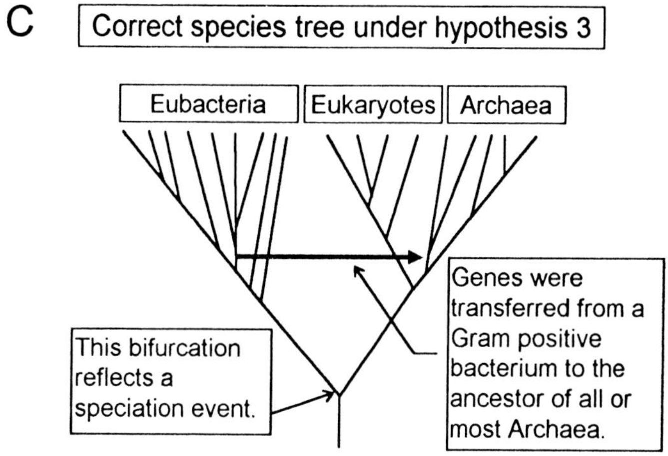
C
Correct species tree under hypothesis 3
Eubacteria
Eukaryotes
Archaea
Genes were transferred from a Gram positive bacterium to the ancestor of all or most Archaea.
This bifurcation reflects a speciation event.

Fig. 5. Schematic diagrams depicting possible relationships between the three cell lineages. Three hypotheses can be used to explain the conflict between those molecular markers that support three distinct domains of life, and those which group the archaea together with the Gram positive eubacteria (see text). If one suggests unrecognized paralogous genes as an explanation (hypothesis 2), then the deepest bifurcation in the ATPase, elongation factor, and aminoacyl-tRNA synthetase genes reflects another round of gene duplication (panel A). The true species phylogeny under the same hypothesis is depicted in panel B. To reconcile this species phylogeny with the observed molecular phylogenies several independent losses of complementary character sets have to be assumed (numbers 2, 3 and 4 in panel B). Assuming horizontal gene transfer to reconcile the conflicting molecular phylogenies (hypothesis 3) results in the species phylogeny depicted in panel C.

following questions arise: Under what conditions did archaea obtain these genes and why was it a selective advantage for archaea to retain them? Did the archaeal ancestor lack these enzymes before the transfer took place? In the example of Fig. 5C, the archaea and the eukaryotes are sister groups. Many (though not all) of the enzymes that were transferred horizontally have counterparts in the eukaryotic nucleocytoplasm. This suggests that these enzymes were already present in the last common ancestor. According to this line of reasoning these enzymes were also present at the bifurcation leading to archaea and the eukaryotic nucleocytoplasm and thus many of the genes transferred from the Gram positives to the archaea would have replaced similar enzymes already present in this lineage.

However, an alternative explanation is possible. An early eubacterial contribution to the eukaryotic nucleocytoplasm has been repeatedly suggested (Zillig, Palm & Klenk, 1992). This contribution, supported by glyceraldehyde-3-phosphate dehydrogenase (GAPDH) phylogeny (Hensel *et al.*, 1989; Markos, Miretsky & Muller, 1993) predated the symbiosis that gave rise to mitochondria and plastids. It has been suggested that an early eubacterial contribution was involved in the evolution of the eukaryotic cytoskeleton (Margulis, 1993). This earlier eubacterial contribution could also have contributed the genes involved in the various anabolic pathways

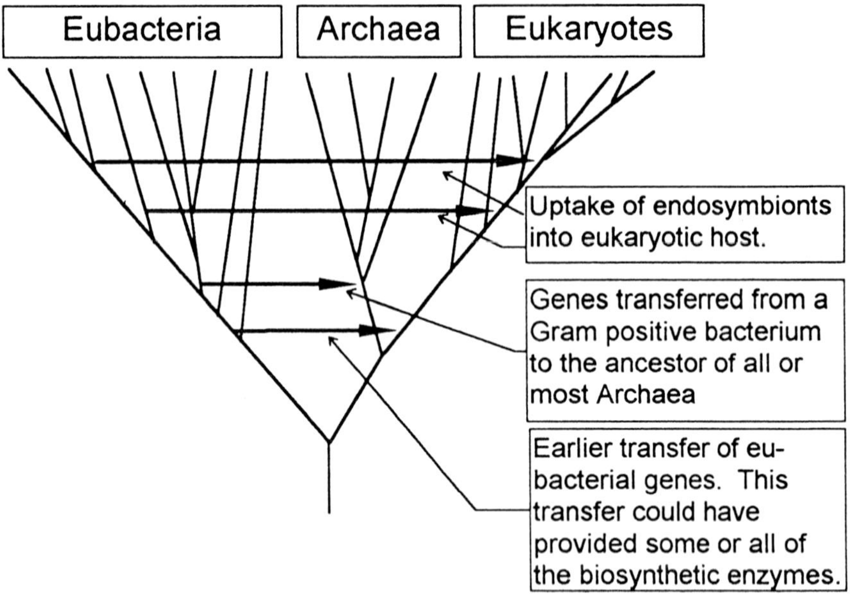

Fig. 6. Hypothetical species phylogeny that is compatible with many anabolic enzymes not being present in the last common ancestor. According to this hypothesis, these enzymes evolved early in the eubacterial lineage and subsequently were transferred into the other two cell lineages. See text for further discussion.

(Fig. 6). According to this alternative scheme these pathways would have evolved only after the last common ancestor in the eubacterial lineage. They first were transferred into the eukaryotic ancestor, and later these genes and others that had since evolved in the eubacterial lineage (e.g. genes encoding nitrogenase) were transferred from the eubacteria to the archaea. This scenario bears some resemblance to Kandler's pre-cell proposal (Kandler, 1994) in that one lineage makes successive inventions that are subsequently passed on into the three domains; the main difference is that in the proposal presented here the last common ancestor already had well functioning genetic and bioenergetic machineries in place.

PROPERTIES OF THE LAST COMMON ANCESTOR

Clearly, a reticulate species phylogeny greatly complicates the inference of characteristics present in the last common ancestor. The fact that a character is present in all three lineages no longer is sufficient to demonstrate that it was already present in that last common ancestor. To infer that a trait was present in the last common ancestor the phylogeny of this trait has to be compared and reconciled with the net-like species phylogeny. In many instances a definitive conclusion remains elusive at present. In the cases discussed above (i.e. glutamate dehydrogenase, carbamoylphosphate synthetase, glutamate dehydrogenase, etc.) there are two possible scenarios:

(i) All these genes were present in the last common ancestor, subsequently these genes were lost early in the archaeal lineage and were replaced by genes obtained via horizontal gene transfer from Gram positive eubacteria.

(ii) These enzymatic functions were not present in the last common ancestor. These genes evolved in the eubacterial lineage and were transferred twice, first to the ancestor of the eukaryotic nucleocytoplasm, and second to the archaeal ancestor (Fig. 6).

At present there is no obvious rationale to decide between these scenarios. Scenario B involves two horizontal transfers and a less anabolically competent last common ancestor; scenario A involves a loss of functions followed by horizontal transfer of the same functions. It also requires a much more complex last common ancestor.

If the tree sketched in Fig. 5C approximates the true species phylogeny, it appears likely that the last common ancestor already possessed effective transcription and translation machineries. Using a structure function relationship derived for extant H^+-ATPases/ATPsynthases, Gogarten and Taiz (1992) concluded that the H^+-ATPases present in the last common ancestor were likely to function as an ATP synthase under physiological conditions. Recently, Castresana, Lübben, and Saraste (1995) studied the phylogeny of redox chain components. Many of these are either small proteins or contain many membrane spanning helices that retained only little phylogenetic information. Figure 7 shows a distance matrix analysis (panel A) and a bootstrapped parsimony analysis (panel B) of the cytochrome c oxidase subunit 1 homologues. The tree depicted in panel A is essentially identical to the phylogeny published by Castresana *et al.* (1994). Although many of the deeper branches are not strongly supported (Fig. 7B), the data suggest that the ancestor of eubacteria and archaea had a cytochrome c oxidase homologue. The archaeal sequences so far obtained suggest at least two types are present in archaea. The halobacterial cytochrome oxidase groups with the majority of all other eubacterial cytochrome oxidase homologues. *Sulfolobus* contains two Sox genes which do not group within the major eubacterial lineage. However, *Thermus*, an eubacterium, contains two cytochrome oxidases, CbaA and CaaB. One of these (CaaB) is clearly eubacterial; the other falls outside the eubacterial lineage, between the two Sox genes. More archeal sequences are desirable to determine if this is a case of paralogy, however, the possibility exists that *Thermus* may have received an archeal-type cytochrome oxidase via horizontal gene transfer.

The cytochrome c oxidase homologues do not reflect a close association between archaea and Gram positives, suggesting that this enzyme was not part of the horizontal transfer discussed previously. The FixN cytochrome

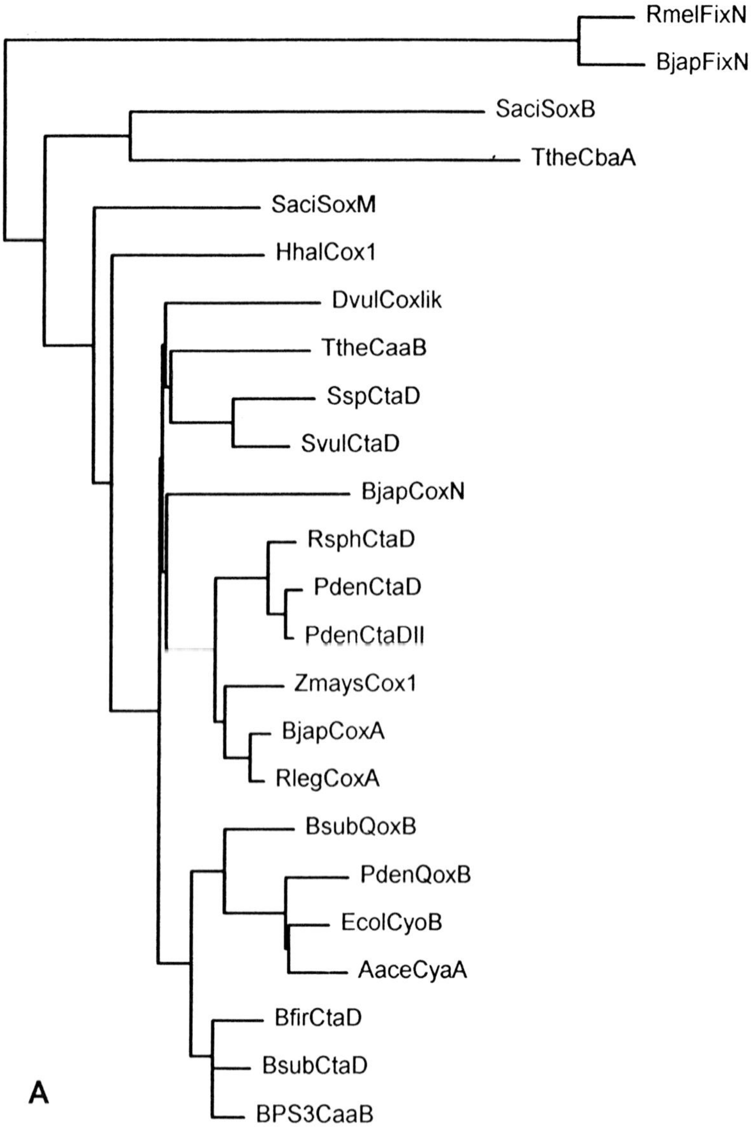

Fig. 7. Molecular phylogenies of cytochrome c oxidase subunit I homologues calculated using distance analyses (A) and protein parsimony (B). Sequences used were retrieved from GenBank (see Castresana *et al.*, 1994 for accession numbers). Analyses were performed as described in Fig. 2. Abbreviations for species are as follows: Rmel, *Rhizobium melliloti*; Bjap, *Bradyrhizobium japonicum*; Saci, *Sulfolobus acidocaldarius*; Tthe, *Thermus thermophilus*; Hhal, *Halobacterium halobium*; Dvul, *Desulfovibrio vulgaris*; Ssp, *Synechocystis* sp.; Svul, *Synechococcus vulcanus*; Rsph, *Rhodobacter sphaeroides*; Pden, *Paracoccus denitrificans*;

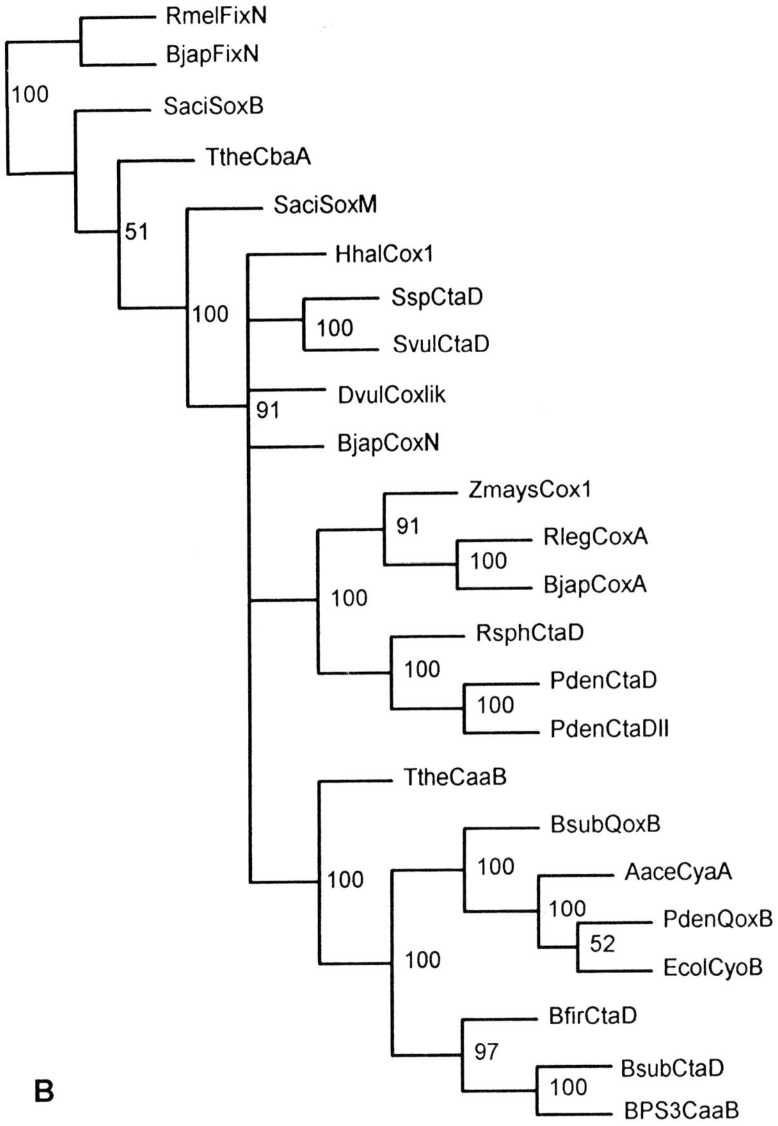

Zmays, *Zea mays*; Rleg, *Rhizobium leguminosarum*; Bsub, *Bacillus subtillus*; Ecol, *Escherichia coli*; Aace, *Acetobacter aceti*; Bfir, *Bacillus firmus*; BPS3, Thermophilic *Bacillus* BPS3; All proteins included in the analyses are members of the heme-copper oxidase family and are homologues of cytochrome c oxidase subunit 1. FixN is found in microaerophilic bacteria; Sox B and Sox M are archaebacterial quinol oxidases; QoxB, CyaA and CyoB are eubacterial quinol oxidases; DvulCoxlik(e) and TtheCbaA are inferred to be cytochrome oxidases based on sequence homology and function; all others are known cytochrome c oxidases.

oxidase found in microaerophilic endosymbionts is homologous to and believed to have evolved from nitric oxide reductases (NorB) (Saraste & Castresana, 1994; van der Oost *et al.*, 1994). The finding that the eubacterial nitrate reducing homologue (FixN) branches close to the archeal forms of oxygen-utilizing cytochrome oxidase homologues, suggests that a redox chain operative in the last common ancestor could have used another substrate instead of oxygen as a terminal electron acceptor. However, the trees depicted by Castresana and Saraste (1994) and in Fig. 7 are unrooted phylogenies. Considering this and the finding that many of the deeper branches remain ill resolved, alternative interpretations of these phylogenies cannot be excluded at present.

CONCLUSIONS

Differences between well resolved molecular phylogenies can be due either to unrecognized gene duplications or to horizontal gene transfer. Two examples that conflict with well documented phylogenetic relationship are the presence of archaeal type H^+-ATPases in *Thermus* and *Enterococcus* and the close association between Gram positive eubacteria and archaea as supported by many different genes. In both cases the best explanation for these conflicts is horizontal gene transfer. Assuming unrecognized paralogies (duplicated genes) as an explanation would necessitate many instances of convergent evolution; furthermore this assumption would also result in species phylogenies that are at odds with two distinct prokaryotic domains (archaea and eubacteria). Many characters reflect the close association between archaea and eubacteria; therefore, it seems likely that a substantial portion of the eubacterial genome participated in this transfer. A net-like species phylogeny complicates inferring the properties of the last common ancestor. Even so, the data strongly indicate that the *last common ancestor* was a cellular organism, with a DNA based genome, and a sophisticated transcription and translation machinery. Furthermore, the analysis of ATPase structure function relationships and the evolution of cytochrome oxidase homologues suggest that the last common ancestor already used chemiosmotic coupling to synthesize ATP, and that electron transport chains could be used to energize the plasma membrane.

ACKNOWLEDGEMENTS

Research in the author's lab was supported by NSF grant BSR-9020868, and through the NASA-Exobiology program. EH was supported in part by a fellowship from the DGAPA-Universidad Nacional Autónoma de México. JPG thanks Henrik Kibak and Lincoln Taiz for numerous stimulating discussions.

REFERENCES

Benachenhou-Lafha, N., Forterre, P. & Labedan, B. (1993). Evolution of glutamate dehydrogenase genes: Evidence for two paralogous protein families and unusual branching patterns of the archaebacteria in the universal tree of life. *Journal of Molecular Evolution*, **36**, 335–46.

Brown, J. R., Masuchi, Y., Robb, F. T. & Doolittle, W. F. (1994). Evolutionary relationships of bacterial and archaeal glutamine synthetase genes. *Journal of Molecular Evolution*, **38**, 566–76.

Brown, J. R. & Doolittle, W. F. (1995). Root of the universal tree of life based on ancient aminoacyl-tRNA synthetase gene duplications. *Proceedings of the National Academy of Sciences, USA*, **92**, 2441–5.

Cammarano, P., Palm, P., Creti, R., Ceccarelli, E., Sanangelantoni, A. M. & Tiboni, O. (1992). Early evolutionary relationships among known life forms inferred from elongation factor EF-2 (EF-G) sequences: Phylogenetic coherence and structure of the archaeal domain. *Journal of Molecular Evolution*, **34**, 396–405.

Castresana, J., Lübben, M. & Saraste, M. (1995). New archaebacterial genes coding for redox proteins: Implications for the evolution of aerobic metabolism. *Journal of Molecular Biology*, **250**, 202–10.

Castresana, J., Lübben, M., Saraste, M. & Higgins, D. G. (1994). Evolution of cytochrome oxidase, an enzyme older than atmospheric oxygen. *European Molecular Biology Organization Journal*, **13**, 2516–25.

Chien, Y. T. & Zinder, S. H. (1994). Cloning, DNA sequencing, and characterization of a nifD-homologous gene from the archaeon *Methanosarcina barkeri* 227 which resembles nifD1 from the eubacterium *Clostridium pasteurianum*. *Journal of Bacteriology*, **176**, 6590–8.

Felsenstein, J. (1988). Phylogenies from molecular sequences: Inference and reliability. *Annual Review of Genetics*, **22**, 521–65.

Felsenstein, J. (1993). Phylogeny Inference Package, version 3.5c. Distributed by the author. Dept. of Genetics, Univ. of Washington, Seattle.

Fitch, W. S. (1970). Distinguishing homologous from analogous proteins. *Systematic Zoology*, **19**, 99–113.

Forterre, P., Benachenhou-Lafha, N., Confalonieri, F., Duguet, M., Elie, C. & Labedan, B. (1993). The nature of the last universal ancestor and the root of the tree of life, still open questions. *BioSystems*, **28**, 15–32.

Gandbhir, M., Rasched, I., Marlière, P. & Mutzel, R. (1995). Convergent evolution of amino acid usage in archaebacterial and eubacterial lineages adapted to high salt. *Research in Microbiology*, **146**, 113–20.

Gogarten, J. P. (1994). Which is the most conserved group of proteins? Homology-orthology, paralogy, xenology and the fusion of independent lineages. *Journal of Molecular Evolution*, **39**, 541–3.

Gogarten, J. P. (1995). The early evolution of cellular life. *Trends in Ecology and Evolution*, **10**, 147–51.

Gogarten, J. P. & Taiz, L. (1992). Evolution of proton pumping ATPases: rooting the tree of life. *Photosynthesis Research*, **33**, 137–46.

Gogarten, J. P., Starke, T., Kibak, H., Fichmann, J. & Taiz, L. (1992). Evolution and isoforms of V-ATPase subunits. *Journal of Experimental Biology*, **172**, 137–47.

Gogarten, J. P., Kibak, H., Dittrich, P., Taiz, L., Bowman, E. J., Bowman, B. J., Manolson, M. F., Poole, R. J., Date, T., Oshima, T., Konishi, J., Denda, K. & Yoshida, M. (1989). Evolution of the vacuolar H^+-ATPase: Implications for the origin of eukaryotes. *Proceedings of the National Academy of Sciences, USA*, **86**, 6661–5.

Gray, G. S. & Fitch, W. M. (1983). Evolution of antibiotic resistance genes: the DNA sequence of a kanamycin resistance gene from *Staphylococcus aureus. Molecular Biology and Evolution*, **1**, 57–66.

Gupta, R. S. & Golding, G. B. (1993). Evolution of HSP70 gene and its implications regarding relationships between archaebacteria, eubacteria and eukaryotes. *Journal of Molecular Evolution*, **37**, 573–82.

Gupta, R. S. & Singh, B. (1992). Cloning of the HSP70 gene from *Halobacterium marismortui*: relatedness of archaebacterial HSP70 to its eubacterial homologs and a model of the evolution of the HSP70 gene. *Journal of Bacteriology*, **174**, 4594–605.

Harvey, W. R. & Nelson, N., eds. (1992). V-ATPases. *Journal of Experimental Biology*, **172**.

Hashimoto, T., Nakamura, Y., Nakamura, F., Shirakura, T., Adachi, J., Goto, N., Okamoto, K.-I. & Hasegawa, M. (1994). Protein phylogeny gives a robust estimation for early divergences of eukaryotes: phylogenetic place of a mitochondria-lacking protozoan, *Giardia lamblia. Molecular Biology and Evolution*, **11**, 65–71.

Hensel, R., Demharter, W., Kandler, O., Kroppenstedt, M. & Stackebrandt, E. (1986). Chemotaxonomic and molecular-genetic studies of the genus *Thermus*: evidence for a phylogenetic relationship of *Thermus aquaticus* and *Thermus ruber* to the genus *Deinococcus. International Journal of Systematic Bacteriology*, **36**, 444–53.

Hensel, R., Zwickl, P., Fabry, S., Lang, J. & Palm, P. (1989). Sequence comparison of glyceraldehyde-3-phosphate dehydrogenases from the three kingdoms: evolutionary implication. *Canadian Journal of Microbiology*, **35**, 81–5.

Hilario, E. & Gogarten, J. P. (1993). Horizontal transfer of ATPase genes–the tree of life becomes a net of life. *BioSystems*, **31**, 111–19.

Hillis, D. M., Huelsenbeck, J. P. & Swofford, D. L. (1994). Hobgoblin of phylogenetics? *Nature*, **3692**, 363–4.

Iwabe, N., Kuma, K.-I., Hasegawa, M., Osawa, S. & Miyata, T. (1989). Evolutionary relationships of archaebacteria, eubacteria and eukaryotes inferred from phylogenetic trees of duplicated genes. *Proceedings of the National Academy of Sciences, USA*, **86**, 9355–9.

Kandler, O. (1994). The early diversification of life. In *Early Life on Earth: Nobel Symposium No. 84*, Bengston, S., ed., pp. 152–160. Columbia University Press, New York.

Kumada, Y., Benson, D. R., Hillemann, D., Hosted, T. J., Rochford, D. A., Thompson, C. J., Wohlleben, W. & Tateno, Y. (1993). Evolution of the glutamine synthase gene, one of the oldest existing and functioning genes. *Proceedings of the National Academy of Sciences, USA*, **90**, 3009–13.

Langer, D., Hain, J., Thuriaux, P. & Zillig, W. (1995). Transcription in archaea: similarity to that in eucarya. *Proceedings of the National Academy of Sciences, USA*, **92**, 5768–72.

Larsen, N., Olsen, G. J., Maidak, B. L., McCaughey, M. J., Overbeek, R., Macke, T. J., Marsh, T. L. & Woese, C. R. (1993). The ribosomal database project. *Nucleic Acids Research*, **21**, 3021–3.

Lipman, D. J. & Pearson, W. R. (1985). Rapid and sensitive protein similarity searches. *Science*, **227**, 1435–41.

Lockhart, P. J., Beanland, T. J., Howe, C. J. & Larkum, A. W. (1992). Sequence of

Prochloron didemni atpBE and the inference of chloroplast origins. *Proceedings of the National Academy of Sciences, USA*, **89**, 2742–6.

Ludwig, W., Neumaier, J., Klugbauer, N., Brockmann, E., Roller, C., Jilg, S., Reetz, K., Schachtner, I., Ludvigsen, A., Bachleitner, M., Fischer, U. & Schleifer, K. H. (1993). Phylogenetic relationships of bacteria based on comparative sequence analysis of elongation factor Tu and ATP-synthase beta-subunit genes. *Antonie Van Leeuwenhoek*, **64**, 285–305.

Margulis, L. (1993). *Symbiosis in Cell Evolution*. 2nd edn. W. J. Freeman and Company, New York.

Markos, A., Miretsky, A. & Muller, M. (1993). A glyceraldehyde-3-phosphate dehydrogenase with eubacterial features in the amitochondriate eukaryote, *Trichomonas vaginalis*. *Journal of Molecular Evolution*, **37**, 631–43.

Marsh, T. L., Reich, C. I., Whitelock, R. B. & Olsen, G. J. (1994). Transcription factor IID in the Archaea: Sequences in the *Thermococcus celer* genome would encode a product closely related to the TATA-binding protein of eukaryotes. *Proceedings of the National Academy of Sciences, USA*, **91**, 4180–4.

Morden, C. W., Delwiche, C. F., Kuhsel, M. & Palmer, J. D. (1992). Gene phylogenies and the endosymbiotic origin of plastids. *Biosystems*, **28**, 75–90.

Needleman, S. B. & Wunsch, C. D. (1970). A general method applicable to the search for similarities in the amino acid sequence of two proteins. *Journal of Molecular Biology*, **48**, 443–53.

Olsen, G. J., Woese, C. R. & Overbeek, R. (1994). The winds of (evolutionary) change: breathing new life into microbiology. *Journal of Bacteriology*, **176**, 1–6.

Pesole, G., Gissi, C., Lanave, C. & Saccone, C. (1995). Glutamine synthetase gene evolution in bacteria. *Molecular Biology and Evolution*, **12**, 189–97.

Puhler, G., Leffers, H., Gropp, F., Palm, P., Klenk, H. P., Lottspeich, F., Garrett, R. A. & Zillig, W. (1989). Archaebacterial DNA-dependent RNA polymerases testify to the evolution of the eukaryotic nuclear genome. *Proceedings of the National Academy of Sciences, USA*, **86**, 4569–73.

Saraste, M. & Castresana, J. (1994). Cytochrome oxidase evolved by tinkering with denitrification enzymes. *Federation of European Biochemical Societies Letters*, **341**, 1–4.

Schofield, J. P. (1993). Molecular studies on an ancient gene encoding for carbamoyl-phosphate synthetase. *Clinical Science*, **84**, 119–28.

Schopf, J. W. (1993). Microfossils of the Early Archean Apex chert: new evidence of the antiquity of life. *Science*, **260**, 640–6.

Smith, M. W., Feng, D. F. & Doolittle, R. F. (1992). Evolution by acquisition: the case for horizontal gene transfer. *Trends in Biochemical Science*, **17**, 489–93.

Sogin, M. L., Gunderson, J. H., Elwood, H. J., Alonso, R. A. & Peattie, D. A. (1989). Phylogenetic meaning of the kingdom concept: an unusual ribosomal RNA from *Giardia lamblia*. *Science*, **243**, 75–7.

Souillard, N., Magot, M., Possot, O. & Sibold, L. (1988). Nucleotide sequence of regions homologous to nifH (nitrogenase Fe protein) from the nitrogen-fixing archaebacteria *Methanococcus thermolithotrophicus* and *Methanobacterium ivanovvi*: evolutionary implications. *Journal of Molecular Evolution*, **27**, 65–76.

Takase, K., Kakinuma, S., Yamato, I., Konishi, K., Igarashi, K. & Kakinuma, Y. (1994). Sequencing and characterization of the ntp gene cluster for vacuolar-type Na(+)-translocating ATPase of *Enterococcus hirae*. *Journal of Biological Chemistry*, **269**, 11037–44.

Thompson, J. D., Higgins, D. G. & Gibson, T. J. (1994). CLUSTAL W: improving the sensitivity of progressive multiple sequence alignment through sequence weighting, position-specific gap penalties and weight matrix choice. *Nucleic Acids Research*, **22**, 4673–80.

Tiboni, O., Cammarano, P. & Sanangelantoni, A. M. (1993). Cloning and sequencing of the gene encoding glutamine synthase I from the Archaeum *Pyrococcus woesei*: anomalous phylogenies inferred from analysis of archaeal and bacterial glutamine synthase I sequences. *Journal of Bacteriology*, **175**, 2961–9.

Tsutsumi, S., Denda, K., Yokoyama, K., Oshima, T., Date, T. & Yoshida, M. (1991). Molecular cloning of genes encoding major two subunits of a eubacterial V-Type ATPase from *Thermus thermophilus. Biochimica et Biophysica Acta*, **1098**, 13–20.

van der Oost, J., de Boer, A. P., de Gier, J. W., Zumft, W. G., Stouthamer, A. H. & van Spanning, R. J. (1994). The heme-copper oxidase family consists of three distinct types of terminal oxidases and is related to nitric oxide reductase. *Federation of European Microbiological Societies Microbiology Letters*, **121**, 1–9.

Woese, C. R. (1987). Bacterial evolution. *Microbiological Reviews*, **51**, 221–71.

Yokoyama, K., Oshima, T. & Yoshida, M. (1990). *Thermus thermophilus* membrane-associated ATPase; Indication of a eubacterial V-type ATPase. *Journal of Biological Chemistry*, **265**, 21946–50.

Zillig, W., Palm, P. & Klenk, H.-P. (1992). A model of the early evolution of organisms: the arisal of the three domains of life from the common ancestor. In *The Origin and Evolution of the Cell*, Hartman, H. & Matsuno, K., eds., pp. 163–182. Singapore, New Jersey, London, and Hong Kong: World Scientific.

Zimniak, L., Dittrich, P., Gogarten, J. P., Kibak, H. & Taiz, L. (1988). The cDNA sequence of the 69kDa subunit of the carrot vacuolar H^+-ATPase. Homology to the beta-chain of F_0F_1-ATPases. *Journal of Biological Chemistry*, **263**, 9102–12.

INDEX